AF411411

New Strategies for Innovative a[n]
Enhanced Meat and Meat Produ[c]

New Strategies for Innovative and Enhanced Meat and Meat Products

Editors

Gonzalo Delgado-Pando
Tatiana Pintado

MDPI • Basel • Beijing • Wuhan • Barcelona • Belgrade • Manchester • Tokyo • Cluj • Tianjin

Editors
Gonzalo Delgado-Pando   Tatiana Pintado
ICTAN   ICTAN
CSIC   CSIC
Madrid   Madrid
Spain   Spain

Editorial Office
MDPI
St. Alban-Anlage 66
4052 Basel, Switzerland

This is a reprint of articles from the Special Issue published online in the open access journal *Foods* (ISSN 2304-8158) (available at: www.mdpi.com/journal/foods/special_issues/New_Strategies_Innovative_Enhanced_Meat_Meat_Products).

For citation purposes, cite each article independently as indicated on the article page online and as indicated below:

LastName, A.A.; LastName, B.B.; LastName, C.C. Article Title. *Journal Name* **Year**, *Volume Number*, Page Range.

ISBN 978-3-0365-4104-4 (Hbk)
ISBN 978-3-0365-4103-7 (PDF)

Contents

About the Editors

Gonzalo Delgado-Pando

Dr. Delgado-Pando graduated in Food Science and Technology in 2006 at Universidad de Burgos (Spain), obtaining merit with distinction. He also studied the first cycle of a Chemistry degree at the same university. He obtained a grant to study a MSc in Food Safety and Biotechnology, completed in 2007 with the qualification of excellent. He finished his PhD in 2013 at Universidad Complutense de Madrid with the thesis entitled: "Design and development of meat products with optimised lipid profile. Evaluation of their functional effect in humans" , obtaining the highest qualification (Excellent cum laude). Dr. Delgado-Pando has had an ample international career as a researcher. After finishing his PhD, he moved to the United Kingdom to Queen's University Belfast to join a European funded FP7 project called STARTEC as work package deputy for two years. Then, he moved to Ireland to work at Teagasc in two projects: PROSSLOW and MTI. In 2020, he moved back to Spain to CSIC-ICTAN to work under a European EIT Food project about consumer attitudes towards meat products and also a CDTI project. Dr Delgado-Pando's lines of research lie within the field of meat science, sensory analysis, data analysis, design of experiments and food quality analysis. He has worked in meat product reformulation with regards to more nutritious and healthier products: improvement of lipid profile, salt reduction strategies and food shelf-life extension with the use of novel food processing technologies such as high pressure processing, continuous microwave or ultrasound.

Dr Delgado-Pando has 70 scientific contributions, including 36 peer-reviewed scientific publications, 3 scientific-technical publications, 15 works submitted to conferences, 3 workshops/seminars, 4 chapters in published books, and 9 pieces in outreach magazines.

Tatiana Pintado

Dr Pintado works as laboral staff in the Spanish National Research Council (CSIC), Institute of Food Science, Technology and Nutrition (ICTAN) from April of 2006. She is part of the Meat and Meat Products group (CARPROCAR), whose objective is the performance of different scientific activities of both basic oriented nature, and technological application aimed at improving the quality and safety of meat and its derivatives, as well as the development of healthier meat products, of which the bioavailability of healthy compounds is studied through "in vitro" assays.

Until 2012, she was taking part in the development of CDTI projects related with the development of healthier meat products, among others, based on the improvement of their lipid content. In October 2017, she read the Doctoral Thesis evaluated as cum laude. During her postdoc research stage, the activity has been focused on the quality, safety and evaluation of traditional and functional meat products and non-meat ingredients; the design, formulation and processing potentially functional healthier meat products (technological, structural, organoleptic and microbiological analysis); development of different strategies to obtain new solid lipid materials based on oil structuring methodologies to be used as animal fat replacers and healthy compound delivery systems; "in vitro" assays to evaluate the bioavailability of healthy compounds and application of traditional and emerging technologies to meat derivatives. These studies were financed by the UE and the Spanish MICIU.

Dr Pintado is coauthor of 32 SCI publications (first author in 9 of them), two book chapters and 14 non-SCI publications.

Preface to "New Strategies for Innovative and Enhanced Meat and Meat Products"

Meat and meat products are an important part of the human diet. Even though non-essential, they provide high amounts of protein, vitamins, and minerals in a concentrated form. However, the consumption of meat and meat products has been associated with an increased risk of health-related problems. Once the harmful components of meat and meat products are elucidated, novel technologies can help in identifying, removing, replacing, and/or minimising their deleterious effects. In addition, meat products can be and are being utilised as carriers of added bioactive compounds due to their processing versatility and high worldwide consumption. New strategies in the field of meat and meat product development are certainly needed in order to overcome not only the health-related problems these products might contribute to, but also from the sustainability and economy perspective. This book compiles ten original studies and two comprehensive reviews that will tackle some of these issues.

Gonzalo Delgado-Pando and Tatiana Pintado
Editors

 **foods**

Editorial

New Strategies for Innovative and Enhanced Meat and Meat Products

Gonzalo Delgado-Pando * and **Tatiana Pintado**

Institute of Food Science, Technology and Nutrition (CSIC), José Antonio Novais 10, 28040 Madrid, Spain; tatianap@ictan.csic.es
* Correspondence: g.delgado.pando@gmail.com

Citation: Delgado-Pando, G.; Pintado, T. New Strategies for Innovative and Enhanced Meat and Meat Products. *Foods* **2022**, *11*, 772. https://doi.org/10.3390/foods 11050772

Received: 21 February 2022
Accepted: 27 February 2022
Published: 7 March 2022

Publisher's Note: MDPI stays neutral with regard to jurisdictional claims in published maps and institutional affiliations.

New strategies in the field of meat and meat product development are certainly needed in order to overcome not only the health-related problems these products might contribute to, but also from the perspectives of sustainability and the economy.

Sustainability is now on the agenda of the United Nations member states, who in 2015 declared the sustainable development goals for 2030. The meat industry should be one of the food industries more concerned about this issue, as the low efficiency of animal production as well as environmental issues make this sector key in advancing towards food sustainability. In this Special Issue, Pintado and Delgado-Pando [1] reviewed the use of meat extenders as a way of contributing towards more sustainable meat products. The use of pulses, cereals, tubers, fruits, mushrooms, food by-products, and insects were evaluated as meat replacers or extenders in several types of meat products. Even though there are many of these ingredients that have been successfully employed for substituting meat content, there is need for further research where not only the product quality but also the consumer acceptance is jointly evaluated. A very interesting manuscript presented by Bakhsh et al. [2] proposed the use of plant-based meat analogues for tackling this sustainability issue. The authors use methylcellulose in different concentrations for the development of beef patty analogues of soy protein isolate and soy-based textured vegetable protein. These analogues were compared to a beef patty control in terms of physicochemical and sensory properties. Promising results were obtained in the patty with soy-based textured vegetable protein and 3% of methylcellulose, although a much lower hardness than the control patty was still an issue to improve. The use of by-products as ingredients in meat products can have a double purpose of increasing sustainability and improving the healthiness. Coffee silver skin (CSS) is a unique by-product of coffee roasting that is usually discarded, contributing to food waste. CSS was used in chicken burgers, contributing to the use of these by-products and improving the nutritional properties of the burgers: more fibre, minerals, and bioactive molecules [3]. Another proposal of using a food by-product was made by Summo et al. [4] that utilised oat hull, a by-product of oat milling, as a fat replacer in low-fat beef burgers. The authors found that a 100% substitution of fat from animal origin by this by-product generated burgers more appealing to the consumers with a higher juiciness and a softer texture.

Meat and meat product consumption has been related to an increased risk of developing certain cancers and cardiovascular disease [5,6]. Saturated fat, naturally present in foods from animal origin, have been targeted as one of the issues towards CVD. Pintado and Cofrades [7] proposed a novel approach of substituting the pork backfat of dried fermented sausage with a mixture of oils from plant origin: olive and chia. The authors manufactured a beeswax oleogel and an emulsion gel as carriers of these oils and found an improved lipid profile and a good oxidative and microbiological status, irrespective of the carrier used. Other components of meat products related to health issues are the additives. An extensive review of clean label alternatives proposed by Delgado-Pando et al. [8] explored the idea of replacing the traditional additives with clean label alternatives. Even though the terminology is not yet properly defined, the ideas of consumers and industry were discussed. The origin of the additive, i.e., natural vs. synthetic (e.g., nitrites from green vegetables vs. synthetic nitrites), could be perceived as a good trait but the health problems associated with some additives do not distinguish if the substance is extracted from nature or synthesised in a laboratory, the chemical component remains the same. Nitrites have been related to the formation of N-nitroso compounds, known as human carcinogenic [8]. Tomovic et al. [9] explored the idea of using the *Juniperus communis* L. essential oil as an alternative for sodium nitrite in dry fermented sausages. The authors found that this essential oil could partially replace the use of nitrite as it provides significant antioxidant activity, maintaining the shelf life. Another additive that is being scrutinised is phosphate. Phosphates are widely used texturisers in meat products and have been related to increased CVD in people with chronic kidney disease. Although not harmful for healthy people, EFSA found that the exposure was higher than the acceptable daily intake for some population groups [8]. Goemaere et al. [10] studied the use of seven protein-based ingredients as phosphate replacers in emulsified meat products. The authors found that blood plasma and soy were superior in phosphate-free cooked sausages compared with sausages containing phosphates, in terms of texture, cooking yield, and stability. However, the authors admit that the meat matrix is important when selecting one ingredient or another as phosphate replacer. On the other hand, restructured ostrich ham was successfully formulated with a partial substitution of phosphates by iota carrageenan [11]. Another clean label alternative was proposed by Mancini et al. [12] who utilised common spices such as salt and garlic powder in rabbit burgers. They observed that these two ingredients played an important role in colour changes during storage and that higher garlic levels should be explored if a bacteriostatic effect is also intended.

This Special Issue was completed with two research articles that will contribute to economic improvement by innovation and consumer information. From the latter, Yang et al. [13] did a thorough study of how providing nutritional information can boost the purchase intention of meat products by consumers in Taiwanese wet markets. This shows that sales could be improved by studying the information the consumer obtains during shopping. In terms of innovation, Hrbek et al. [14] proposed a technique for the authentication of meat and meat products, as well as meat adulteration, by using triacylglycerol profiling and DNA analysis. The authors proposed a direct analysis in real time, coupled with high-resolution mass spectrometry and combined with a multiplex polymerase chain reaction.

Author Contributions: G.D.-P. and T.P. conceived and wrote this editorial. All authors have read and agreed to the published version of the manuscript.

Funding: This research received no external funding.

Conflicts of Interest: The authors declare no conflict of interest.

References

1. Pintado, T.; Delgado-Pando, G. Towards More Sustainable Meat Products: Extenders as a Way of Reducing Meat Content. *Foods* **2020**, *9*, 1044. [CrossRef] [PubMed]
2. Bakhsh, A.; Lee, S.-J.; Lee, E.-Y.; Sabikun, N.; Hwang, Y.-H.; Joo, S.-T. A Novel Approach for Tuning the Physicochemical, Textural, and Sensory Characteristics of Plant-Based Meat Analogs with Different Levels of Methylcellulose Concentration. *Foods* **2021**, *10*, 560. [CrossRef] [PubMed]
3. Martuscelli, M.; Esposito, L.; Mastrocola, D. The Role of Coffee Silver Skin against Oxidative Phenomena in Newly Formulated Chicken Meat Burgers after Cooking. *Foods* **2021**, *10*, 1833. [CrossRef] [PubMed]
4. Summo, C.; De Angelis, D.; Difonzo, G.; Caponio, F.; Pasqualone, A. Effectiveness of Oat-Hull-Based Ingredient as Fat Replacer to Produce Low Fat Burger with High Beta-Glucans Content. *Foods* **2020**, *9*, 1057. [CrossRef] [PubMed]
5. IARC. *Monographs on the Evaluation of Carcinogenic Risks to Humans: Red Meat and Processed Meat*; IARC: Lyon, France, 2018; Volume 114, ISBN 978-92-832-0180-9.
6. Cocking, C.; Walton, J.; Kehoe, L.; Cashman, K.D.; Flynn, A. The Role of Meat in the European Diet: Current State of Knowledge on Dietary Recommendations, Intakes and Contribution to Energy and Nutrient Intakes and Status. *Nutr. Res. Rev.* **2020**, *33*, 181–189. [CrossRef] [PubMed]

7. Pintado, T.; Cofrades, S. Quality Characteristics of Healthy Dry Fermented Sausages Formulated with a Mixture of Olive and Chia Oil Structured in Oleogel or Emulsion Gel as Animal Fat Replacer. *Foods* **2020**, *9*, 830. [CrossRef] [PubMed]
8. Delgado-Pando, G.; Ekonomou, S.I.; Stratakos, A.C.; Pintado, T. Clean Label Alternatives in Meat Products. *Foods* **2021**, *10*, 1615. [CrossRef] [PubMed]
9. Tomović, V.; Šojić, B.; Savanović, J.; Kocić-Tanackov, S.; Pavlić, B.; Jokanović, M.; Đorđević, V.; Parunović, N.; Martinović, A.; Vujadinović, D. New Formulation towards Healthier Meat Products: Juniperus Communis L. Essential Oil as Alternative for Sodium Nitrite in Dry Fermented Sausages. *Foods* **2020**, *9*, 1066. [CrossRef] [PubMed]
10. Goemaere, O.; Glorieux, S.; Govaert, M.; Steen, L.; Fraeye, I. Phosphate Elimination in Emulsified Meat Products: Impact of Protein-Based Ingredients on Quality Characteristics. *Foods* **2021**, *10*, 882. [CrossRef] [PubMed]
11. Schutte, S.; Marais, J.; Muller, M.; Hoffman, L.C. Replacement of Sodium Tripolyphosphate with Iota Carrageenan in the Formulation of Restructured Ostrich Ham. *Foods* **2021**, *10*, 535. [CrossRef] [PubMed]
12. Mancini, S.; Mattioli, S.; Nuvoloni, R.; Pedonese, F.; Dal Bosco, A.; Paci, G. Effects of Garlic Powder and Salt on Meat Quality and Microbial Loads of Rabbit Burgers. *Foods* **2020**, *9*, 1022. [CrossRef] [PubMed]
13. Yang, S.-H.; Suhandoko, A.A.; Chen, D. Impact of Nutritional Information on Consumers' Willingness to Pay for Meat Products in Traditional Wet Markets of Taiwan. *Foods* **2020**, *9*, 1086. [CrossRef] [PubMed]
14. Hrbek, V.; Zdenkova, K.; Jilkova, D.; Cermakova, E.; Jiru, M.; Demnerova, K.; Pulkrabova, J.; Hajslova, J. Authentication of Meat and Meat Products Using Triacylglycerols Profiling and by DNA Analysis. *Foods* **2020**, *9*, 1269. [CrossRef] [PubMed]

Review

Towards More Sustainable Meat Products: Extenders as a Way of Reducing Meat Content

Tatiana Pintado [1] and **Gonzalo Delgado-Pando** [2,*]

[1] Institute of Food Science, Technology and Nutrition (CSIC), José Antonio Novais 10, 28040 Madrid, Spain; tatianap@ictan.csic.es

[2] Teagasc Ashtown Food Research Centre, Dublin, D15 DY05, Ireland

*   Correspondence: g.delgado.pando@gmail.com

Received: 3 July 2020; Accepted: 31 July 2020; Published: 3 August 2020

Abstract: The low efficiency of animal protein (meat products) production is one of the main concerns for sustainable food production. However, meat provides high-quality protein among other compounds such as minerals or vitamins. The use of meat extenders, non-meat substances with high protein content, to partially replace meat, offers interesting opportunities towards the reformulation of healthier and more sustainable meat products. The objective of this review is to give a general point of view on what type of compounds are used as meat extenders and how they affect the physicochemical and sensory properties of reformulated products. Plant-based ingredients (pulses, cereals, tubers and fruits) have been widely used to replace up to 50% of meat. Mushrooms allow for higher proportions of meat substitution, with adequate results in reduced-sodium reformulated products. Insects and by-products from the food industry are novel approaches that present an opportunity to develop more sustainable meat products. In general, the use of meat extenders improves the yield of the products, with slight sensory modifications. These multiple possibilities make meat extenders' use the most viable and interesting approach towards the production of healthier meat products with less environmental impact.

Keywords: meat extenders; meat products; meat substitutes; sustainability; plant-based proteins; insects; by-products; pulses; mushrooms

---

1. Introduction

In 2015, all United Nation Member States adopted "The 2030 Agenda for Sustainable Development" [1]. In this agenda, the countries agreed to 17 Sustainable Development Goals (SDG) to be achieved by the end of 2030. Sustainable food production is one of the main pillars of the document, where foods needs to be sufficient, safe, affordable and nutritious, as well as part of a sustainable production system. The world population growth and industrial development are causing an expansion of food production and an increased demand for animal protein [2]. One of the main concerns is the low efficiency of animal protein production. It is estimated that 7 kg of food from plant origin (animal feed) yields 1 kg of milk or meat for human consumption [3]. In addition, animal production is believed to use around 30% of the global land surface, contributing to deforestation and the loss of biodiversity [4]. However, the environmental impact of livestock production goes further than biodiversity loss: important greenhouse gas emissions, vast use of fertilisers and the deterioration of water quality due to effluents [4,5]. Westhoek et al. [6] estimated that "halving the consumption of meat, dairy products and eggs in the European Union would achieve a 40% reduction in nitrogen emissions, 25–40% reduction in greenhouse gas emissions and 23% per capita less use of cropland for food production". However, meat represents an important source of energy, high-quality protein and micronutrients such as iron, zinc, selenium, vitamin B12 and vitamin D [5,7]. Meat and meat products currently provide one-sixth of the total energy intake of a European adult and widely contribute to total protein, vitamin D and iron consumptions up to 40%, 30% and 23%, respectively [7]. Hence, meat and meat products should not be disregarded in the diet, as they contribute to the avoidance of essential nutrient deficiencies and can also protect against malnutrition in countries where access to other types of highly nutritious products is limited [5,8]. Deficiencies of iron and vitamin D are of high prevalence around the world [9,10]. A suboptimal vitamin B12 status occurs in 30–60% of the population, mainly in less-developed countries [11]. A recent study by Vatanparast et al. [12] found that decreasing by 50% the red and processed meat consumption and increasing by 100% the consumption of plant-based alternatives in Canadian individuals improved the overall nutritional diet value but adversely affected the intake of protein, zinc and vitamin B12. However, not only undeveloped or developing countries are affected. Rippin et al. [8] detected deficiencies in these micronutrients for certain segments of the European population. There is no unique food alternative to meat or meat products with similar nutritional profiles, and even a combination of several foods does not assure the same nutritional intake. Vitamin B12 is only present in foods of animal origin, which makes people following vegan and vegetarian diets in need of supplementations to achieve the dietary reference intake (DRI) for this micronutrient [13]. Furthermore, non-meat foods contain only 20–60% the protein density of that of the meat, and the digestibility and bioavailability of some micronutrients from these sources are known to be lower [14]. Even though highly desirable, a vast improvement of meat production efficiency and sustainability in the near future is not likely. Current strategies should focus on limiting the environmental impact of our diets without risking nutrition deficiencies.

Meat products are inherent to food culture and are widely consumed all around the world. Imamura et al. [15] estimated a global consumption of processed meat from 3.9 g/day (first quintile) to 34 g/day (fifth quintile). Even though their consumption has been linked with the burden of chronic diseases like coronary heart disease, type 2 diabetes and certain types of cancer [16,17], a number of gaps still exist (such as the underlying mechanisms of cancer development and the role of cooking) that could offer room for mitigation during their processing [16,18]. Versatility, more attractive products, waste reduction opportunities and higher shelf lives are some core characteristics that differentiate meat products from fresh meat. Therefore, the reformulation of meat products to produce healthier and more sustainable versions seems like a robust strategy in-line with the SDG. The research on the development of healthier meat products started in the 1990s but still comprises a big proportion of the current research in this field. Two strategies are primarily followed: the reduction of harmful components to appropriate amounts and the incorporation of potentially health-enhancing ingredients [19]. The former is focused on the reduction of harmful saturated fatty acids [20], salt [21], cholesterol [22] and additives such as nitrite [23] or phosphates [24], whereas the latter studies the incorporation of the so-called "functional ingredients", mainly from plant origin, that provide healthier characteristics to the product [25–27].

In the last decade, meat substitutes or analogues have received much interest as plant-based similar-in-properties alternatives to conventional meat products [28]. However, most of these analogues are produced under heavy processing manufacture and, thus, limiting the environmental sustainability gain and losing the healthier prerogative that they were originally based on. Meat reduction arises as a more meaningful alternative to a complete elimination of meat from the diet and, sometimes, a more sustainable option than meat substitutes [29,30]. The integration of plant-based ingredients into meat dishes has been proven as a successful and consumer-accepted strategy [29] and has opened the way to a different approach towards the reformulation of healthier and more sustainable meat products: meat substitutions with plant-based ingredients. Although originally devised to reduce costs, the use of meat extenders presents an opportunity to reduce the meat content while incorporating some healthier ingredients to the meat product. Meat extenders are non-meat substances with high protein contents that can also modify some of the product's properties, such as water-holding capacity (WHC), texture, palatability and appearance [31].

In this review, we aim to evaluate the use of extenders as meat substitutes and how they affect the physicochemical and sensory properties of the meat products. The review has been structured in different sections, where meat extenders have been grouped based on their origin. In addition, a section discussing the consumer perspective about the acceptance of novel and more sustainable meat products has also been included.

2. Meat Extenders

2.1. Pulses as Meat Extenders

"Pulses are edible dry seeds of plants belonging to the *Leguminosae* family" [32]. Pulses' protein contents range from the 18.4% of the Bambara bean to the 34.1% of the lupin. They not only contain a great amount of protein, but they also present the highest protein digestibility score among the plant-origin proteins. In addition, pulses are also a rich source of micronutrients such as iron, zinc and B-vitamins. Even though iron from plant origin is less absorbed by the human tract, when combined with meat, the absorption increases substantially [32]. Therefore, from a nutritional point of view, pulses are a great candidate as meat substitutes, providing high quantities of proteins and similar micronutrients to the ones in meat. Hence, several studies have analysed their role as meat extenders in the past fifteen years (Table 1). Even though soybean is not a pulse as per the definition, it is a legume, and for this reason, a study with texturised soy granules as a meat extender has also been included in Table 1.

Table 1. Use of pulses as meat substitutes (extenders).

Ingredient Used	Meat Product	Meat Substitution (%)	Effect on Properties	References
Green pea flour emulsion	Pork patties	10.1–44.6	Higher yield, lower redness values, increased yellowness at higher substitution levels, increased hardness at all levels	[33]
Chickpea flour emulsion	Pork patties	10.1–44.6	Higher yield, lower redness values, increased yellowness at higher substitution levels, increased hardness at lower substitution levels	[33]
Lentil flour emulsion	Pork patties	10.1–44.6	Higher yield, increased hardness at lower substitution levels	[33]
Bean flour emulsion	Pork patties	10.1–44.6	Higher yield, lower redness values, increased hardness at lower substitution levels	[33]
Texturised soy granules	Dehydrated chicken ring meat	5	Lower meat flavour, lower yellowness hue and chroma	[34]
Cowpea	Chicken seekh kababs	15	Sensory properties not affected, lower TBARS * values, higher microbial counts	[35]
Green gram	Chicken seekh kababs	15	Sensory properties not affected, lower TBARS values, higher microbial counts	[35]
Black bean	Chicken seekh kababs	10	Sensory properties not affected, higher microbial counts	[35]
23 different pulses	Beef patties	35–50	Higher yield (highest for yellow split bean), colour values not affected with most of pulses, texture not different from control on black-eyed pea, baby lima bean, purple hull pea and crowder pea patties	[36]
23 different pulses	Pork sausage patties	35–50	Higher yield (highest for small red), colour values not affected with most of pulses, texture not different from control on black bean, lentil, black-eyed pea, green split pea and baby lima bean	[36]
Bengal gram flour	Quail meat rolls	3–9	Higher yield, sensory not affected at 3–6%, lower protein	[37]
Bengal gram flour/Pea flour	Chicken patties	5–10	Pea flour higher yield, stability and sensory scores than gram flour at higher levels of substitution	[38]
Blackeye bean flour	Meatballs	10	Higher yield and overall palatability, lower yellowness and tougher compared with Rusk	[39]
Chickpea flour	Meatballs	10	Higher yield, lower yellowness and tougher compared with Rusk	[39]
Lentil flour	Meatballs	10	Higher yield, lightness and overall palatability, lower yellowness and tougher compared with Rusk	[39]

* TBARS: thiobarbituric acid reactive substances.

Pulses in different forms have been used in meat product reformulations as binders or to increase their nutritional and healthier properties [40–43]. The starch, fibre and protein contents make pulses great binders, as they can form complex gel networks with meat proteins. These networks can trap the water and other compounds, forming stronger bonds between them and, thus, helping to achieve a higher retention in the meat matrix during processing [44]. Aslinah et al. [45] used adzuki bean flour as a fat and corn flour replacer in meatballs due to its water-holding capacities. Soy protein has been also widely used when developing reduced fat meat products due to its gelling properties [46,47]. The type and quantity of the pulse utilised, as well as the type of product, will determine the overall effect on the product stability in terms of WHC. In this regard, Nagamallika et al. [38] used two different pulses, Bengal gram flour and pea flour, to replace the meat content in chicken patties at two levels:

5% and 10%, yielding a higher stability at the higher level of substitution. Pea flour proved to yield a significantly lower cooking loss (9.5% vs. 30.3%) and higher emulsion stability (4.2% vs. 2.2%) and WHC (64.3% vs. 30.8%) when used at the 10% level compared to the Bengal gram flour patties. Nonetheless, at the 5% level of substitution, the emulsion stability and cooking loss were significantly lower for the patties with gram flour, whereas the WHC was significantly higher (19.8% vs. 47.8%). Yadav and Yadava [37] observed an increase in the yield and emulsion stability with an increasing level of substitution (3–9%) with gram flour in quail meat rolls. In a comprehensive analysis with 23 different type of pulses as meat substitutes in chicken and beef patties, this variation among the pulse types was also found [36]. The authors observed that cooking losses in beef patties ranged from 8.0% in the yellow split pea patty samples to 15.1% in the patties with pink beans, whereas the control had a 37.9% cooking loss. In the case of the pork patties, the control had a cooking loss of 22.9%, and the substitution improved the yield in all cases, with the cooking loss ranging from 5.6% to 10.7%, the lowest being the patties with small red beans and the highest for the ones with speckled butter beans. In an interesting study by Serdaroğlu et al. [39], three different pulse flours (lentil, chickpea and blackeye bean) were used in low-fat meatballs, replacing not the meat but the rusk used in the control samples. Lower cooking losses and an increased WHC were found in the meatballs reformulated with pulses. This gives an idea that not only the starch (on higher quantities in the rusk) but the protein and fibre contents (much higher in the pulses) have big impacts on the water-holding capacities of meat products. The pulses with higher protein contents, blackeye beans and lentils, gave significantly higher yields to the meatballs. The substitution percentage also determined the effect of the extender on the product yield. Argel, Ranalli et al. [33] evaluated four different pulses (chickpea, lentil, green pea and bean) as extenders in pork patties with six different levels of meat contents. At the lower level of substitution (10.1%), the patties manufactured with bean flour had the highest cooking yields, followed by lentil and green pea, significantly different from the ones with chickpea flour. However, at the highest level of meat substitution (44.6%), the bean flour had the lowest cooking yield among the four pulses; the reformulated patties had higher cooking yields at all substitution levels than a commercial one. The different compositions of these flours might explain this, as the chickpea flour had the lowest protein and fibre contents but the highest fat levels. On the other hand, no significant yield changes were observed in dehydrated chicken ring meat using soy as a meat extender at a 5% level of substitution [34].

Another property closely related to the water-holding properties of meat products is the texture [48]. Texture is usually evaluated using a texturometer by means of a texture profile analysis (TPA) or a measurement of the hardness with the shear force value. A TPA analysis of pork patties substituted with pulse flour showed that the hardness and chewiness increased when compared to the control and commercial ones, but that this difference disappeared when the substitution level was above 35% and added water was at its highest level [33]. In the same study, the authors found that cohesiveness was lower in all the extended pork patties and that the bean flower had the lowest hardness among the four pulses studied. In addition to the level of substitution, the type of pulse will also affect the textural properties. An increase in hardness was observed when the rusk used in low-fat meatballs was replaced (10%) by the flour of three different pulses, being the meatballs with chickpea flour the ones with significantly higher hardness, followed by black bean and lentil flour [39]. Out of 23 varieties of pulses, only four pulses did not affect the shear force value when substituting 50% of the meat in beef patties and five pulses when pork patties were prepared instead [36]. The overall mean shear force was lower for the majority of the pulses used. As the substitution values for this study ranged between 35% and 50%, this agrees with the aforementioned results. The type of meat product will also affect how the substitution alters the textural parameters, as the networks formed in the matrix will be different depending on the degree of comminution and the quantity of fat, water and proteins. A great example can be observed in the 23-varieties study where the beef patties with green northern beans as the extender had the lowest hardness value, whereas in pork sausage patties, the hardness was one of the highest for this same pulse.

Colour is perhaps the attribute most difficult to mask when substituting meat with pulses, as not many of them have similar colour to meat. In addition, cooking of the meat product can also affect the colour changes generated by the use of pulses as extenders. Any colour comparison should be mainly addressed on the product at the state it is going to be purchased, although extra analyses can also be taken into consideration. Lightness was not affected by pulses as extenders in pork and beef patties with varying levels of substitution [33,36]. However, the use of some pulses as extenders in a variety of meat products have significantly affected the redness and yellowness values [33,34,36,39].

The product appearance is the first attribute the consumer observes before purchasing the product, and even though the instrumental colour measurements are correlated with the appearance, the consumer might not be able to detect the differences as the instrument does, or they could like better the colour change. In the same way, texture results from instrumental measurements and those from sensory panels differ substantially. Serdaroğlu et al. [39] found that general appearance scores for meatballs with pulses as extenders did not significantly differ when compared to the rusk, but instrumental colour values for the same products showed significant changes in the yellowness value. In the same study, the meatballs with chickpea flour had the harder texture value, and it was scored lower by the panellists, but the one with the highest score was not the one with the softer texture but the second-hardest. These sensory analyses were done by trained panellists on a nine-point hedonic scale, and even though this practise should be avoided—hedonic analyses should always be carried out by non-trained panellists—it can give an idea of the sensorial properties of the product. When black beans, green grams and cowpeas were used as extenders in chicken seekh kababs, the sensory properties remained unaltered throughout storage, with no significant differences among the pulse varieties [35]. Yadav and Yadava [37] found that gram flour substituting meat in quail meat rolls at levels 3% and 6% did not affect the sensory properties, although, at 9%, they observed a significant decrease in the colour and flavour scores by the panellists. The use of texturised soy granules in dehydrated chicken meat only affected the meat flavour intensity, according to a sensory panel [34]. Argel et al. [33] found that pork patties where the meat was substituted (37%) with chickpea, lentil, green pea and bean flour emulsions had acceptable sensory properties, with no significant differences among the pulse types.

The use of pulses as meat extenders has been researched mainly in patties and similar meat products, but no work on comminuted ones, although some studies with pulses as binders can be found for these types of products [40–42]. In general, pulses seem to be an adequate ingredient to be used as a meat replacer, as they have a very similar nutritional composition and do not affect extremely the physicochemical properties of the finished product. Unfortunately, a limitation from pulses and legumes as extenders can be found on the allergenic potential of some proteins contained in soybean and peanuts that would restrict population access to these products (people with allergies) and would need proper labelling [49].

2.2. Other Meat Extenders of Plant Origin: Cereals, Tubers and Fruits

Other plants such as cereals, tubers and fruits have also been used in meat product formulations. The main reason of using these food products as ingredients in meat products has been the healthy properties they possess: high fibre contents, vitamins and minerals, important proportions of phytochemicals and antioxidants and void of cholesterol, among others [50,51]. Apart from their nutritional properties, some of these plants also have good functional and technological properties, such as improved water-binding and yield properties, fat emulsifiers, increased flavour, etc. [52]. Even though their main usage has been for the development of functional meat products [53–56], there has been also some research about the use of these ingredients as meat substitutes/extenders. Research about the use of cereals, tubers and fruits as meat extenders in the last thirteen years is summarised in Table 2.

Table 2. Use of cereals, tubers and fruits as meat substitutes (extenders).

Ingredient(s) Used	Meat Product	Meat Substitution (%)	Effect on Properties	References
Blend of potato, soy protein, oat meal, barley flour, whey protein concentrate	Restructured spent hen	23.5–25.5	Sensory properties not affected and higher yield. Softer texture and increased chroma values	[57]
Dried pumpkin pulp and seed	Beef patties	2.8–6.9	Increased water-holding capacity (WHC), lower redness, no changes in texture and sensory	[58]
Olive cake powder	Beef patties	2.6–7.9	Lower sensory scores, higher protein and yield, increased yellowness	[59]
Different blends of fibre, carrageenan and pork rind	Beef and chicken sausage	35–50	Decreased hardness, similar flavour to control but loss of general sensory quality, with the exception of a few blends	[60]
Rice flour	Dehydrated chicken ring meat	10	Sensory properties not affected, higher yield, lower iron	[34]
Barnyard millet flour	Dehydrated chicken ring meat	10	Higher yield, multiplied iron content, lower meat flavour	[34]
Blend of lentil flour, sorghum, potato and water chestnut flour	Restructured chicken meat blocks	15	Higher yield, similar texture properties, lower sensory scores	[61]
Plum puree	Beef patties	5.1–15.4	10% substitution best sensory results with no detrimental effects on physicochemical properties	[52]
Corn flour	Quail meat rolls	3–9	Higher yield, sensory not affected at 6%, lower protein	[37]
Several cereals, tubers and plants	Meat cubes	10	Pearl millet, carrot and cabbage showed highest-ranking scores in sensory properties	[62]
Melon flour from kernels	Beef sausages	10–40	Higher yield, no changes in sensory attributes up to 20% substitution. Lower TBARS values.	[63,64]
Sorghum flour	Chicken patties	5	Lower TBARS at end of storage, sensory properties not significantly different	[65]
Barley flour	Chicken patties	10	Lower TBARS at end of storage, sensory properties not significantly different	[65]
Pressed rice flour	Chicken patties	5	Lower TBARS at end of storage, sensory properties not significantly different	[65]

Fruits and their by-products have been used as ingredients in meat products to improve the shelf lives and provide meat with antioxidants, fibre and other phytochemicals [66]. However, their role as meat extenders is yet to be explored, with only a few studies in the scientific literature. Melon flour, from defatted melon kernels, was used to substitute meat in beef sausages at levels 10–40% [63]. The authors found an increased yield, WHC and better sensory properties with the increasing levels of substitution. No significant differences with control on the overall acceptability and appearance were found at the 20% substitution level. The same authors found that, after two and four weeks of storage, the thiobarbituric acid reactive substance (TBARS) values were significantly lower for the sausages with substitution levels above 20% [64]. Low-fat beef patties where the meat was substituted with plum puree (5–15%) showed an increase in the cooking yield and redness of the patties but a decrease in WHC, lightness and yellowness [52]. The TBARS values of the extended patties with plum were lower at the end of the storage period, irrespective of the substitution level. In addition, the sensory properties were improved at the 10% and 15% levels of substitution, being the former the one with the best scores in overall acceptability, flavour, texture and juiciness. An increased cooking yield has been also found in beef patties extended with olive cake powder at levels 2.6–7.9% [59]. The olive cake powder also increased the amount of polyphenols and the antioxidant activity of the patties, but the instrumental colour was also affected, with a decrease of the lightness and an increase of the yellowness with increasing levels of substitution. The sensory properties were negatively affected, with significantly lower values at the higher levels of substitution. When using plum puree as an extender in beef patties (2.8–6.9% substitutions), the cooking yield and sensory attributes remained unaltered, but the WHC increased with the increasing levels, the redness dropped and the hardness increased [58]. All of these studies proved that fruits can be used as meat extenders, but further research is needed on different meat products (not only patties) and with different types of fruits and substitution levels.

Cereals are crops of the family *Gramineae*, which comprises nine species: corn, barley, millet, oat, rice, rye, sorghum, triticale and wheat. They are an important source of proteins (ranging from 7–18% dry matter) and vitamins (B group and E) [67]. Chicken patties where the meat was substituted by sorghum (5%), pressed rice (5%) and barley flour (10%) showed a significant decrease on the extract release volume and lower TBARS values at the end of storage, with no significant impact on the sensory properties [65]. Mishra et al. [34] found that rice flour at a 10% substitution level in dehydrated chicken ring meat did not affect the sensory properties, whereas a 5% meat substitution with barnyard millet flour decreased the sensory perception of the meat flavour intensity while not affecting any of the other sensory attributes. The same authors also observed that the yield was improved by these two extenders without affecting the instrumental colour. Both cereals also significantly reduced the cholesterol content and increased the manganese; the millet chicken meats had also a 10-fold increase of their iron contents, while the meats with rice had lower iron contents when compared to the control. Corn flour used as a meat extender in quail meat rolls increased the yield and emulsion stability with the increasing level of substitutions (3–9%) [37]. However, the sensory perceptions of colour and flavour were impacted on the rolls where meat was substituted at a 9% level but remained unaffected at the lower substitution levels. A screening of a combination of five different cereals and six plants and tubers as meat extenders (10%) in sheep meat cubes was performed using a Plackett-Burman design [62]. The authors found that millets, carrots and cabbages gave the cubes the most desirable sensory characteristics and that further research with these ingredients should be guaranteed. Malav et al. [61] analysed the use of a blend of sorghum with potato, lentil and water chestnut flours as extenders (15%) in restructured chicken. The blend of extenders exerted higher yields and similar texture attributes but lower sensory scores. Another study where cereals were combined with other ingredients as meat extenders in the same type of product was done by Gupta and Sharma [57]. Wheat, oat and barley were blended with potato, whey and texturised soy protein in three different combinations that were compared to a control. The three blends increased the cooking yield and decreased the hardness, but only one of them did not differ in the overall acceptability of the product; the other two had lower scores for flavour. With regards to the instrumental colour, the redness was not affected, but the yellowness increased in all the reformulated samples. However, the sensory appearance was higher for the sample with the highest chroma value. Cereals proved to be important and successful ingredients when used as meat extenders, but their behaviours in meat products different than restructured meat and chicken are still unknown. It is also important to highlight that cereals containing gluten (wheat, rye, barley and oats) have allergenic potentials that must be declared in the labelling.

2.3. By-Products of the Food Industry as Meat Extenders

The food industry (from vegetables or animal products) produces high amounts of residues and/or by-products that are edible compounds with high percentages of proteins and/or fibres. In today's global scenario, the use of these compounds—in many cases, undervalued—could be an opportunity to replace meat for manufacturing more sustainable meat products [68]. Furthermore, many of these residues are a source of polyphenols, organic acids and fatty acids, among others, which are underutilised, providing added value to the products in which they are included [69]. In this regard, some studies have assayed the use of residues from the agri-food industry as meat extenders (Table 3).

Table 3. Use of by-products of the food industry as meat substitutes (extenders).

Ingredient Used	Meat Product	Meat Substitution (%)	Effect on Properties	References
Okara	Beef patties	7.5–37.5	Cholesterol reduces for raw (6–56%) and cooked (9–42%). Higher cooking yield, pH, lightness and yellowness. Sensory attributes valued negatively with 37.5% of meat replacements.	[70]
Okara	Beef burger	5–25	Increase lipid and moisture contents. Higher luminosity and dimmed during storage. Changes in the brown colour	[71]
Okara	Beef sausages	10–40	Carbohydrate, ash and fibre contents increased, while moisture, fat and protein contents decreased. Improved WHC but decreased textural parameters	[72]
Okara	Beef burgers	6 and 12	Sixty percent less calories. Increased hardness but decreased cohesiveness, chewiness and springiness. Lower sensory scores with 12% of substitutions	[73]
Okara	Pork meat gels	3–27	Higher cooking yield. Increased in lightness, hardness, chewiness and breaking force of gels but decreased in cohesiveness. Higher storage (G′) and loss (G″) modulus by heating.	[74]

Table 3. *Cont.*

Ingredient Used	Meat Product	Meat Substitution (%)	Effect on Properties	References
Cashew apple residue powder	Hamburgers	7.1–14.3	Reduced 35% of the lipid content and increased of up to 7.6% of the fibre. Lower humidity but sensorial acceptable with 7.1 and 10.7% of meat replacements	[75]
Apple pomace	Buffalo emulsion-based sausage	2–8	Increased fibre content and improve cooking yield and emulsion stability	[76]
Enoki (*Flammulina velutipes*) mushroom stem waste powder	Goat nuggets	2–6	Increased dietary fibre, ash and phenolics compounds. Improved the emulsion stability, WHC, oxidative stability and shelf-life. Slight hardness decrease. No negative effects in the sensory attributes.	[77]
Textured whey proteins (TWP)	Beef Patties	0–50	Higher cooking yields. Patties containing up to 40% of hydrated TWP obtained similar sensory evaluations than all-beef patties	[78]
Protein concentrates from porcine blood	Irish breakfast sausage	15 and 30	Higher protein contents in raw samples. Decreased fat levels in cooked samples. Higher cooking yield and WHC for 15% of replacements. Decreased hardness and chewiness with 30% of meat substitutions	[79]
Protein concentrates from pork hams exudates	Irish breakfast sausage	15 and 30	Lower fat contents in raw samples. Higher protein contents with 30% of meat replacements. Decreased WHC. Decreased hardness and chewiness values with 30% of replacements	[79]
Protein concentrate from residues of edible fat processing	Irish breakfast sausage	15 and 30	Decreased fat contents. Similar WHC and cooking yield. Decrease hardness and chewiness values with 30% of meat replacements	[79]
Protein concentrate from brine solutions	Irish breakfast sausage	15 and 30	Higher protein contents. Higher cooking losses. Decreased redness in raw samples but increased when they are cooked	[79]

Okara is a by-product with low commercial value that is generated in massive volumes (about two to three tons for each ton of soybean processed) during the manufacturing of soymilk and tofu [80]. This component presents solvent-binding properties, making it an ideal low-cost ingredient to increase yields in meat products (Table 3). Moreover, okara contains valuable components such as fibre and high-quality protein (40% on a dry weight basis) due to the presence of a good essential amino acids profile and its digestibility [80]. In that sense, okara has been applied to extend meat contents both in fresh and cooked emulsion-based meat products (Table 3). In beef burgers, lean meat has been replaced by wet okara in different quantities, up to 37.5% (Table 3). In general, it was observed an increase of the moisture content and a decrease of the protein level in the reformulated burgers [70,71,73]. Moreover, Tie Su et al. [73] obtained beef burgers with 60% less calories than commercial products when 12% of okara was added. The use of okara as a meat extender improved the cooking yields of the samples [70]. Tie Su et al. [73] noticed that, as the percentage of okara increases, an increase in hardness occurs, while Strada de Oliveira et al. [71] observed an improvement in tenderness with respect to the control samples. The effect of wet okara on the sensory properties was significant, and higher scores for overall acceptability were recorded for products with approximately 20% added okara [70,73]. In cooked emulsion-based sausages, contrary to those observed in fresh meat products, the moisture content was increased with an okara addition [72]. Water and oil-holding capacities were improved as a consequence of okara additions, and in that sense, the cooking yield was improved [72]. For textural properties in cooked emulsion-based products, the incorporation of okara presented contradictory behaviours. The same authors observed an increase for the texture parameters with up to 40% of okara added to beef sausages, while a decrease of the hardness, chewiness and breaking force occurred when okara was incorporated in pork meat batters [74]. On the other hand, the overall acceptability of the samples decreased with the okara incorporation [72].

The residue obtained from the production of cashew apple juice (skin and the husk) has been used to extend beef meat in the formulation of hamburgers. With increasing the concentrations of the residues, the samples showed lower moisture, protein and lipid levels, while their fibre contents were higher. Hamburgers with improved yields and similar flavours than the control were observed with additions up to 10.70% of the residue [75]. Apple pomace powder was employed (2–8%) to replace buffalo meat in emulsion-based sausages by increasing the fibre contents. Moreover, the cooking yield and emulsion stability got enhanced [76].

Mushroom by-products are described as a good source of protein, dietary fibre and phenolic components, with the potential to be strong antioxidants [77]. In that sense, the use of different amounts (2%, 4% and 6%) of enoki (*Flammulina velutipes*) mushroom stem wastes as meat extenders in nuggets enhanced their composition (Table 3). The inclusion of meat extenders improved the oxidative stability and shelf-lives of treated nuggets without impacting the sensory attributes of reformulated nuggets.

Whey is a by-product of the dairy industry, which is generated in massive quantities during the manufacture of cheeses, yogurts and other dairy products [81]. Its great content of high biological value proteins offers interesting possibilities to be used during the processing and manufacturing of meat products. Hale et al. [78] extruded a dry whey protein concentrate (80% protein) to obtain an ingredient that they used to substitute from 0% to 50% of beef in the elaboration of patties. Samples containing up to 40% of whey extrudes were as acceptable to a consumer panel as all-beef patties. Moreover, the cooking yield was improved, and these patties suffered less diameter reductions and less water and fat losses by the cooking process.

The meat industry also generates compounds that hold strong potential for higher-value techno-functional applications due to their high-quality protein contents (Table 3). However, their use as meat extenders in meat products is very limited. For example, Álvarez et al. [79] extracted protein concentrates from different residues of the meat industry to be used as meat replacers in the elaboration of an Irish breakfast-type sausage: blood plasma, exudates generated from ham elaboration, brine solutions and water produced during edible fat processing. Two levels were assayed: 15% and 30% (Table 3). Regarding the composition, raw products showed lower fat contents and higher protein levels than the control ones. However, the technological properties were conditioned by the type of protein used and the level of meat substitution. In general, for all types of protein, the 15% meat replacement offered products with a better overall final product quality. Regarding the type of protein, plasma proteins at both replacement levels had the most positive effect on the technological properties, whereas the use of protein concentrates from brine solutions to substitute meat resulted in sausages with lower fat and water-binding properties and redness values (Table 3).

Based on the foregoing, it seems that the use of residues or by-products from the food industry as meat replacers endows products with compounds that offer positive effects on health without being a detriment to their technological properties. In addition, this strategy offers multiple advantages to maintaining a more sustainable world by both using industrial residues and reducing meat productions.

2.4. Mushrooms as Meat Extenders

Fungi have been used in human foods for a long time. Of more than 14.000 species of mushrooms, at least 2000 of them have various degrees of edibility [82]. Mycoprotein is fungal in origin, and it is utilised as a high-protein, low-fat, health-promoting food ingredient [49]. Mycoproteins could be obtained by the continuous-flow fermentation of *F. venenatum* on a glucose substrate, and it is used to elaborate meat analogues. However, in the development of more sustainable meat products, some studies were carried out adding mushrooms directly to meat products (Table 4), replacing different proportions of meat proteins by mycoproteins.

Table 4. Use of mushrooms as meat substitutes (extenders).

Ingredient Used	Meat Product	Meat Substitution (%)	Effect on Properties	References
Mushroom (*Agaricus bisporus*)	Beef Patties	10–50	Allows reduced sodium patties (1.5% NaCl). Increasing mushroom extender level; samples perform similar to an all-meat control in yield, lightness and redness; increase moisture and yellowness and decrease mechanical properties, sodium and fat contents.	[83]
White mushroom (*Agaricus bisporus*)	Beef taco blend	50 and 80	Enhancement of overall flavour and mitigated salt reduction.	[84]
White mushroom (*Agaricus bisporus*)	Carne Asada	50	Allows reduced sodium samples. No alterations on the overall flavour strength.	[84]
White jelly mushroom (*Tremella fuciformis*)	Pork Patties	10–30	Improve cooking yield and increase lightness and yellowness. Ten percent substitution improved the sensory acceptance, while 30% decreased the approval of patties.	[85]

Table 4. *Cont.*

Ingredient Used	Meat Product	Meat Substitution (%)	Effect on Properties	References
Lentinula edodes	Pork sausage	25–100	Increased moisture, fibre, essential amino acids and total phenolic content. Higher cooking yield and antioxidant activity. Decreased protein, energy ash, pH and texture parameters. Twenty-five percent substitutions are the best sensory acceptance.	[86]
Pleurotus sajor-caju	Beef patties	25 and 50	Increased insoluble fibre content, mainly β-glucan. Decreased fat retention during the cooking process. Best cooking yield with 25% of substitutions. No differences in sensory attributes.	[87]
Pleurotus sajorcaju	Chicken frankfurters	2–6	Decreased fat content. Enhancement of dietary fibres up to 6.20% and β-glucan up to 14.30%. Hardness was decreased.	[88]

Mushrooms are a good source of dietary fibre, where approximately one-third is chitin and two-thirds β-1,3 glucan and 1,6 glucan. Chitin is a modified polysaccharide with an analogous structure to cellulose and considered an insoluble fibre with potential prebiotic properties in gut microbiota [89]. In addition, mushrooms are also a source of proteins; essential amino acids; vitamins (such as thiamine, riboflavin and niacin) and essential minerals (such as Ca, P, Mg, Cu, Se and Zn). Moreover, these products are low in calories, fat and sodium [90]. In that sense, the application of mushrooms as meat extenders could also be an opportunity to improve the presence of health-promoting bioactive components in meat products.

White mushrooms (*Agaricus bisporus*), the most cultivated edible mushroom, poses a dual opportunity as a meat extender by reducing the meat content while also imparting flavours that can complemented and enhance the saltiness perception [82]. Wong et al. [83] compared two meat extenders, a traditional one (textured soy) and *Agaricus bisporus*, to replace 10% to 50% of meat in the development of beef patties (Table 4). Increasing textured soy improved the cooking yield of patties but did not affect their colour or textural properties. However, increasing the level of mushroom extenders performed statistically similar to an all-meat control in yield, lightness and redness, while decreasing the mechanical properties. Additionally, meat extension using mushrooms yielded sensory liking scores more similar to the all-meat formulations than textured soy in reduced sodium samples. In the same way, white mushrooms were used to replace meat in two meat based dishes, carne asada and beef taco blends, whose sodium contents were reduced [84]. In carne asada, the beef substitution (50%) with mushrooms did not alter the overall flavour strength of the dish, but the replacement of 50% or 80% of meat in the beef taco blend enhanced its overall flavour. The ability of mushrooms to mitigate sodium reductions in terms of the overall flavour has been attributed to the fact that mushrooms contain umami tastants [82]. White jelly mushroom (*Tremella fuciformis*) is another type of edible mushroom that has been used as a meat extender in pork meat patties (Table 4) [85]. In this case, higher mushroom quantities (30%) decreased the sensory acceptance of patties because of the mushroom flavour. However, patties containing 10% of mushrooms improved significantly the sensory affections due to their oil-holding capacities. Furthermore, this ability, along with its capacity to bind water, allowed improving the cooking yield of patties formulated with white jelly mushroom [85]. In pork sausages, *Lentinula edodes* has been used as meat extender to replace 25%, 50% and 100% of the meat (Table 4). Regarding sensory acceptability, all samples were satisfactory. Although those with 25% of substitutions showed the highest scores for sensory attributes. From a technological point of view, the presence of mushrooms improves the oxidation stability and the cooking yield of sausages [86].

The use of *Pleurotus sajor-caju* as a meat extender (25% and 50% of meat substitutions) in beef patties and in lower proportions (2% to 6%) to replace chicken meat in the formulation of frankfurters produced an increase of their fibre contents. It should be noted that this fibre was insoluble mainly based on β-glucans (0.78 g/100 g in the case of patties and 1.43 g/100 g in frankfurters) [87,88]. As with other mushrooms, the use of *Pleurotus sajor-caju* as a meat replacer improved the cooking yield of the products. The hardness values of the reformulated products were lower. However, the sensory analysis scores indicated that the products were accepted by the panellists [87,88].

Mushrooms seem to be an adequate ingredient to be utilised as a meat replacer. The use of mushrooms allows for the development of healthier meat products with higher fibre and less salt contents (as they have the potential to increase saltiness perceptions) without affecting much the physicochemical properties.

2.5. Insects as Meat Extenders

Entomophagy, or the practise of eating insects, is a long-time practise and an important nutritional source (high-quality protein, lipids, carbohydrates, mineral elements and certain vitamins) for many cultures, mainly located in Africa, Asia and Latin America [91]. More than 40 years ago, Meyer-Rochow [92] already suggested that insects could supplement traditional animal protein sources. Currently, there is a growing interest in edible insects as a novel source of protein due to their high contents, as well as their functionalities, which have been described similar to conventional proteins (included meat proteins) [91]. However, probably due to insect food neophobia in Western countries, there are only a few studies using insects as meat extenders, and the majority are from Eastern Asian countries (Table 5). With the aim to decrease this well-known food neophobia related to insects, Caparros Megido et al. [93] decided to test the level of sensory-liking of patties in which beef was replaced (53%) by mealworms, allowing them to hide insects and to present them in a familiar way. The authors concluded that insect integration into Western food culture could be feasible, as the taste and appearance of burgers were rated higher than neutral scores, positioning them between a fully meat burger and a fully vegetable burger.

Table 5. Use of insects as meat substitutes (extenders).

Ingredient Used	Meat Product	Meat Substitution (%)	Effect on Properties	References
Mealworm (*Tenebrio molitor* L.)	Pork patties	10–60	Improved cooking yield. Higher fat content. Decreased moisture and protein content. Lower lightness but higher force shear. No sensory characteristics affected	[94]
Mealworm larvae (*Tenebrio molitor*)	Burger patties	53	The appearance of insect-based burgers was preferred by men. In terms of overall liking, meat substitution by insects was better valuated than by legumes	[93]
Mealworm larvae (*Tenebrio molitor* L.)	Frankfurter	10–60	Decreased moisture and fat content while increased protein level. Decreased lightness and textural parameters. Greater replacement than 15%. decreased emulsion stability. Less sensory acceptance	[95]
Mealworm larvae (*Tenebrio molitor*)	Emulsion sausage	10	Increased protein and mineral contents but decreased moisture. Improved cooking yield. Samples with more lightness but with lower values for textural parameters.	[96]
Silkworm pupae (*Bombyx mori*)	Emulsion sausage	10	Increased protein and mineral contents but decreased moisture. Improved cooking yield. Samples more lightness but with lower values for textural parameters	[96]
House Cricket (*Acheta domesticus*)	Emulsion sausage	5 and 10	Increased protein and minerals (P, K and Mg), no negative impacts on cooking yield and textural properties	[97]

The incorporation of mealworms as meat replacers was also studied to evaluate their effects in the composition and technological properties of new products. Ju-Hye et al. [94] studied the effects of different replacement ratios (10% to 60%) of pork meat in the development of patties (Table 5). The addition of mealworms conditioned significantly the composition of the samples, decreased protein contents and increased fat levels. The cooking yield was improved with the presence of insects. There were no significant differences in the sensory characteristics of burgers, although the shear force was reduced and the lightness was increased with the replacement of meat by insects.

In emulsion-based meat products, mealworms (*Tenebrio molitor* L.) have been used to replace 10–60% of pork meat (Table 5). Reformulated samples had increased protein and fat contents when the meat was replaced at the 10% level [95,96]. However, Choi, Kim, Choi, Park, Sung, Jeon, Paik and Kim [95], who assayed higher levels of extended meat (up to 60%), observed that frankfurters with a higher meat replacement by mealworms increased the protein content but decreased the fat content approximately to 30% in respect to all-pork meat samples. Moreover, the incorporation of edible insects increased the mineral contents of emulsion sausages [96]. The cooking yield was improved with a substitution of meat of 10%; extended higher meat decreased the cooking yield [95,96]. Additionally, replacing pork meat with up to 10% mealworms successfully maintained the sensory quality of frankfurters.

Silkworm pupae (*Bombyx mori*) and the House cricket (*Acheta domesticus*) are two other types of edible insects used as meat extenders (Table 5). Kim, Setyabrata, Lee, Jones and Kim [96] added freeze-dried Silkworm pupae (*Bombyx mori*) to replace 10% of the pork meat in an emulsion-based meat product. They assayed three strategies to incorporate the insects: ground, defatted and acid-hydrolysed. The inclusion of insects had no impact on the protein solubility of emulsion sausages. The protein contents of sausages were increased for all the treatments; however, the fat contents only were increased when insects were ground. Additionally, the mineral content was improved when ground and defatted Silkworm pupae was incorporated [96]. The replacement of pork meat with house cricket flour within a 10% level could fortify the product with proteins and some micronutrients (phosphorus, potassium and magnesium) without a negative impact on the cooking yield and textural behaviours [97].

Edible insects possess the necessary physical properties to be used as an alternative nonmeat ingredient for incorporation within fresh or emulsified meat products, which could be further promoting to improve the image that the consumers have of them. Moreover, the addition of invisible insects in food preparations helps to reduce insect food neophobia [93].

3. Meat Products' Sustainability from the Consumer Perspective

As stated before, protein production has a large impact on the climate change, with proteins from meat being much less sustainable than plant-based proteins [98]. It seems logical to think that the daily choice of food has a high impact on the environment, and therefore, acting to change consumer preferences seems an appropriate strategy to reduce the negative impact that food production may have [99].

Some alternatives for meat products made entirely of vegetable components (e.g., tofu) can be already found in the supermarkets, although the market shares of these products are still very low compared to meat and meat products. The lower penetration of these products in households could be partially explained by the lack of texture and taste reported for some of them [100]. In addition, the heavy processing conditions to obtain these products and, in consequence, the multiple additives that they contain are sometimes neglected; besides, they can have a really high carbon footprint [28].

Complex external cues (perceived healthfulness, animal welfare, environmental impact and sustainability) are increasingly taken into account in our preference for meat [101]. However, despite a seemingly close match between the consumers' image of a sustainable, healthy and a plant-based diet [102], there is actually low consumer awareness of the environmental impact of meat production, as well as a low willingness to change meat consumption behaviours in terms of reducing or substituting meat in Europe. It is therefore relevant to determine the opportunities and barriers for consumers to adopt such alternative meat protein sources in their diets [100]. Preconceptions towards vegetarian diets, habits and prices and a lack of familiarity with meat substitutes, among others, are barriers to changing meat consumption behaviours [103]. Despite all of the above, it must be taken into account that the complete elimination of meat from our diet is impractical and might even have negative societal consequences [104].

The challenge of developing healthier foods with high consumer appeal underscores the need for integrated culinary, sensory and consumer research in this area [105]. Although Hoek et al. [99] concluded that, for the development of new foods, more emphasis is needed on consumer evaluation instead of on the sensory properties of the individual product. In that regard, studies that also take consumer behaviours into consideration could be an alternative to standard consumer sensory analyses. A recent alternative method called Mind Genomics has been applied on meat analogues, with promising results [106]. In addition, in order to increase the acceptance of novel products, it is necessary to obtain knowledge about the demographics, the consumption patterns and the sensory drivers of consumers [107]. In Western countries, vegetable proteins have a high level of acceptance and are consumed regularly. However, the same does not occur with the inclusion of nonconventional meats, insects or food by-products in our diet.

An alternative to conventional meat production is the use of more sustainable species like rats or other pest rodents [108,109]. Although rats are a regular staple in some Asian regions, the mere suggestion of its consumption in Western countries generates a big consumer rejection. Caparros Megido et al. [93] concluded that insect-tasting sessions are important to decrease their neophobia, because they observed that people with previous entomophagy experience gave globally higher ratings to meat products that contained insects-based proteins. In addition, Meyer-Rochow and Hakko [110] concluded that the acceptability of insect consumption would be higher if they were presented in flours or pastes. The inclusion of food by-products or residues from the meat industry can also present a challenge to consumer acceptance. Even though this practise presents a double opportunity towards healthier and more sustainable meat products, their acceptance is quite limited. Some of the reasons are related with consumer perceptions of these by-products as actual waste and, thus, unhealthy, but even if healthiness would be proven, consumers would also reject some of these reformulated products due to "ideational" reasons [111]. This concept is linked to the sensation of disgust some products produce in consumers just because of their origin (e.g., insects, by-products, etc.) and bad taste.

Meat eating is a habitual behaviour that is difficult to change; there is an unwillingness to reduce or substitute meat among the vast majority of consumers in various European countries [100]. In search of new alternatives, it is necessary to know how different food-related attitudes and behaviours (food choice motives, food fussiness, etc.) and socio-demographics (gender, age, education, etc.) influence the consumption of such protein sources [103]. In that sense, although some studies concluded that there is an urgent need for meat moderation campaigns that provide a broad spectrum of measures and habit-breaking interventions—including the promotion of vegetarian options [112]—the use of extenders to reduce animal proteins in the development of meat products could help to minimise their environmental impact without having to give up entirely the meat products in our diet.

4. Conclusions

A global demand for high-protein foods is on the rise. Meat and meat products are an important protein source in our diets but also great contributors to environment degradation through the far-from-sustainable production and increased carbon footprint of the finished products. Alternatives to more sustainable protein productions fall into two categories: mitigation of the negative impact and the use of more sustainable protein sources. With the use of meat extenders in meat products, we would be mitigating their negative impact by reducing the meat content, but we would also be maintaining the nutritional properties (i.e., protein and minerals) by using more sustainable sources. Even though pulses are the main extenders we should be looking at—similar nutritional profiles to meat—there are other extenders worth exploring. Apart from mushrooms, cereals, tubers and fruits that can be a great choice for some types of meat products, novel approaches such as insects and by-products from the food industry present an opportunity to develop healthier and more sustainable meat products. However, there is a need to devise strategies to increase consumer awareness and acceptance of these types of products. The plethora of sources and possibilities make the use of meat extenders the most viable and interesting approach towards the production of more sustainable meat products.

Author Contributions: Conceptualisation: G.D.-P., occurrence: T.P., writing—original draft preparation: G.D.-P. and T.P., writing—editing: T.P. and review: G.D.-P. All authors have read and agreed to the published version of the manuscript.

Funding: This research received no external funding.

Acknowledgments: We would like to thank the reviewers for their time, effort, insightful comments and suggestions that helped improve this manuscript.

References

1. UN. *Transforming Our World: The 2030 Agenda for Sustainable Development*; General Assembly of the United Nations: New York, NY, USA, 2015.
2. Fasolin, L.H.; Pereira, R.N.; Pinheiro, A.C.; Martins, J.T.; Andrade, C.C.P.; Ramos, O.L.; Vicente, A.A. Emergent food proteins—Towards sustainability, health and innovation. *Food Res. Int.* **2019**, *125*, 108586. [CrossRef]
3. Nadathur, S.R.; Wanasundara, J.P.D.; Scanlin, L. Chapter 1—Proteins in the diet: Challenges in feeding the global population. In *Sustainable Protein Sources*; Nadathur, S.R., Wanasundara, J.P.D., Scanlin, L., Eds.; Academic Press: San Diego, CA, USA, 2017; pp. 1–19. [CrossRef]
4. Stoll-Kleemann, S.; O'Riordan, T. The sustainability challenges of our meat and dairy diets. *Environ. Sci. Policy Sustain. Dev.* **2015**, *57*, 34–48. [CrossRef]
5. Salter, A.M. The effects of meat consumption on global health. *Revue Sci. Tech.* **2018**, *37*, 47–55. [CrossRef]
6. Westhoek, H.; Lesschen, J.P.; Rood, T.; Wagner, S.; De Marco, A.; Murphy-Bokern, D.; Leip, A.; van Grinsven, H.; Sutton, M.A.; Oenema, O. Food choices, health and environment: Effects of cutting Europe's meat and dairy intake. *Glob. Environ. Chang.* **2014**, *26*, 196–205. [CrossRef]
7. Cocking, C.; Walton, J.; Kehoe, L.; Cashman, K.D.; Flynn, A. The role of meat in the European diet: Current state of knowledge on dietary recommendations, intakes and contribution to energy and nutrient intakes and status. *Nutr. Res. Rev.* **2020**, 1–9. [CrossRef]
8. Rippin, H.L.; Hutchinson, J.; Jewell, J.; Breda, J.J.; Cade, J.E. Adult nutrient intakes from current national dietary surveys of European populations. *Nutrients* **2017**, *9*, 1288. [CrossRef]
9. Lopez, A.; Cacoub, P.; Macdougall, I.C.; Peyrin-Biroulet, L. Iron deficiency anaemia. *Lancet* **2016**, *387*, 907–916. [CrossRef]
10. Van Schoor, N.; Lips, P. Global overview of vitamin d status. *Endocrinol. Metab. Clin. N. Am.* **2017**, *46*, 845–870. [CrossRef]
11. Smith, A.D.; Warren, M.J.; Refsum, H. Vitamin B(12). *Adv. Food Nutr. Res.* **2018**, *83*, 215–279. [CrossRef]
12. Vatanparast, H.; Islam, N.; Shafiee, M.; Ramdath, D.D. Increasing plant-based meat alternatives and decreasing red and processed meat in the diet differentially affect the diet quality and nutrient intakes of Canadians. *Nutrients* **2020**, *12*, 2034. [CrossRef]
13. Rizzo, G.; Laganà, A.S.; Rapisarda, A.M.; La Ferrera, G.M.; Buscema, M.; Rossetti, P.; Nigro, A.; Muscia, V.; Valenti, G.; Sapia, F.; et al. Vitamin B12 among vegetarians: Status, assessment and supplementation. *Nutrients* **2016**, *8*, 767. [CrossRef]
14. Bohrer, B.M. Review: Nutrient density and nutritional value of meat products and non-meat foods high in protein. *Trends Food Sci. Technol.* **2017**, *65*, 103–112. [CrossRef]
15. Imamura, F.; Micha, R.; Khatibzadeh, S.; Fahimi, S.; Shi, P.; Powles, J.; Mozaffarian, D. Dietary quality among men and women in 187 countries in 1990 and 2010: A systematic assessment. *Lancet. Glob. Health* **2015**, *3*, e132–e142. [CrossRef]
16. Domingo, J.L.; Nadal, M. Carcinogenicity of consumption of red meat and processed meat: A review of scientific news since the IARC decision. *Food Chem. Toxicol.* **2017**, *105*, 256–261. [CrossRef]
17. Micha, R.; Michas, G.; Mozaffarian, D. Unprocessed red and processed meats and risk of coronary artery disease and type 2 diabetes–an updated review of the evidence. *Curr. Atheroscler Rep.* **2012**, *14*, 515–524. [CrossRef]
18. De Smet, S.; Vossen, E. Meat: The balance between nutrition and health. A review. *Meat Sci.* **2016**, *120*, 145–156. [CrossRef]
19. Jiménez-Colmenero, F.; Carballo, J.; Cofrades, S. Healthier meat and meat products: Their role as functional foods. *Meat Sci.* **2001**, *59*, 5–13. [CrossRef]
20. Özer, C.O.; Çelegen, Ş. Evaluation of quality and emulsion stability of a fat-reduced beef burger prepared with an olive oil oleogel-based emulsion. *J. Food Process. Preserv.* **2020**. [CrossRef]
21. Delgado-Pando, G.; Allen, P.; Kerry, J.P.; O'Sullivan, M.G.; Hamill, R.M. Optimising the acceptability of reduced-salt ham with flavourings using a mixture design. *Meat Sci.* **2019**, *156*, 1–10. [CrossRef]
22. Do Amaral, D.S.; Cardelle-Cobas, A.; Do Nascimento, B.M.S.; Monteiro, M.J.; Madruga, M.S.; Pintado, M.M.E. Development of a low fat fresh pork sausage based on chitosan with health claims: Impact on the quality, functionality and shelf-life. *Food Funct.* **2015**, *6*, 2768–2778. [CrossRef]

23. Zhu, Y.; Guo, L.; Yang, Q. Partial replacement of nitrite with a novel probiotic Lactobacillus plantarum on nitrate, color, biogenic amines and gel properties of Chinese fermented sausages. *Food Res. Int.* **2020**, *137*. [CrossRef]

24. Câmara, A.K.F.I.; Vidal, V.A.S.; Santos, M.; Bernardinelli, O.D.; Sabadini, E.; Pollonio, M.A.R. Reducing phosphate in emulsified meat products by adding chia (Salvia hispanica L.) mucilage in powder or gel format: A clean label technological strategy. *Meat Sci.* **2020**, *163*. [CrossRef]

25. Pintado, T.; Herrero, A.M.; Jiménez-Colmenero, F.; Ruiz-Capillas, C. Strategies for incorporation of chia (Salvia hispanica L.) in frankfurters as a health-promoting ingredient. *Meat Sci.* **2016**, *114*, 75–84. [CrossRef]

26. Alirezalu, K.; Pateiro, M.; Yaghoubi, M.; Alirezalu, A.; Peighambardoust, S.H.; Lorenzo, J.M. Phytochemical constituents, advanced extraction technologies and techno-functional properties of selected Mediterranean plants for use in meat products. A comprehensive review. *Trends Food Sci. Technol.* **2020**, *100*, 292–306. [CrossRef]

27. Barrón-Ayala, C.G.; Valenzuela-Melendres, M.; Camou, J.P.; Sebranek, J.G.; Dávila-Ramírez, J.L.; Cumplido-Barbeitia, G. Pork frankfurters prepared with hydrolyzed whey: Preliminary product quality aspects and inhibitory activity of the resulting peptides on angiotensin-converting enzyme. *Meat Sci.* **2020**, *166*. [CrossRef]

28. Bohrer, B.M. An investigation of the formulation and nutritional composition of modern meat analogue products. *Food Sci. Hum. Wellness* **2019**, *8*, 320–329. [CrossRef]

29. Lang, M. Consumer acceptance of blending plant-based ingredients into traditional meat-based foods: Evidence from the meat-mushroom blend. *Food Qual. Prefer.* **2020**, *79*, 103758. [CrossRef]

30. Van der Weele, C.; Feindt, P.; Jan van der Goot, A.; van Mierlo, B.; van Boekel, M. Meat alternatives: An integrative comparison. *Trends Food Sci. Technol.* **2019**, *88*, 505–512. [CrossRef]

31. Mills, E. Additives | Extenders. In *Encyclopedia of Meat Sciences*, 2nd ed.; Dikeman, M., Devine, C., Eds.; Academic Press: Oxford, UK, 2014; pp. 1–6. [CrossRef]

32. Rawal, V.; Navarro, D.K. *The Global Economy of Pulses*; FAO: Rome, Italy, 2019.

33. Argel, N.S.; Ranalli, N.; Califano, A.N.; Andrés, S.C. Influence of partial pork meat replacement by pulse flour on physicochemical and sensory characteristics of low-fat burgers. *J. Sci. Food Agric.* **2020**. [CrossRef]

34. Mishra, B.P.; Chauhan, G.; Mendiratta, S.K.; Sharma, B.D.; Desai, B.A.; Rath, P.K. Development and quality evaluation of dehydrated chicken meat rings using spent hen meat and different extenders. *J. Food Sci. Technol.* **2015**, *52*, 2121–2129. [CrossRef]

35. Bhat, Z.F.; Pathak, V.; Fayaz, H. Effect of refrigerated storage on the quality characteristics of microwave cooked chicken seekh kababs extended with different non-meat proteins. *J. Food Sci. Technol.* **2013**, *50*, 926–933. [CrossRef]

36. Holliday, D.L.; Sandlin, C.; Schott, A.; Malekian, F.; Finley, J.W. Characteristics of meat or sausage patties using pulses as extenders. *J. Culin. Sci. Technol.* **2011**, *9*, 158–176. [CrossRef]

37. Yadav, S.; Yadava, R. Effect of using bengal gram flour and corn flour as extenders on physicochemical and nutritional quality of quail meat rolls. *Vet. Pract.* **2008**, *9*, 150–153.

38. Nagamallika, E.; Prabhakara Reddy, K.; Masthan Reddy, P. Development of chicken meat patties with different extenders. *J. Food Sci. Technol.* **2005**, *42*, 529–531.

39. Serdaroğlu, M.; Yildiz Turp, G.; Abrodímov, K. Quality of low-fat meatballs containing Legume flours as extenders. *Meat Sci.* **2005**, *70*, 99–105. [CrossRef]

40. Purohit, A.S.; Reed, C.; Mohan, A. Development and evaluation of quail breakfast sausage. *LWT-Food Sci. Technol.* **2016**, *69*, 447–453. [CrossRef]

41. Pietrasik, Z.; Janz, J.A.M. Utilization of pea flour, starch-rich and fiber-rich fractions in low fat bologna. *Food Res. Int.* **2010**, *43*, 602–608. [CrossRef]

42. Sanjeewa, W.G.T.; Wanasundara, J.P.D.; Pietrasik, Z.; Shand, P.J. Characterization of chickpea (Cicer arietinum L.) flours and application in low-fat pork bologna as a model system. *Food Res. Int.* **2010**, *43*, 617–626. [CrossRef]

43. Marti-Quijal, F.J.; Zamuz, S.; Tomašević, I.; Rocchetti, G.; Lucini, L.; Marszałek, K.; Barba, F.J.; Lorenzo, J.M. A chemometric approach to evaluate the impact of pulses, Chlorella and Spirulina on proximate composition, amino acid, and physicochemical properties of turkey burgers. *J. Food Sci. Technol.* **2019**, *99*, 3672–3680. [CrossRef]

44. Farooq, Z.; Boye, J.I. Novel food and industrial applications of pulse flours and fractions. In *Pulse Foods: Processing, Quality and Nutraceutical Application*; Tiwari, B.K., Gowen, A., McKenna, B., Eds.; Academic Press: London, UK, 2001.

45. Aslinah, L.N.F.; Mat Yusoff, M.; Ismail-Fitry, M.R. Simultaneous use of adzuki beans (Vigna angularis) flour as meat extender and fat replacer in reduced-fat beef meatballs (bebola daging). *J. Food. Sci. Technol.* **2018**, *55*, 3241–3248. [CrossRef]

46. Paglarini, C.D.S.; Martini, S.; Pollonio, M.A.R. Using emulsion gels made with sonicated soy protein isolate dispersions to replace fat in frankfurters. *LWT* **2019**, *99*, 453–459. [CrossRef]

47. Nasonova, V.V.; Tunieva, E.K. A comparative study of fat replacers in cooked sausages. In Proceedings of the IOP Conference Series: Earth and Environmental Science, Kopaonik, Serbia, 22–25 September 2019.

48. Haque, M.A.; Timilsena, Y.P.; Adhikari, B. Food proteins, structure, and function. In *Reference Module in Food Science*; Elsevier: Amsterdam, The Netherlands, 2016. [CrossRef]

49. Asgar, M.A.; Fazilah, A.; Huda, N.; Bhat, R.; Karim, A.A. Nonmeat protein alternatives as meat extenders and meat analogs. *Compr. Rev. Food Sci. Food Saf.* **2010**, *9*, 513–529. [CrossRef]

50. Mireles-Arriaga, A.I.; Ruiz-Nieto, J.E.; Juárez-Abraham, M.; Mendoza-Carrillo, M.; Martínez-Loperena, R. Functional restructured meat: Applications of ingredients derived from plants. *Vitae* **2017**, *24*, 196–204. [CrossRef]

51. Hygreeva, D.; Pandey, M.C.; Radhakrishna, K. Potential applications of plant based derivatives as fat replacers, antioxidants and antimicrobials in fresh and processed meat products. *Meat Sci.* **2014**, *98*, 47–57. [CrossRef]

52. Yildiz-Turp, G.; Serdaroglu, M. Effects of using plum puree on some properties of low fat beef patties. *Meat Sci.* **2010**, *86*, 896–900. [CrossRef]

53. Fernández-López, J.; Lucas-González, R.; Viuda-Martos, M.; Sayas-Barberá, E.; Ballester-Sánchez, J.; Haros, C.M.; Martínez-Mayoral, A.; Pérez-Álvarez, J.A. Chemical and technological properties of bologna-type sausages with added black quinoa wet-milling coproducts as binder replacer. *Food Chem.* **2020**, *310*. [CrossRef]

54. Ghafouri-Oskuei, H.; Javadi, A.; Saeidi Asl, M.R.; Azadmard-Damirchi, S.; Armin, M. Quality properties of sausage incorporated with flaxseed and tomato powders. *Meat Sci.* **2020**, *161*. [CrossRef]

55. Peiretti, P.G.; Gai, F.; Zorzi, M.; Aigotti, R.; Medana, C. The effect of blueberry pomace on the oxidative stability and cooking properties of pork patties during chilled storage. *J. Food Process. Preserv.* **2020**, e14520. [CrossRef]

56. Pereira, J.; Hu, H.; Xing, L.; Zhang, W.; Zhou, G. Influence of rice flour, glutinous rice flour, and tapioca starch on the functional properties and quality of an emulsion-type cooked sausage. *Foods* **2020**, *9*, 9. [CrossRef]

57. Gupta, S.; Sharma, B.D. Quality characteristics of functional restructured spent hen meat slices developed by utilizing different binders and extenders. *Food Sci. Technol. Res.* **2018**, *24*, 241–247. [CrossRef]

58. Serdaroğlu, M.; Kavuşan, H.S.; İpek, G.; Öztürk, B. Evaluation of the quality of beef patties formulated with dried pumpkin pulp and seed. *Korean J. Food Sci. Anim. Resour.* **2018**, *38*, 1–13. [CrossRef] [PubMed]

59. Hawashin, M.D.; Al-Juhaimi, F.; Ahmed, I.A.M.; Ghafoor, K.; Babiker, E.E. Physicochemical, microbiological and sensory evaluation of beef patties incorporated with destoned olive cake powder. *Meat Sci.* **2016**, *122*, 32–39. [CrossRef] [PubMed]

60. González Rodríguez, D.M.; Giraldo Lopera, E.; Restrepo Molina, D.A. Effect of postproduction heating on the texture properties of a standard sausage that contains a chicken paste meat extender. *Rev. Fac. Nac. de Agron. Medellín* **2015**, *68*, 7713–7720. [CrossRef]

61. Malav, O.P.; Sharma, B.D.; Talukder, S.; Mendiratta, S.K.; Kumar, R.R. Quality characteristics and storage stability of restructured chicken meat blocks extended with different combinations of vegetative extenders. *J. Food Sci. Technol. -Mysore* **2015**, *52*, 391–398. [CrossRef]

62. Modi, V.K.; Prakash, M. Quick and reliable screening of compatible ingredients for the formulation of extended meat cubes using Plackett–Burman design. *LWT-J. Food Sci. Technol.* **2008**, *41*, 878–882. [CrossRef]

63. Igyor, M.A.; Ankeli, J.A.; Badifu, G.I.O. Quality of beef sausages as affected by defatted melon (Citrullus vulgaris Schrad) kernel flour supplementation and refrigerated storage. *J. Food Sci. Technol.* **2008**, *45*, 173–176.

64. Igyor, M.A.; Ankeli, J.A.; Badifu, G.I.O. Effect of defatted melon (citrulllus vulgaris schrad.) kernel flour supplementation on the storage stability and microbiological quality of refrigerated beef-based sausages. *J. Food Process. Preserv.* **2008**, *32*, 143–158. [CrossRef]

65. Kumar, R.R.; Sharma, B.D.; Kumar, M.; Chidanandaiah; Biswas, A.K. Storage quality and shelf life of vacuum-packaged extended chicken patties. *J. Muscle Foods* **2007**, *18*, 253–263. [CrossRef]
66. Peiretti, P.G.; Gai, F. Fruit and pomace extracts: Applications to improve the safety and quality of meat products. In *Fruit and Pomace Extracts: Biological Activity, Potential Applications and Beneficial Health Effects*; Nova Science Publishers: Hauppauge, NY, USA, 2015; pp. 1–28.
67. Papageorgiou, M.; Skendi, A. Introduction to cereal processing and by-products. In *Sustainable Recovery and Reutilization of Cereal Processing By-Products*; Galanakis, C.M., Ed.; Woodhead Publishing: Cambridge, UK, 2018; pp. 1–25. [CrossRef]
68. Calderon-Oliver, M.; Lopez-Hernandez, L.H. Food vegetable and fruit waste used in meat products. *Food Rev. Int.* **2020**, 1–27. [CrossRef]
69. Ben-Othman, S.; Joudu, I.; Bhat, R. Bioactives from Agri-Food Wastes: Present insights and future challenges. *Molecules* **2020**, *25*, 510. [CrossRef]
70. Turhan, S.; Temiz, H.; Sagir, I. Utilization of wet okara in low-fat beef patties. *J. Muscle Foods* **2007**, *18*, 226–235. [CrossRef]
71. Strada de Oliveira, R.B.; Della Lucia, F.; Ferreira, E.B.; Evangelista de Oliveira, R.M.; Pimenta, C.J.; de Sousa Gomes Pimenta, M.E. Quality of beef burger with addition of wet okara along the storage. *Cienc. E Agrotecnologia* **2016**, *40*, 706–717. [CrossRef]
72. Noriham, A.; Muhammad Ariffaizuddin, R.; Noorlaila, A.; Faris Zakry, A.N. Potential use of okara as meat replacer in beef sausage. *J. Teknol.* **2016**, *78*, 13–18.
73. Tie Su, S.I.; Pedroso Yoshida, C.M.; Contreras-Castillo, C.J.; Quinones, E.M.; Venturini, A.C. Okara, a soymilk industry by-product, as a non-meat protein source in reduced fat beef burgers. *Food Sci. Technol.* **2013**, *33*, 52–56.
74. Chang, T.; Wang, C.J.; Wang, S.L.; Shi, L.; Yang, H.; Cui, M. Effect of okara on textural, color and rheological properties of pork meat gels. *J. Food Qual.* **2014**, *37*, 339–348. [CrossRef]
75. Xerez Pinho, L.; Rodrigues Amorim Afonso, M.; Beserra Carioca, J.O.; Correia da Costa, J.M.; Mota Ramos, A. The use of cashew apple residue as source of fiber in low fat hamburgers. *Cienc. Tecnol. Aliment.* **2011**, *31*, 941–945. [CrossRef]
76. Younis, K.; Ahmad, S. Waste utilization of apple pomace as a source of functional ingredient in buffalo meat sausage. *Cogent Food Agric.* **2015**, *1*, 1119397. [CrossRef]
77. Kumar Banerjee, D.; Arun, K.D.; Rituparna, B.; Mirian, P.; Kumar Nanda, P.; Yogesh, P.G.; Subhasish, B.; McClements, D.J.; Lorenzo, J.M. Application of enoki mushroom (Flammulina Velutipes) stem wastes as functional ingredients in goat meat nuggets. *Foods* **2020**, *9*, 432. [CrossRef]
78. Hale, A.B.; Carpenter, C.E.; Walsh, M.K. Instrumental and consumer evaluation of beef patties extended with extrusion-textured whey proteins. *J. Food Sci.* **2002**, *67*, 1267–1270. [CrossRef]
79. Álvarez, C.; Drummond, L.; Mullen, A.M. Protein recovered from meat co-products and processing streams as pork meat replacers in Irish breakfast sausages formulations. *LWT* **2018**, *96*, 679–685. [CrossRef]
80. Colletti, A.; Attrovio, A.; Boffa, L.; Mantegna, S.; Cravotto, G. Valorisation of by-products from soybean (Glycine max (L.) Merr.) processing. *Molecules* **2020**, *25*, 2129. [CrossRef] [PubMed]
81. Ahmad, T.; Aadil, R.M.; Ahmed, H.; Rahman, U.U.; Soares, B.C.V.; Souza, S.L.Q.; Pimentel, T.C.; Scudino, H.; Guimarães, J.T.; Esmerino, E.A.; et al. Treatment and utilization of dairy industrial waste: A review. *Trends Food Sci. Technol.* **2019**, *88*, 361–372. [CrossRef]
82. Zhang, Y.; Venkitasamy, C.; Pan, Z.; Wang, W. Recent developments on umami ingredients of edible mushrooms—A review. *Trends Food Sci. Technol.* **2013**, *33*, 78–92. [CrossRef]
83. Wong, K.M.; Corradini, M.G.; Autio, W.; Kinchla, A.J. Sodium reduction strategies through use of meat extenders (white button mushrooms vs. textured soy) in beef patties. *Food Sci. Nutr.* **2019**, *7*, 506–518. [CrossRef]
84. Miller, A.M.; Mills, K.; Wong, T.; Drescher, G.; Lee, S.M.; Sirimuangmoon, C.; Schaefer, S.; Langstaff, S.; Minor, B.; Guinard, J.X. Flavor-enhancing properties of mushrooms in meat-based dishes in which sodium has been reduced and meat has been partially substituted with mushrooms. *J. Food Sci.* **2014**, *79*, S1795–S1804. [CrossRef]
85. Cha, M.-H.; Heo, J.-Y.; Lee, C.; Lo, Y.M.; Moon, B. Quality and sensory characterization of white jelly mushroom (*Tremella fuciformis*) as a meat substitute in pork patty formulation. *J. Food Process. Preserv.* **2014**, *38*, 2014–2019. [CrossRef]

86. Wang, L.; Guo, H.; Liu, X.; Jiang, G.; Li, C.; Li, X.; Li, Y. Roles of Lentinula edodes as the pork lean meat replacer in production of the sausage. *Meat Sci.* **2019**, *156*, 44–51. [CrossRef]

87. Rosli, W.I.W.; Solihah, M.A. Effect on the addition of Pleurotus sajor-caju (PSC) on physical and sensorial properties of beef patty. *Int. Food Res. J.* **2012**, *19*, 993–999.

88. Rosli, W.W.I.; Maihiza, N.M.S.; Raushan, M. The ability of oyster mushroom in improving nutritional composition, beta-glucan and textural properties of chicken frankfurter. *Int. Food Res. J.* **2015**, *22*, 311–317.

89. Stull, V.J.; Finer, E.; Bergmans, R.S.; Febvre, H.P.; Longhurst, C.; Manter, D.K.; Patz, J.A.; Weir, T.L. Impact of edible cricket consumption on gut microbiota in healthy adults, a double-blind, randomized crossover trial. *Sci. Rep.* **2018**, *8*, 10762. [CrossRef]

90. Dikeman, C.L.; Bauer, L.L.; Flickinger, E.A.; Fahey, G.C. Effects of stage of maturity and cooking on the chemical composition of select mushroom varieties. *J. Agric. Food Chem.* **2005**, *53*, 1130–1138. [CrossRef] [PubMed]

91. De Carvalho, N.M.; Madureira, A.R.; Pintado, M.E. The potential of insects as food sources—A review. *Crit. Rev. Food Sci. Nutr.* **2019**, 1–11. [CrossRef] [PubMed]

92. Meyer-Rochow, V.B. Can insects help to ease the problem of world food shortage? *Search* **1975**, *6*, 261–262.

93. Caparros Megido, R.; Gierts, C.; Blecker, C.; Brostaux, Y.; Haubruge, É.; Alabi, T.; Francis, F. Consumer acceptance of insect-based alternative meat products in Western countries. *Food Qual. Prefer.* **2016**, *52*, 237–243. [CrossRef]

94. Ju-Hye, C.; Hae In, Y.; Su-Kyung, K.; Tae-Kyung, K.; Yun-Sang, C. The quality characteristics of pork patties according to the replacement of mealworm (Tenebrio molitor L.). *Korean J. Food Cook. Sci.* **2019**, *35*, 441–449. [CrossRef]

95. Choi, Y.-S.; Kim, T.-K.; Choi, H.-D.; Park, J.-D.; Sung, J.-M.; Jeon, K.-H.; Paik, H.-D.; Kim, Y.-B. Optimization of replacing pork meat with yellow worm (Tenebrio molitor L.) for frankfurters. *Korean J. Food Sci. Anim. Resour.* **2017**, *37*, 617–625. [CrossRef]

96. Kim, H.-W.; Setyabrata, D.; Lee, Y.J.; Jones, O.G.; Kim, Y.H.B. Pre-treated mealworm larvae and silkworm pupae as a novel protein ingredient in emulsion sausages. *Innov. Food Sci. Emerg. Technol.* **2016**, *38*, 116–123. [CrossRef]

97. Kim, H.-W.; Setyabrata, D.; Lee, Y.; Jones, O.G.; Kim, Y.H.B. Effect of house cricket (Acheta domesticus) flour addition on physicochemical and textural properties of meat emulsion under various formulations. *J. Food Sci.* **2017**, *82*, 2787–2793. [CrossRef]

98. Harwatt, H. Including animal to plant protein shifts in climate change mitigation policy: A proposed three-step strategy. *Clim. Policy* **2019**, *19*, 533–541. [CrossRef]

99. Hoek, A.C.; van Boekel, M.A.J.S.; Voordouw, J.; Luning, P.A. Identification of new food alternatives: How do consumers categorize meat and meat substitutes? *Food Qual. Prefer.* **2011**, *22*, 371–383. [CrossRef]

100. Hartmann, C.; Siegrist, M. Consumer perception and behaviour regarding sustainable protein consumption: A systematic review. *Trends Food Sci. Technol.* **2017**, *61*, 11–25. [CrossRef]

101. Frank, D.; Oytam, Y.; Hughes, J. Chapter 27—Sensory perceptions and new consumer attitudes to meat. In *New Aspects of Meat Quality*; Purslow, P.P., Ed.; Woodhead Publishing: Cambridge, UK, 2017; pp. 667–698. [CrossRef]

102. Van Loo, E.J.; Hoefkens, C.; Verbeke, W. Healthy, sustainable and plant-based eating: Perceived (mis)match and involvement-based consumer segments as targets for future policy. *Food Policy* **2017**, *69*, 46–57. [CrossRef]

103. Grasso, A.C.; Hung, Y.; Olthof, M.R.; Verbeke, W.; Brouwer, I.A. Older consumers' readiness to accept alternative, more sustainable protein sources in the European Union. *Nutrients* **2019**, *11*, 1904. [CrossRef]

104. Hartmann, C.; Siegrist, M. Our daily meat: Justification, moral evaluation and willingness to substitute. *Food Qual. Prefer.* **2020**, *80*. [CrossRef]

105. Circus, V.E.; Robison, R. Exploring perceptions of sustainable proteins and meat attachment. *Br. Food J.* **2019**, *121*, 533–545. [CrossRef]

106. Gere, A.; Harizi, A.; Bellissimo, N.; Roberts, D.; Moskowitz, H. Creating a mind genomics wiki for non-meat analogs. *Sustainability* **2020**, *12*, 5352. [CrossRef]

107. Guinard, J.-X.; Myrdal Miller, A.; Mills, K.; Wong, T.; Lee, S.M.; Sirimuangmoon, C.; Schaefer, S.E.; Drescher, G. Consumer acceptance of dishes in which beef has been partially substituted with mushrooms and sodium has been reduced. *Appetite* **2016**, *105*, 449–459. [CrossRef] [PubMed]

108. Meyer-Rochow, V.B.; Megu, K.; Chakravorty, J. Rats: If you can't beat them eat them! (Tricks of the trade observed among the Adi and other North-East Indian tribals). *J. Ethnobiol. Ethnomed.* **2015**, *11*, 1–12. [CrossRef] [PubMed]

109. Gruber, K. Rodent meat—A sustainable way to feed the world? Using rodents as food has a long tradition in many parts of the world. *EMBO Rep.* **2016**, *17*, 630–633. [CrossRef]

110. Meyer-Rochow, V.B.; Hakko, H. Can edible grasshoppers and silkworm pupae be tasted by humans when prevented to see and smell these insects? *J. Asia-Pac. Entomol.* **2018**, *21*, 616–619. [CrossRef]

111. Henchion, M.; McCarthy, M.; O'Callaghan, J. Transforming beef by-products into valuable ingredients: Which spell/recipe to use? *Front. Nutr.* **2016**, *3*, 53. [CrossRef] [PubMed]

112. De Boer, J.; Schösler, H.; Aiking, H. Towards a reduced meat diet: Mindset and motivation of young vegetarians, low, medium and high meat-eaters. *Appetite* **2017**, *113*, 387–397. [CrossRef] [PubMed]

 **foods**

MDPI

Article

A Novel Approach for Tuning the Physicochemical, Textural, and Sensory Characteristics of Plant-Based Meat Analogs with Different Levels of Methylcellulose Concentration

Allah Bakhsh [1], Se-Jin Lee [1], Eun-Yeong Lee [1], Nahar Sabikun [1], Young-Hwa Hwang [2] and Seon-Tea Joo [1,2,*]

[1] Division of Applied Life Science (BK21 Four), Gyeongsang National University, Jinju 52852, Korea; drallahbakhsh1@yahoo.com (A.B.); sejinlee1994@gmail.com (S.-J.L.); ley9604@gmail.com (E.-Y.L.); seturajjo@yahoo.com (N.S.)

[2] Institute of Agriculture & Life Science, Gyeongsang National University, Jinju 52852, Korea; philoria@hanmail.com

* Correspondence: stjoo@gnu.ac.kr; Tel.: +82-55-772-1943

Abstract: This study assessed the effects of Methylcellulose (MC) at different concentrations on plant-based meat analog (PBMA) patties, comprised of commercial texture vegetable protein (C-TVP) and textured isolate soy protein (T-ISP) as key ingredients, and compared to beef patty control. A significantly higher difference was observed in moisture content in control with increasing MC concentration than the C-TVP and T-ISP patties. However, protein varied significantly among three different protein sources, with control had higher protein content than PBMA patties. Crude fiber content recorded higher values in C-TVP as compared to control. Significantly lower pH values were recorded in control than C-TVP and T-ISP respectively. Regardless, with the addition of MC or ingredient PBMA and control patties tend to reduce lightness (L*) and redness (a*) value after cooking. Although control sample before cooking exhibits lighter and redder than PBMA patties (C-TVP and T-ISP). Likewise, water holding capacity (WHC) decreases as the concentration of MC increases (1.5–4%) in control and PBMA patties. Warner-Bratzler shear force (WBSF) and texture profile analysis (TPA), including hardness, chewiness, and gumminess of control, were significantly higher than C-TVP and T-ISP. Consequently, panelists' in the sensory analysis presented that C-TVP patties containing 3% of MC had better sensory properties than T-ISP. Hence, PBMA patties with C-TVP and incorporation of 3% MC are considered ideal for manufacturing of meat analog as related to control (beef).

Keywords: plant-based meat analog; commercial texture vegetable protein; texture soy isolate protein; methylcellulose

Citation: Bakhsh, A.; Lee, S.-J.; Lee, E.-Y.; Sabikun, N.; Hwang, Y.-H.; Joo, S.-T. A Novel Approach for Tuning the Physicochemical, Textural, and Sensory Characteristics of Plant-Based Meat Analogs with Different Levels of Methylcellulose Concentration. *Foods* **2021**, *10*, 560. https://doi.org/10.3390/foods 10030560

Academic Editors: Gonzalo Delgado-Pando and Tatiana Pintado

Received: 18 January 2021
Accepted: 4 March 2021
Published: 8 March 2021

Publisher's Note: MDPI stays neutral with regard to jurisdictional claims in published maps and institutional affiliations.

1. Introduction

The term "meat analog" denotes food products that are not made from red meat exclusively, commonly known as meat alternatives, meat substitutes, fake, mock, and imitation meat [1]. However, it possesses texture, mouth-feel, taste, and nutritional qualities that resemble meat [2]. Meat contributes to the food industry by supplying specific functionalities and has its attraction on consumers for its organoleptic features. Meat proteins are responsible for their characteristic appearance, textural and functional properties [3]. However, mimicking these meat protein characteristics by any other source of protein is difficult. Moreover, recently the International Agency for Research on Cancer, the cancer agency of WHO (World Health Organization), has classified the consumption of red meat (particularly processed meat) as carcinogenic to humans [4]. Furthermore, Food and Agricultural Organization (FAO) reports have been critical of the ecological impact of high levels of meat consumption and potentially transmissible diseases [5,6]. To mask these disadvantages of red meat, meat analogs are just one example of a variety of products recently demanded by a substantial portion of the population, especially those concerned about red meat's potential health effects [7]. Additionally, the projections for the increasing demand for animal protein in the coming decades are distressing, while extensive livestock production is also causing severe environmental and ecological imbalance [8]. Consequently, the research community is targeting refining the current production systems, searching for efficient novel technologies, while at the same time focusing on the improvement of consumption habits and food cultures [9].

In the current study, soy-based texture vegetable protein (TVP), and textured isolate soy protein (T-ISP) have been used as a meat replacer with many economic and functional benefits [3]. Soy-based TVP_S are plant-based protein products with low saturated fat, a high concentration of essential amino acids, and is cholesterol-free [10]. The manufacturing process of TVP involves a high-pressure extrusion process and a final spinning or extraction of the finishing product, which can then be used in meat analogs [7]. The low/intermediate moisture TVP has advantages in handling, storage, and shelf stability but requires time to hydrate before consumption. Upon hydration, it presents a spongy textured, fibrous structure mimicking meat [7]. Furthermore, numerous investigators have reported that by using soy protein and wheat gluten as TVP constituents, the final product could mimic the texture, appearance, taste, smell, and functionality of red meat [5].

In red meat, textural and taste parameters are important to the consumers and represent high economic value as some cuts bring exorbitant prices. In contrast, meat analogs lack these features and are generally regarded as substandard to cheaper meats. Numerous plant proteins, including cereal, oilseed, legume, and soy proteins (textured, flour, concentrate, and isolate), are recommended additions to the meat analogs. These elements have appropriate functional properties (e.g., water and oil absorption capacity, emulsification), which allow them to create numbers of distinctive meat substitutes [5,11].

The binding ability of the different ingredients in plant-based meat is of significant importance as non–adhesive behavior of varying plant ingredients can significantly affect the final analogy. Earlier binding agents such as egg solids, hydrocolloids, starch, and milk protein have been used in various commercial products [12]. In the present study, Methylcellulose (MC) has been used as a binder. Quality characteristics of MC include binding abilities and moisture retention, boil-out control, increase volume, and texture improvement in several types of meat analogs and processed meat [13]. Through synthetic modification, the naturally occurring polymer cellulose is converted to hypromellose or MC and is considered safe for consumption by humans [13]. Moreover, MC is classified as GRAS (generally recognized as safe) by the FDA (21 CFR 182.1480) and is also allowed in USDA regulated meat patties at concentrations up to 0.15% (9 CFR 3 t 8.7). Previously, the use of binding agents in meat analogs has been widely investigated, although no such attempt has been made to study the effects of MC on quality characteristics of Plant-based meat analog (PBMA) patties. Therefore, the objective of the present study was to evaluate the effects of MC on quality characteristics of PBMA with the incorporation of different texturized soy vegetable proteins.

2. Materials and Methods

2.1. Materials

Commercial texture vegetable protein (C-TVP) (Anthony's goods, Glendale, CA, USA) and ISP (isolate soy protein) (Shandong, China) were as the base for PBMA and MC (high viscosity, Modernist Pantry, Eliot, ME, USA) was incorporated as a binder. Other ingredients, including molasses, yeast seasoning, umami seasoning, coconut oil, canola oil, garlic powder, and pepper were used in the formulation (Table 1).

Table 1. Treatment and formulation of plant-based meat analogs.

Ingredients % Concentration (MC)	Treatments								
	Control (Beef)			C-TVP			T-ISP		
	1.5%	3%	4%	1.5%	3%	4%	1.5%	3%	4%
Lean beef C-TVP T-ISP	82.1	80.82	80.00	76.08	74.90	74.13	70.08	74.90	74.13
Methylcellulose	1.5	3.00	4.00	1.50	3.00	4.00	1.50	3.00	4.00
Garlic powder				2.28	2.25	2.22	2.28	2.25	2.22
Yeast extract				2.28	2.25	2.22	2.28	2.25	2.22
Black pepper				1.52	1.50	1.49	1.52	1.50	1.49
Mushroom				2.28	2.25	2.22	2.28	2.25	2.22
Salt				1.14	1.11	1.11	1.14	1.11	1.11
Beef back fat	16.4	16.18	16.00						
Coconut oil				3.80	3.75	3.71	3.80	3.75	3.71
Canola oil				3.80	3.75	3.71	3.80	3.75	3.71
Beet juice				3.04	3.00	2.96	3.04	3.00	2.96
Molasses				1.52	1.50	1.49	1.52	1.50	1.49
Umami seasoning				0.76	0.74	0.74	0.76	0.74	0.74

C-TVP: Commercial textured vegetable protein. T-ISP: Textured isolate soy protein. MC: Methylcellulose.

2.2. Sample Preparation and Processing

The flow diagram for processing meat analog is described in Figure 1. For meatless patties, C-TVP and texture isolate soy protein (T-ISP) were used as the base. The texturization of ISP was carried out by mixing ISP powder with water at a ratio of 1:6 (w/v). The mixture was stirred continuously over a lower flame until it forms a thickened paste. Subsequently, the paste was heated in an oven for two hours at a temperature of 120 °C T-ISP was a secondary option for comparing the quality characteristics of the created meatless patties to C-TVP. A total of three hundred g of each C-TVP and T-ISP were mixed with water separately (2 times in volume) and allowed to hydrate for "1 h" at 4 °C for a single concentration of MC with three repetitions and two formulations having raw and cooked patties respectively. After that, the hydrated C-TVP and T-ISP were mixed with the ingredients listed in Table 1 using a Kitchen Aid (Classic Plus Stand Mixer, St Joseph, MI, USA). Subsequently from the whole mixture, 50 g of the mixture was then shaped into patties using a patty press maker. The current experiment had three different concentrations of MC (1.5%, 3%, and 4%), from every single concentration of MC three patties (repetition) were prepared with one control and two treatments. In total, for one control, two treatments, and two formulations, 27 raw and 27 cooked patties were prepared. Therefore, in total, 54 patties were shaped. Eighteen patties were allocated for each control and two treatments separately.

A beef patty was used for the control formulated as describe in Table 1. The patties were cooked by dry heat, cooking on a non-stick pan at 150 °C for 5 min per side. They were flipped three times or until the internal temperature reached 75 °C as measured by a probe thermometer. Patties were allowed to cool at ambient temperature for 30 min before measuring the physicochemical and sensory attributes.

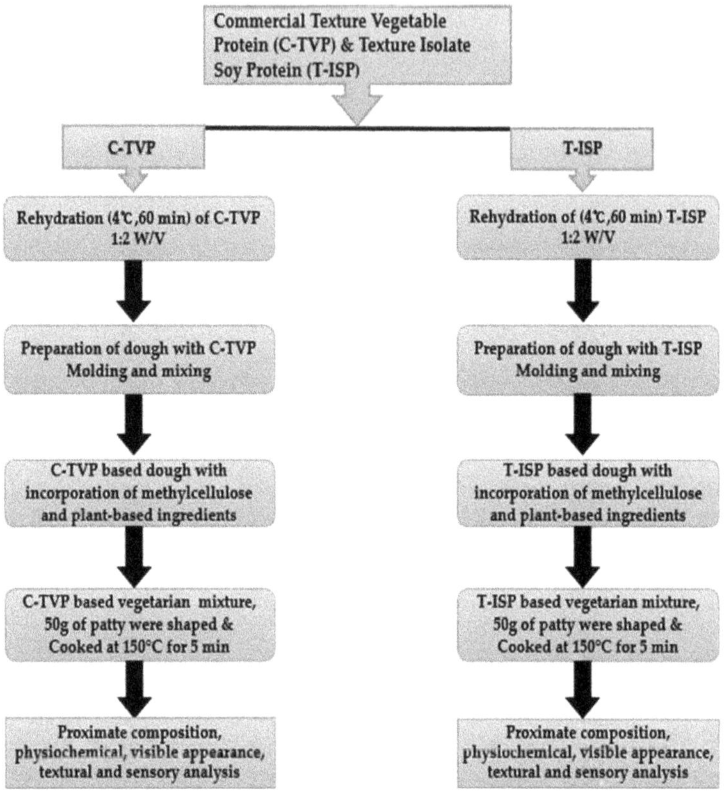

Figure 1. Flow diagram for manufacturing the meat analog. C-TVP: Commercial textured vegetable protein. T-ISP: Textured isolate soy protein. PBMA: Plant-based meat analog.

2.3. Proximate Analysis

Moisture, protein, fat, and ash contents were determined based on the standard AOAC [14]. Moisture content was quantified by the oven (BioFree, BF-150C, Buchen Korea), drying 5 g samples at 105 °C for 16 h. Protein was determined by the established procedure of Kjeldahl assay N analyzer (B-324, 412, 435 and 719 S Titrino, BUCHI, Flawil, Switzerland) ($N \times 6.25$) using 0.1 g of sample. The crude protein was determined by using the following formula.

$$\%N = \frac{[V(1) - V(B1)].F.c.F.M(N) \times 100}{M.1000} \tag{1}$$

$$\% P = \% N \times PF \tag{2}$$

V(1): consumption of titrant, sample (mL)
V(BI): average consumption of titrant, blank (mL)
F: molar reaction factor (1 = HCl, 2 = H_2SO_4)
c: concentration of titrant [mol/L]
M(N): molecular weight of N (14,007 (g/mol))
M: sample weight (g)
1000: conversion factor (mL in L)
PF: protein factor
% N: % of weight of N
% P: % of weight of protein

Crude fat was measured with 2 g samples by extraction in a Soxhlet apparatus (MS-EAM9203-06, Seoul Korea) by using petroleum ether as a solvent. The crude fat content was calculated by using the following formula.

$$\%Crudefat = (W2 - W1) \times \frac{100}{S} \tag{3}$$

Weight of empty flask (g) = W1
Weight of flask and extracted fat (g) = W2
Weight of sample = S

Ash was determined after incineration of 2 g of sample in a furnace (CFMD2, Changsin, Korea) at 500 °C. Crude fiber determination was estimated using an Ankom 200 Fiber Analyzer (Ankom Technology, Macedon, NY, USA) by digesting 0.5 g with H_2SO_4 and NaOH. The loss of weight resulting from ashing (2 h at 600 ± 15 °C) was collected to calculate the crude fiber content [15].

2.4. Physicochemical Analysis

The pH values of raw and cooked patties were measured with a digital pH meter (Mettler Toledo, MP230, Schwerzenbach, Switzerland) using 3 g of sample homogenized with 20 mL of distilled water.

The color of raw and cooked patties was measured using a Konica Minolta Colorimeter (Chroma meter, CR-300, Japan). The apparatus was standardized through a white ceramic plate (Y = 93.5, X = 0.3132, y = 0.3198), and lightness (L*), redness (a*), and yellowness (b*) values were recorded.

Release water percentage (RW%) was measured based on a method described by Joo [16]. The cooking loss (CL%) was determined as a percentage method adopted by Biswas et al. [17] using the following formula: Cooking loss (%) = (Weight of the patties after cooking/Weight of the patties before cooking) × 100.

Warner-Bratzler shear force (WBSF) was determined on the cooked sample using the established AMSA procedure [18]. The shrinkage percentage of the patties' diameter was measured at four different locations both before and after cooking. A total of 18 (nine raw and nine cooked) patties were allocated for physiochemical analysis.

2.5. Visible Appearance

The appearance of the control and PBMA patties were assessed by adding the different concentrations of MC (1.5%, 3%, and 4%) respectively. The external and internal appearance were photographed using a digital camera (EOS 700D, Canon, Tokyo, Japan), and various features were distinguished. In total, 18 (nine raw and nine cooked) patties were used for visible appearance.

2.6. Texture Profile Analysis

Samples were uniformly cut into $1 \times 1 \times 1$ cm, and they were axially compressed using a Sun Rheometer (Compact-100 II, Sun Scientific Co., LTD., Tokyo, Japan) with a flat pressure adaptor of 25 mm in diameter (No. 1). The samples were compressed at a crosshead speed of 60 mm/min at a final strain of 60% through a 2-cycle sequence with a load cell of 10 kg [19]. The following parameters were determined: hardness, cohesiveness, springiness, gumminess, and chewiness. A total of nine patties were assigned for the determination of texture profile analysis.

2.7. Sensory Evaluation

A 10-member trained panel from the laboratory of meat science Gyeongsang National University Korea, with 20 members of the untrained panel, includes students and researchers from the Department of Animal Sciences at Gyeongsang National University, Republic of Korea, assessed sensory characteristics of prepared patties. The panelist assortment was approved according to Lawless and Heymann [20], modified by Rahman et al. [21]. Small pieces of different samples (2 cm × 2 cm × 2 cm) were prepared and marked, random coding was allotted on pre-positioned glass container (Pyrex, Charleroi, PA, USA), and the pieces of samples were permitted to rest for 30 min at room temperature and then disseminated among the panelists. For judging each sample in a triplicates way, fluorescent light was used. For every sensory evaluation procedure, the panelist was provided with drinking water for washing the mouth for every new sample evaluation. Sensory traits that were recorded included appearance, shape, firmness, color, and overall acceptability. The samples were judged using a 9-point hedonic scale ranging from extreme dislike (score = 1) to extreme like (score = 9). A total nine number of PBMA patties were assigned for sensory evaluation.

2.8. Statistical Analysis

The results of PBMA based on C-TVP and T-ISP content are represented as the mean plus/minus standard error of the mean (SEM). The effect of main ingredients and concentration of MC on the variation of proximate composition, physicochemical properties, and visible appearance was described as mean and standard error of mean (SEM). Analysis of variance (factorial ANOVA) was carried out using SPSS version 23 (IBM Corp., Armonk, NY, USA). For multiple mean comparisons, the Tukey's test was run at the level of 5%.

3. Results and Discussion

3.1. Proximate Analysis

The proximate composition of control and PBMA patties are presented in Table 2. Moisture content prepared from a lower MC concentration (1.5%) was not statistically different among treatments. However, control beef patties with a higher concentration of MC (3–4%) expressing a higher moisture content than patties prepared from soy-based C-TVP and T-ISP. The ability of MC in reducing the loss of moisture content was due to the thermal gelation of MC. During heating, MC formed an adhesive layer, which acted as a barrier to prevent moisture loss [22].

Table 2. Proximate chemical composition of plant-based meat and control (beef) with different concentration of methylcellulose.

Ingredient	Beef (Control)			C-TVP			T-ISP			SEM	P Ingre-dient	P Concen-tration	P ing * P Conc.
Concentration	1.5%	3%	4%	1.5%	3%	4%	1.5%	3%	4%				
Moisture	57.49 [b,c,d]	59.46 [a,b]	62.64 [a]	57.43 [b,c,d]	51.54 [e,f]	48.32 [f]	57.80 [b,c]	53.77 [d,e]	54.28 [c,d,e]	1.21	<0.001	0.026	<0.001
Protein	20.56 [b]	21.19 [b]	21.18 [b]	16.48 [b]	16.77 [b]	16.93 [b]	16.96 [b]	16.95 [b]	16.07 [b]	0.71	0.079	0.650	0.345
Fat	18.38 [a]	19.12 [a]	18.85 [a]	14.28 [a]	15.05 [a]	15.83 [a]	17.25 [a]	16.13 [a]	15.28 [a]	2.17	0.126	0.997	0.945
Ash	2.92 [a,b]	2.22 [b]	2.64 [a,b]	3.11 [a]	3.24 [a]	2.84 [a,b]	2.65 [a,b]	2.83 [a,b]	2.65 [a,b]	0.25	<0.001	0.857	0.831
Crude fibre	1.38 [c]	1.56 [c]	1.69 [c]	6.04a [b]	6.87 [a]	7.82 [a]	3.15 [c]	3.66 [bc]	3.70 [b,c]	0.90	<0.001	0.499	0.932

[a–f] Different superscript letters within the same row mean significantly different between treatments ($p < 0.05$). SEM: standard error of mean; *: interaction between ingredient and concentration. C-TVP: Commercial texture vegetable protein; T-ISP: Texture isolate soy protein

The mechanism by which MC gelation is achieved between meat protein and plant-based protein is still unclear. One standard theory is that when in solution, hydrophobic methyl groups along the methylcellulose polymers are surrounded by cage-like structures of water molecules [23]. With increasing temperature, the cage structure is disrupted, and the polymers gradually lose their hydrated water. At the gelation point, polymers' association occurs due to extensive hydrophobic associations between exposed hydrophobic segments [24]. Elevated temperatures highly favor the hydrophobic associations, and strong gels can form [25].

However, in the current study protein belongs to a heterogeneous mixture of different sources. Therefore, purifying the protein following different sources will result in different protein profiles, quality, and functionality [26]. The protein content of three types of patties' varied significantly between various protein sources, with control (beef) indicated higher protein than PBMA patties. Although at any concentration of MC, there was no significant difference detected between C-TVP and T-ISP.

Consequently, PBMA patties with different MC concentrations exhibit no major ($p > 0.05$) difference in fat content among control (beef), C-TVP, and T-ISP respectively. It has been reported that the fat content of plant-based meat is rationally varied as compared to traditional patties [27], however, the fat content of the present study was within the range of Bohrer [27]. Generally, meat analogs are considered low in fat and protein content; however, the new generation of meat analogs products contain substantially greater fat and protein content than traditional meat analog products [9]. Therefore, our argument regarding the average level of fat and protein in meat analog was supported by Ahirwar et al. [28] who described that ready-to-eat meat analog has a good percentage of protein and average fat content manufactured from vegetable and cereal sources.

Irrespective with an application of different concentrations of MC or C-TVP and T-ISP ash content showed no difference. As expected, fiber content for PBMA patties was recorded higher than the control sample, with C-TVP represents the highest value. Similar results were also reported by Bohrer [27] in modern meat analogs. The higher fiber in PBMA patties was probably due to the plants and polysaccharides incorporated into the plant-based patties recipe. The fibrous nature of meat alternatives gives good textural and sensory sensation. Additionally, dietary fiber has been considered to play an essential role in preventing large bowel disease, ischaemic heart disease, and diabetes mellitus [29].

3.2. Physicochemical Analysis

The physiochemical indicators, including pH and colorimetric evaluation, are given in Table 3. There was a significant difference in pH between meat analogs and control (beef patties). The lower pH value of control was likely due to the regular glycolytic changes in meat [30]. However, C-TVP and T-ISP showed a pH of more than 6. The higher pH of PBMA could be due to the slight alkalinity of TVP (pH 7.42–7.43) [31]. Consistent with the current study, Bell and Shelef [32] recorded the pH of minced meat containing vegetable protein had higher pH than as compared to control, while Ahmad et al. [33] also determined that integration of soy protein isolate at 25% expressively increase the pH in meat sausage, which is similar to the outcomes of the present study.

Table 3. Physiochemical characteristics of plant-based meat and control (beef) with different concentration of methylcellulose.

Ingredient	Beef (Control)			C-TVP			T-ISP			SEM	P Ingredient	P Concentration	P ing * P Conc.
Concentration	1.5%	3%	4%	1.5%	3%	4%	1.5%	3%	4%				
pH before	5.54 [d]	5.35 [d]	5.52 [d]	6.34 [c]	6.50 [b,c]	6.35 [c]	6.28 [c]	6.68 [b]	7.08 [a]	0.08	<0.001	0.002	<0.001
pH after	5.71 [e]	5.63 [e,f]	5.51 [f]	6.43 [c]	6.20 [d]	6.77 [a,b]	6.15 [d]	6.61 [b]	6.88 [a]	0.05	<0.001	<0.001	<0.001
L before	46.47 [a]	47.83 [a]	46.93 [a]	45.91 [a]	39.95 [b]	38.88 [b,c]	34.70 [d]	36.63 [c,d]	40.16 [b]	0.85	<0.001	0.456	<0.001
a before	17.83 [a]	16.05 [b]	16.25 [b]	12.39 [d]	14.09 [c]	11.68 [d,e]	12.94 [c,d]	10.83 [e,f]	9.90 [f]	0.41	<0.001	<0.001	0.001
b before	11.87 [c,d]	11.24 [d]	11.55 [d]	17.50 [a]	13.37 [b,c]	13.76 [b]	12.04 [c,d]	11.02 [d]	11.85 [c,d]	0.54	<0.001	0.001	0.011
L after	38.48 [a]	35.46 [a,b]	32.20 [b,c,d]	29.15 [d]	31.15 [c,d]	30.45 [c,d]	31.33 [c,d]	30.71 [c,d]	32.93 [b,c]	1.10	<0.001	0.470	0.009
a after	8.59 [b,c]	8.73 [b,c]	7.75 [c]	10.08 [a,b]	9.77 [a,b]	10.01 [a,b]	10.62 [a]	9.60 [a,b]	8.52 [b,c]	0.53	0.004	0.089	0.359
b after	11.36 [a,b]	12.34 [a,b]	10.19 [a,b]	14.30 [a,b]	12.89 [a,b]	11.49 [a,b]	15.61 [a,b]	12.60 [a,b]	15.56 [a,b]	1.57	0.061	0.539	0.476

[a–f] Different superscript letters within the same row mean significantly different between treatments ($p < 0.05$). SEM: standard error of mean; *: interaction between ingredient and concentration. C-TVP: Commercial texture vegetable protein; T-ISP: Texture isolate soy protein.

Likewise, pH and calorimetric measurements are interconnected with each other. The color coordinates are considered to be one of the essential physical properties in determining consumer acceptance of the product. All patties tended to decrease in lightness (L*) and redness (a*) after cooking. The results show that the control sample before cooking was lighter and redder than PBMA patties (C-TVP and T-ISP). However, our results were in contrast with the reported results of the literature on L* and a* values. Deliza et al. [34] reported an increase in the textured soy protein concentration in beef patties increased the L* values, but a* values were not statistically different. Hidayat et al. [35] also found a similar observation on the beef sausage. The variation of L* and a* values in the present study compared to other studies could be due to the substitution of plant-based proteins (100% substitution) in the formulation. The small globules from meat, such as water and fat, can cause more light reflection, which will probably contribute to higher lightness [36].

The a* values of control before cooking were higher than C-TVP and T-ISP treatments due to the myoglobin pigment in red meat. However, an increase in myoglobin denaturation can be shown by the lower a* values after applying the heat treatment [37], which in tandem with our result in Table 3. The cooking did not affect the redness values of textured soybean protein incorporated samples [34]. Similar effects were noticed in the raw and cooked samples incorporated with either C-TVP or T-ISP as described in Table 3. The b* values of C-TVP and T-ISP before cooking were higher than control. The yellowish coloration of PBMA patties can be associated with the yellow color of soy protein ingredients. Subsequently, the yellowish-brown color initially, affecting the final products' [9]. However, MC's concentration at different percentages only plays a minor role in reducing the b* values of raw and cooked patties.

In the current study, WHC is expressed in two parts, RW and CL, shown in Table 4. The concentration of MC had a significant effect on the RW and CL. An increase in MC concentration from 1.5% to 4.0% lowered the RW and CL in all treatments. These findings were similar to the result reported by Hill and Prusa [38] for beef patties. They described that cellulose hydrocolloids bind moisture in product formulation, and it can gel upon heating. According to Hill and Prusa [38], surface moisture probably would not be affected by gum addition; therefore, evaporative losses were not affected by treatment. Consequently, the present data shows that MC's incorporation did not increase cooked moisture content, but it generally reduced total cook loss. Previously Arora et al. [3] proved that carrageenan and xanthan gum types binding agents had a higher yield than protein-based binding agents. At the same time, they concluded that WHC depends upon protein binding properties, which consequently agreed with our results.

Table 4. Water-holding capacity and tenderness related measurement of plant-based meat and control (beef) with different concentration of methylcellulose.

Ingredient	Beef (Control)			C-TVP			T-ISP			SEM	P Ingredient	P Concentration	P ing * P Conc.
Concentration	1.5%	3%	4%	1.5%	3%	4%	1.5%	3%	4%				
Release water (%)	3.9 ^{a}	1.8b ^{c,d}	1.8 ^{b,c,d}	4.06 ^{a}	2.21 ^{b,c}	1.44 ^{c,d}	3.47 ^{a}	2.43 ^{b}	1.77 ^{d}	0.20	0.980	<0.001	0.044
Cooking loss (%)	7.91 ^{c,d}	6.55 ^{d,e,f}	5.36 ^{f}	9.98 ^{b}	8.69 ^{b,c}	7.12d ^{e}	12.01 ^{a}	7.30 ^{c,d,e}	6.11 ^{e,f}	0.48	<0.001	<0.001	0005
WBSF (N)	3.6 ^{b,c}	3.80 ^{b}	4.26 ^{a}	2.14 ^{f,g}	2.74 ^{e}	3.20 ^{d}	2.41 ^{e,f}	2.41 ^{e,f}	3.29 ^{c,d}	0.12	<0.001	<0.001	0.044
Diameter before	14.99 ^{a,b}	15.83 ^{a,b}	15.52 ^{a,b}	16.63 ^{a,b}	15.47 ^{a,b}	15.83 ^{a,b}	16.30 ^{a,b}	15.37 ^{a,b}	15.70 ^{a,b}	0.38	0.256	0.406	0.121
Diameter after	11.59 ^{d}	12.30 ^{c,d}	12.81 ^{c,d}	14.96 ^{a}	13.57 ^{b,c}	15.16 ^{a}	15.08 ^{a}	14.29 ^{a,b}	14.81 ^{a,b}	0.41	<0.001	0.308	0.038

$^{a-f}$ Different superscript letters within the same row mean significantly different between treatments ($p < 0.05$). SEM: standard error of mean; *: interaction between ingredient and concentration. C-TVP: Commercial texture vegetable protein; T-ISP: Texture isolate soy protein; WBSF: Warner-Bratzler shear force.

Subsequently, WBSF for control represents the highest value, and there is no significant difference between C-TVP and T-ISP treatments. The softer textural properties of C-TVP and T-ISP affect their shear force values. Ruiz de Huidobro et al. [39] reported that shear force value was significantly correlated to hardness, springiness, and chewiness. The shear force in meat is a good measure of initial bite tenderness, which can cause changes during the cooking process are related to heat-induced alteration of myofibrillar proteins and connective tissue, as solubilizes the connective tissue leading to meat tenderization. In contrast, the denaturation of myofibrillar proteins causes meat toughening [40]. The finding of the current study aligns with the Danowska-Oziewicz [41], who detected lower values shear force for the samples containing soy isolate protein as likened to control (pork patties).

Diameter before and after cooking of control and PBMA patties are presented in Table 4. The degree of shrinkage (diameter after) was ranged from about 17.46–22.68% for control,

4.23–12.28% for C-TVP, and 3.64–8.98% for T-ISP. Control represents higher shrinkage due to the connective tissue denaturation and fluid (moisture and fat) loss Table 4. The substitution with plant-based protein reduces the shrinkage markedly in T-ISP, although no difference to C-TVP. According to Gujral et al. [42] the addition of fibers and non-meat protein ingredients may reduce diameter shrinkage and weight loss. Similarly, in the current study, the increase in MC concentration (4%) decreases shrinkage of all patties (control: 17.46%, TVP: 4.23%, and T-ISP: 3.64%).

3.3. Visible Appearance

The external and internal appearance of meat analogs before and after cooking has been presented in Figure 2. The external appearance before cooking showed no difference in observation at different concentrations of MC. However, MC's effect can be seen after thermal treatment, in which the higher concentration (4%) can maintain the structure of patties. MC is essentially incorporated in some modern meat analog due to product consistency and binds all ingredients together to be more intact and stable [9]. MC is a useful binder, especially on the meat analog that does not require pre-heat for gel formation due to its unique thermal gelling and right emulsifier properties [43].

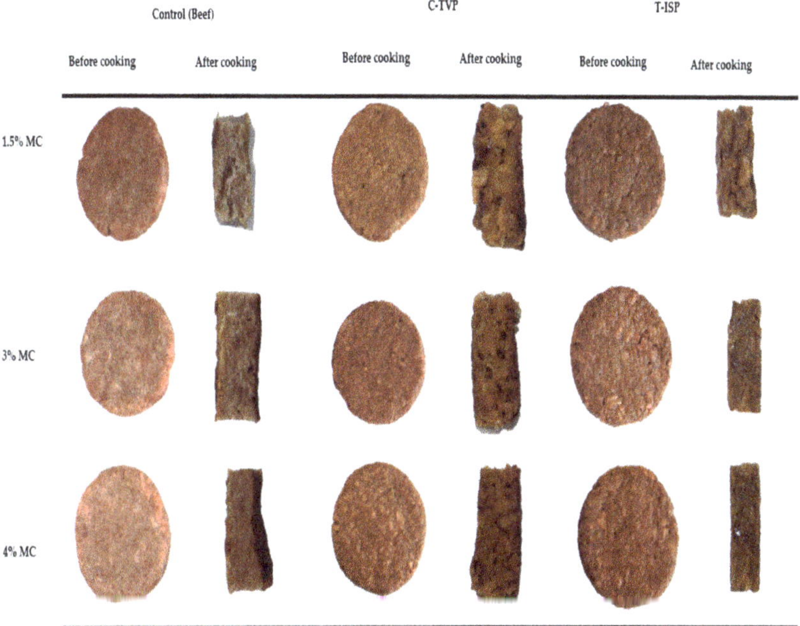

Figure 2. The external-internal appearance of cooked and uncooked plant-based meat patties.

The drawback of using TVP and T-ISP is that we can see the patties' surface's granular appearance. The internal appearance of all patties appeared more homogenous and cohesive with a higher concentration of MC. This proved that the addition of MC could bind well all the ingredients. The interior of C-TVP and T-ISP patties show a rough with intact and no crack appearance. Nevertheless, the interior of C-TVP shows more finely structure than T-ISP patties. The probable reason could be due to adequate hydration of C-TVP during the preparation of the dough. Earlier, MC's phenomena as a binder have been reported, which confirmed that MC helps maintain product shape and firm texture in various commercially available products, i.e., impossible burgers and beyond burgers [27].

3.4. Texture Profile Analysis

The textural properties are crucial for developing meatless patties because in meat analog's texture is an essential factor in mimicking the organoleptic taste of muscle. Figure 2 illustrates the textural parameters, including hardness, chewiness, gumminess cohesiveness, and meat analogs' springiness with different concentrations of MC%. The hardness, chewiness, and gumminess of control were significantly higher in comparison to C-TVP and T-ISP. The higher hardness in control was expected due to the muscle proteins denaturation phenomenon, which led to hardness in the meat system [19]. It is evident from the shrinkage percentage shown in Table 4, whereby meat protein has a higher degree of shrinkage than plant-based proteins. An increase in MC concentration from 1.5% to 4% increases 'hardness of all patties. The current result was consistent with the results reported by Arora et al. [3], who described that when the binding agent increased, the hardness, chewiness, gumminess, and compression values increased proportionally. Similarly, Ayadi et al. [44] reported that incorporating carrageenan at higher concentration (0.5% to 1.5%) increased hardness of sausage products. The reason for lower hardness values in TVP and T-ISP treatments were due to extensive hydration of textured protein with water at the early stage of the processing phase, ultimately causes the PBMA patties to be softer. According to, Ruiz de Huidobro et al. [39] hardness, chewiness, and springiness are instrumental parameters for assessing meat texture.

However, in the present study, only springiness values of PBMA patties (TVP and T-ISP) showing marginally or no difference to the control. The hardness and chewiness values were showed a substantial difference between treatments and control. As we mentioned earlier (introduction), to mimic conventional beef patties' textural properties is the most challenging part in the development of meat analogs.

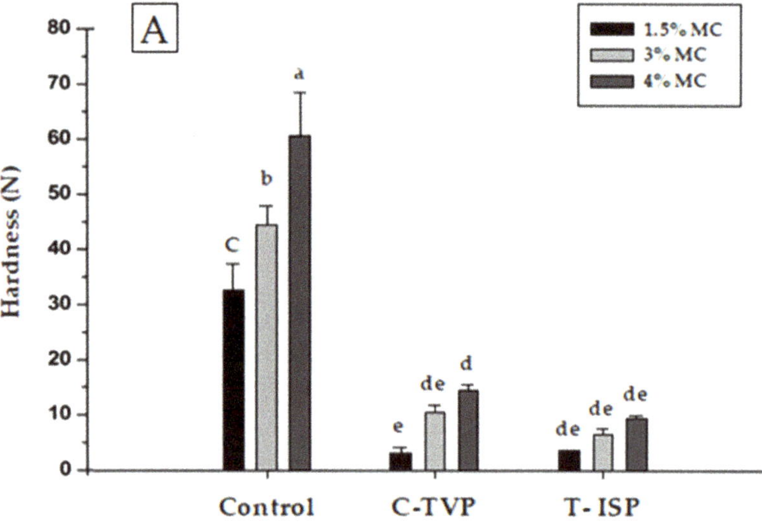

Figure 2. *Cont.*

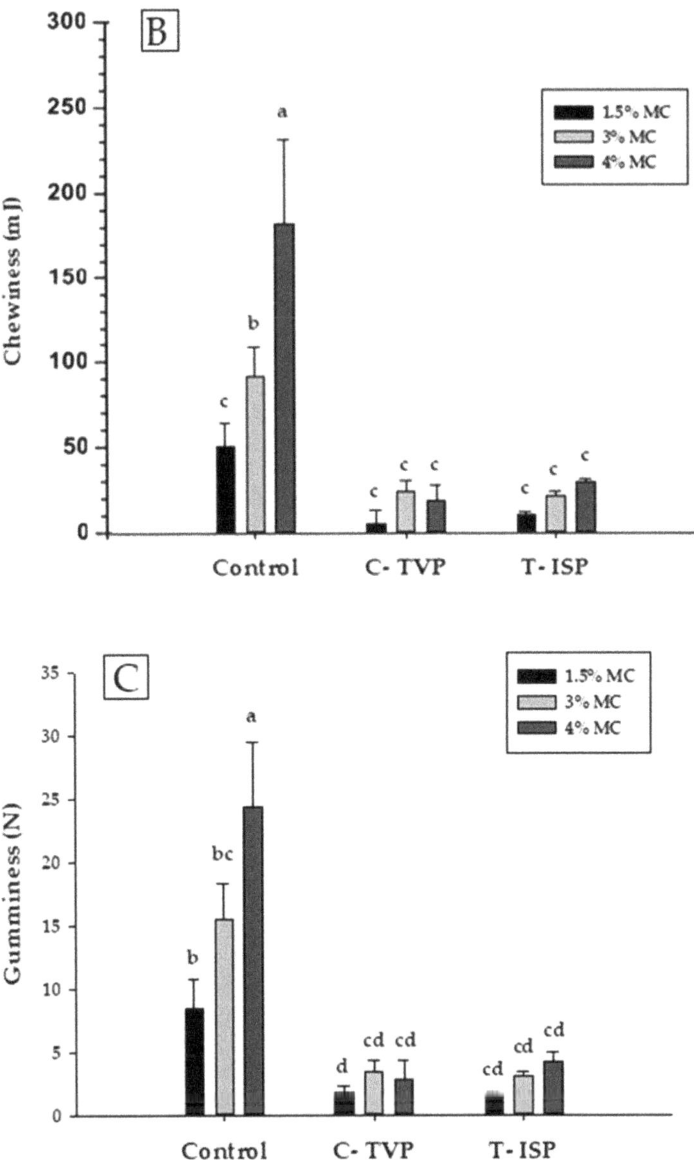

Figure 2. *Cont.*

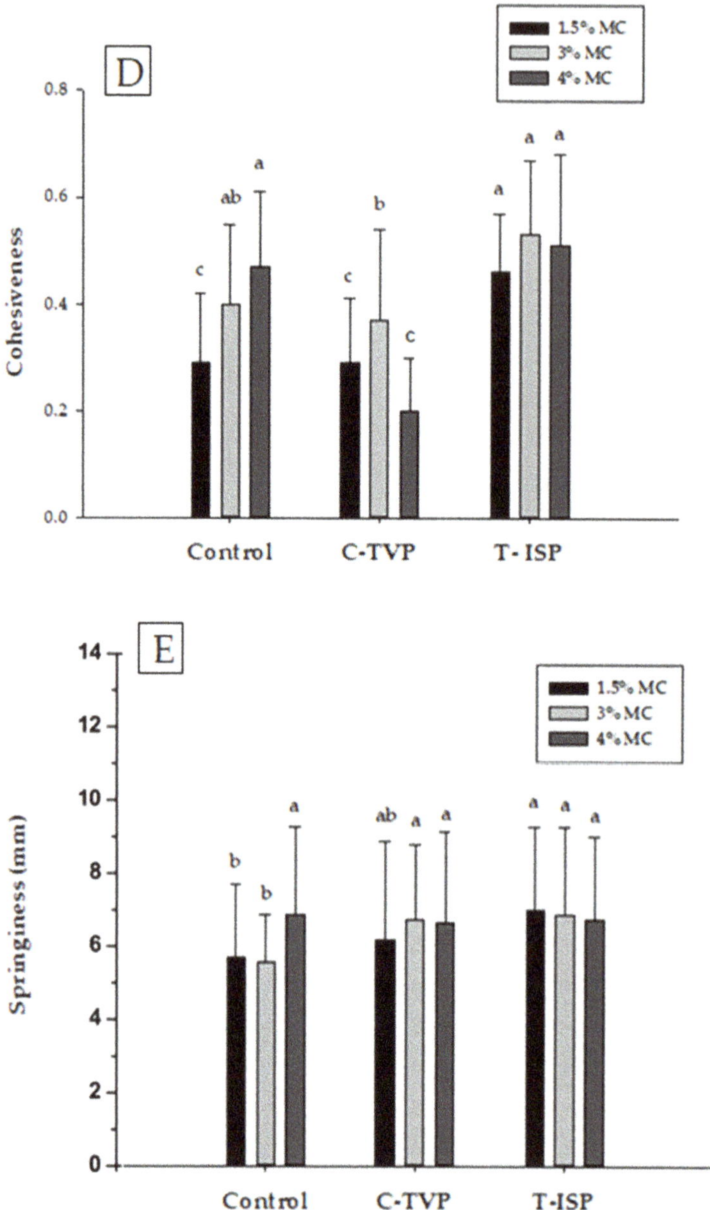

Figure 2. The texture profile analysis (TPA) parameters of plant-based meat on the type of texture soy isolate protein and different methylcellulose concentrations. (**A**) Hardness (N); (**B**) Chewiness (mJ); (**C**) Gumminess (N); (**D**) Cohesiveness; (**E**) Springiness (mm). Those are just different concentrations of MC (Methylcellulose) and the concentration is there on the top right.

3.5. Sensory Evaluation

Sensory parameters are a chief concern for the development of PBMA patties using MC as a binder. The sensory traits for control (beef), C-TVP and T-ISP are presented in Figure 3. Based on the percentage of MC, the control patties expressing higher values in

4% MC for shape, firmness and color, although panelists scored higher, appearance and overall acceptability with 1.5% MC respectively. C-TVP patties obtained the highest score for appearance, shape, firmness, color and overall acceptability with 3% and 4% of MC concentration. Though, T-ISP samples incorporating 3% MC performed well than 1.5% and 4% MC concentration. The subjective evaluation demonstrated a clear preference towards 3% MC in PBMA patties (C-TV and T-ISP). Samples with the integration of 1.5 and 4% of MC were the least preferred on sensory evaluation basis. In contrast to our study, Imkyung et al. [45] described that with hydroxypropyl methylcellulose application as an animal fat replacer for meat patties, there is no significant difference in color, flavor and taste; however, tenderness, juiciness and overall acceptability show a statistically significant difference.

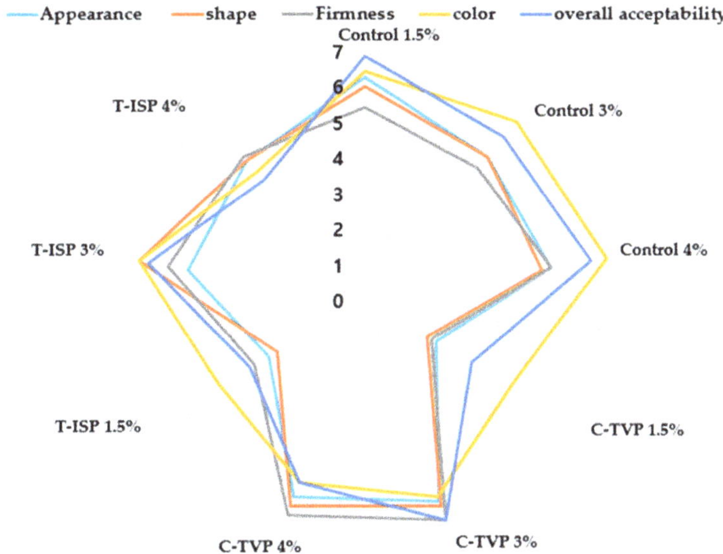

Figure 3. Sensory profile of plant-based meat-based with different soy isolate protein and methylcellulose percentage.

Based on previous literature, it has been reported that the application of water alone in ground beef patties in control without methylcellulose and hydroxypropyl methylcellulose did not satisfy the sensory panel; they further reported that color and aroma of ground patties were least affected by the application of methylcellulose [46]. The vast variability of PBMA patties as compared to control could be due to plant derived proteins (soy and wheat protein) in meat analogs expressing more elastic, rubbery and chewy sensation and poor mouth feel due to their agglomeration properties.

Moreover, previous literature confirmed that incorporating a different type of soy family (soy paste, soy protein isolate or texture soy protein) generates a unique beany essence in meat products and downgrade sensory scores [41]. Remarkably, in the current study, no beany essence was noticed. The possible reason might be due to various types of plant-based ingredients (Table 1) used to mask the beany flavor in PBMA patties successfully. Furthermore, due to natural differences between muscle and plant materials, i.e., structure and size of protein molecules, amino acid composition, peptide sequence, and the chemical composition of both intracellular and extracellular materials, it is difficult to reproduce the complex and delicate sensory profile of animal meat products.

4. Conclusions

The present study assessed the physicochemical, textural, and sensory properties of PBMA patties with two types of texturized soy isolate protein (C-TVP and T-ISP) and incorporation of different concentrations of binding agent (MC). The addition of MC significantly affected the quality characteristics of C-TVP and T-ISP-based PBMA patties. C-TVP with 3% MC showed promising results, with adequate physicochemical, textural parameters, and with satisfactory patty visible appearance, thereby improving the comprehensive process yield compared to T-ISP. Although samples with 4% MC also exhibit similar results compared to 3 % MC, they failed to satisfy the sensory panelist in C-TVP and T-ISP. Using beef as a control, it can be concluded that C-TVP with a 3% MC (binding agent) is recommended to prepare acceptable PBMA patties with good physicochemical, textural, and sensory acceptability.

Author Contributions: All of the authors contributed significantly to the research. A.B. wrote the manuscript. Contribution to experimental work by A.B., S.-J.L., E.-Y.L. and N.S., S.-T.J. designed experimental work, and the manuscript was reviewed and revised by Y.-H.H. All authors have read and agreed to the published version of the manuscript.

Funding: This research was supported by the National Research Foundation of Korea (NRF) grant funded by the Korea government (MSIT) (No. 2020R1I1A206937911).

Conflicts of Interest: The authors declare that they have no conflict of interest.

References

1.   Ismail, I.; Hwang, Y.-H.; Joo, S.-T. Meat analog as future food: A review. *J. Anim. Sci. Technol.* **2020**, *62*, 111–120. [CrossRef]
2.   Wild, F. *Manufacture of Meat Analogues through High Moisture Extrusion*; Elsevier: Amsterdam, The Netherlands, 2016.
3.   Arora, B.; Kamal, S.; Sharma, V. Effect of Binding Agents on Quality Characteristics of Mushroom Based Sausage Analogue. *J. Food Process.* **2017**, *41*, e13134. [CrossRef]
4.   Apostolidis, C.; McLeay, F. Should we stop meating like this? Reducing meat consumption through substitution. *Food Policy* **2016**, *65*, 74–89. [CrossRef]
5.   Asgar, M.; Fazilah, A.; Huda, N.; Bhat, R.; Karim, A. Nonmeat protein alternatives as meat extenders and meat analogs. *Compr. Rev. Food Sci. Food Saf.* **2010**, *9*, 513–529. [CrossRef] [PubMed]
6.   Tubiello, F.; Salvatore, M.; Cóndor Golec, R.; Ferrara, A.; Rossi, S.; Biancalani, R.; Federici, S.; Jacobs, H.; Flammini, A. *Agriculture, Forestry and Other Land Use Emissions by Sources and Removals by Sinks*; FAO Statistics Devision: Rome, Italy, 2014.
7.   Sadler, M.J. Meat alternatives—market developments and health benefits. *Trends Food Sci. Technol.* **2004**, *15*, 250–260. [CrossRef]
8.   de Bakker, E.; Dagevos, H. Reducing meat consumption in today's consumer society: Questioning the citizen-consumer gap. *J. Agric. Environ.* **2012**, *25*, 877–894. [CrossRef]
9.   Kyriakopoulou, K.; Dekkers, B.; van der Goot, A.J. Plant-based meat analogues. In *Sustainable Meat Production and Processing*; Elsevier: Amsterdam, The Netherlands, 2019; pp. 103–126.
10. Ranasinghesagara, J.; Hsieh, F.H.; Yao, G. An image processing method for quantifying fiber formation in meat analogs under high moisture extrusion. *J. Food Sci.* **2005**, *70*, 450–454. [CrossRef]
11. Malav, O.; Talukder, S.; Gokulakrishnan, P.; Chand, S. Meat analog: A review. *Crit. Rev. Food Sci. Nutr.* **2015**, *55*, 1241–1245. [CrossRef]
12. Garcıa, M.; Dominguez, R.; Galvez, M.; Casas, C.; Selgas, M. Utilization of cereal and fruit fibres in low fat dry fermented sausages. *Meat Sci.* **2002**, *60*, 227–236. [CrossRef]
13. Al-Tabakha, M.M. HPMC capsules: Current status and future prospects. *J. Pharm. Pharm. Sci.* **2010**, *13*, 428–442. [CrossRef]
14. AOAC. *Official Methods of Analysis*, 16th ed.; Association of Official Analytical Chemists: Arlington, TX, USA, 2002.
15. ANKOM. *Crude Fiber Analysis in Feeds by Filter Bag Technique*; ANKOM: Macedon, NY, USA, 2008.
16. Joo, S.T. Determination of water-holding capacity of porcine musculature based on released water method using optimal load. *Korean J. Food Sci. An.* **2018**, *38*, 823–828.
17. Biswas, S.; Chakraborty, A.; Sarkar, S. Comparison Among the Qualities of Patties Prepared from Chicken Broiler, Spent Hen and Duck Meats. *J. Poult. Sci.* **2006**, *43*, 180–186. [CrossRef]
18. AMSA. *Research Guidelines for Cookery*; American Meat Science Association in Cooperation with National Live Stock and Meat Board: Chicago, IL, USA, 1995; p. 8.
19. Ismail, I.; Hwang, Y.-H.; Joo, S.-T. Interventions of two-stage thermal sous-vide cooking on the toughness of beef semitendinosus. *Meat Sci.* **2019**, *157*, 107882. [CrossRef]
20. Lawless, H.T.; Heymann, H. *Sensory Evaluation of Food: Principles and Practices*; Springer Science & Business Media: Berlin/Heidelberg, Germany, 2013.

21. Rahman, M.S.; Seo, J.-K.; Zahid, M.A.; Park, J.-Y.; Choi, S.-G.; Yang, H.-S. Physicochemical properties, sensory traits and storage stability of reduced-fat frankfurters formulated by replacing beef tallow with defatted bovine heart. *Meat Sci.* **2019**, *151*, 89–97. [CrossRef] [PubMed]
22. Amboon, W.; Tulyathan, V.; Tattiyakul, J. Effect of hydroxypropyl methylcellulose on rheological properties, coating pickup, and oil content of rice flour-based batters. *Food Bioprocess Tech.* **2012**, *5*, 601–608. [CrossRef]
23. Sarkar, N. Thermal gelation properties of methyl and hydroxypropyl methylcellulose. *J. Appl. Polym. Sci.* **1979**, *24*, 1073–1087. [CrossRef]
24. Sarkar, N.; Walker, L. Hydration—Dehydration properties of methylcellulose and hydroxypropylmethylcellulose. *Carbohydr. Polym.* **1995**, *27*, 177–185. [CrossRef]
25. Walstra, P. *Physical Chemistry of Foods*; CRC Press: Boca Raton, FL, USA, 2002.
26. Ismail, B.P.; Senaratne-Lenagala, L.; Stube, A.; Brackenridge, A. Protein demand: Review of plant and animal proteins used in alternative protein product development and production. *Anim. Front.* **2020**, *10*, 53–63. [CrossRef]
27. Bohrer, B.M. An investigation of the formulation and nutritional composition of modern meat analogue products. *Food Sci. Hum. Well.* **2019**, *8*, 320–329. [CrossRef]
28. Ahirwar, R.; Jayathilakan, K.; Reddy, K.; Pandey, M.; Batra, H. Development of mushroom and wheat gluten based meat analogue by using response surface methodology. *IJAR* **2015**, *3*, 923–930.
29. Trowell, H. Dietary fibre, ischaemic heart disease and diabetes mellitus. *Proc. Nutr. Soc.* **1973**, *32*, 151–157. [CrossRef]
30. Chauhan, S.S.; LeMaster, M.N.; Clark, D.L.; Foster, M.K.; Miller, C.E.; England, E.M. Glycolysis and pH Decline Terminate Prematurely in Oxidative Muscles despite the Presence of Excess Glycogen. *Meat Muscle Biol.* **2019**, *3*. [CrossRef]
31. Anjum, F.M.; Naeem, A.; Khan, M.I.; Nadeem, M.; Amir, R.M. Development of texturized vegetable protein using indigenous sources. *Pak. J. Food Sci.* **2011**, *21*, 33–44.
32. Bell, W.; Shelef, L. Availability and microbial stability of retail beef-soy blends. *J. Food Sci.* **1978**, *43*, 315–333. [CrossRef]
33. Ahmad, S.; Rizawi, J.; Srivastava, P. Effect of soy protein isolate incorporation on quality characteristics and shelf-life of buffalo meat emulsion sausage. *J. Food Sci. Technol.* **2010**, *47*, 290–294. [CrossRef] [PubMed]
34. Deliza, R.; SALDIVAR, S.S.; Germani, R.; Benassi, V.; Cabral, L. The effects of colored textured soybean protein (TSP) on sensory and physical attributes of ground beef patties. *J. Sens. Stud.* **2002**, *17*, 121–132. [CrossRef]
35. Hidayat, B.; Wea, A.; Andriat, N. Physicochemical, sensory attributes and protein profile by SDS-PAGE of beef sausage substituted with texturized vegetable proteins. *Food Res.* **2018**, *2*, 20–31. [CrossRef]
36. Barbut, S.; Marangoni, A. Organogels use in meat processing–Effects of fat/oil type and heating rate. *Meat Sci.* **2019**, *149*, 9–13. [CrossRef]
37. Sen, A.R.; Naveena, B.M.; Muthukumar, M.; Vaithiyanathan, S. Colour, myoglobin denaturation and storage stability of raw and cooked mutton chops at different end point cooking temperature. *J. Food Sci. Technol.* **2014**, *51*, 970–975. [CrossRef] [PubMed]
38. Hill, S.; Prusa, K. Physical and sensory properties of lean ground beef patties containing methylcellulose and hydroxypropyl methylcellulose. *J. Food Qual.* **1988**, *11*, 331–337. [CrossRef]
39. de Huidobro, F.R.; Martínez, V.C.; Gómez, S.L.; Velasco, S.; Pérez, C.; Onega, E. Sensory characterization of meat texture in sucking lamb: Methodology. *Investig. Agrar. Prod. Sanid. Anim.* **2001**, *16*, 245–256.
40. Laakkonen, E.; Wellington, G.; Sherbon, J. Low-temperature, long-time heating of bovine muscle 1. Changes in tenderness, water-binding capacity, pH and amount of water-soluble components. *J. Food Sci.* **1970**, *35*, 175–177. [CrossRef]
41. Danowska-Oziewicz, M. Effect of soy protein isolate on physicochemical properties, lipid oxidation and sensory quality of low-fat pork patties stored in vacuum, MAP and frozen state. *J. Food Process.* **2014**, *38*, 641–654. [CrossRef]
42. Gujral, H.S.; Kaur, A.; Singh, N.; Sodhi, N.S. Effect of liquid whole egg, fat and textured soy protein on the textural and cooking properties of raw and baked patties from goat meat. *J. Food Eng.* **2002**, *53*, 377–385. [CrossRef]
43. Sanz, T.; Fernández, M.; Salvador, A.; Munoz, J.; Fiszman, S. Thermogelation properties of methylcellulose (MC) and their effect on a batter formula. *Food Hydrocoll.* **2005**, *19*, 141–147. [CrossRef]
44. Ayadi, M.; Kechaou, A.; Makni, I.; Attia, H. Influence of carrageenan addition on turkey meat sausages properties. *J. Food Eng.* **2009**, *93*, 278–283. [CrossRef]
45. Oh, I.; Lee, J.; Lee, H.G.; Lee, S. Feasibility of hydroxypropyl methylcellulose oleogel as an animal fat replacer for meat patties. *Food Res. Int.* **2019**, *122*, 566–572. [CrossRef]
46. Raharjo, S.; Sofos, J.; Schmidt, G. Performance of Methylcellulose and Hydroxypropyl-methylcellulose as Moisture Retention and Gelling Additives in Low-Fat Ground Beef Patties. *Indones. Food Nutr. Prog.* **1996**, *3*, 1.

Article

The Role of Coffee Silver Skin against Oxidative Phenomena in Newly Formulated Chicken Meat Burgers after Cooking

Maria Martuscelli *, Luigi Esposito and Dino Mastrocola

Faculty of Bioscience and Technology for Food, Agriculture and Environment, University of Teramo, Via R. Balzarini 1, 64100 Teramo, Italy; lesposito2@unite.it (L.E.); dmastrocola@unite.it (D.M.)
* Correspondence: mmartuscelli@unite.it; Tel.: +39-0861-266-912

Abstract: Coffee Silver Skin (CSS) is the unique by-product discarded after the roasting of coffee beans. This research aimed to evaluate the effect of two levels of CSS (1.5% and 3%) added as a natural ingredient in new formulations of chicken meat burgers. This is one of the first studies proposing a "formulation approach" to control the emergence of off flavours after meat cooking. Physical, chemical, and sensory analyses were carried out, within the CSS content and the evolution of volatile organic compounds in different samples. Newly formulated chicken burgers could limit food waste, while also becoming a source of fibres, minerals, and bioactive molecules. CSS limited weight losses (after cooking process) to 10.50% (1.5% addition) and 11.05% (3% addition), significantly lower ($p < 0.01$) than the control (23.85%). In cooked burgers, the occurrence of hexanal was reduced from 55.1% (CTRL T_0) to 11.7% (CSS T_0 1.5%) to 0 (CSS T_0 3%). As for the limitation of off-flavours, CSS also showed good activity, contrasting with the emergence of octanal, alcohols and other markers of lipid oxidation. From the sensory test carried out, the volatile profile of CSS does not seem to impair the flavour of burgers, though at higher percentages hydrocarbons and pyrazines are traceable. The thiobarbituric acid reactive substances (TBARS assay confirmed the protective effect of CSS against oxidation.

Keywords: coffee by-products; chicken burger; meat formulation; cooking yield; volatile compounds; warmed off-flavours

Citation: Martuscelli, M.; Esposito, L.; Mastrocola, D. The Role of Coffee Silver Skin against Oxidative Phenomena in Newly Formulated Chicken Meat Burgers after Cooking. *Foods* **2021**, *10*, 1833. https://doi.org/10.3390/foods10081833

Academic Editors: Gonzalo Delgado-Pando and Tatiana Pintado

Received: 1 July 2021
Accepted: 5 August 2021
Published: 8 August 2021

Publisher's Note: MDPI stays neutral with regard to jurisdictional claims in published maps and institutional affiliations.

1. Introduction

Minimally processed raw by-products are available in large quantities and their re-utilization might be enhanced to recover bioactive compounds, on top of their promising technological properties [1]. Many valuable molecules such as phenolic acids, carotenoids, and flavonoids can mitigate oxidation occurrence, so there is an increased demand for new methods and technologies to recover and use these [2]. Many publications attest the positive role of by-products' addition in meat formulations to limit oxidation occurrence [3,4].

Oswell et al. [5] explain how some unprocessed food components can help reducing the list of ingredients of a formulation, supporting trends towards the clean and green label. In fact, many by-products have technological properties, acting as additives and ingredients [6].

Coffee Silver Skin (CSS) is a thin layer tightly adherent to coffee seeds, present in all coffee species and impossible to separate when seeds are unroasted [7]. Among all the by-products of the coffee industry, CSS is unique in being discarded immediately after the roasting step [8].

Common features of CSS are high content in fibres (both soluble and insoluble), in minerals such as Calcium and Potassium, and in capability as an adsorbing material [2].

To our knowledge, there are no studies of the inclusion of untreated coffee silver skin (CSS) in meat-based foods such as chicken products. It is well known that poultry meat is easily oxidised; its content of monounsaturated fatty acids (MUFAs) and polyunsaturated fatty acids (PUFAs) (including phospholipids that are distributed in muscles and cellular membranes) make it the elective substrate for lipid oxidation [3]. Feeding of animals is responsible for many qualitative characteristics of meat, progressing to the eventual development of off-flavours (after slaughtering) [9,10]. In chicken products, clove, oregano, thyme, and sage were successfully used as natural substitutes of synthetic antioxidants [11]. Kim et al. [12] used residues of coffee brewing (spent coffee) as antioxidants against meat oxidation in raw and cooked samples. Recently Delgado-Ospina et al. [13] added cocoa pod husk flour, discovering an interesting application for quality improvement of frankfurters. Cooked and refrigerated meat products develop undesired rancidity and a great variety of off flavours known as warmed over flavours (WOF). These defects can also come about by heating ready to eat foods or high-processed meat-based items and can be present in many products where food remains exposed to light, oxygen, and heat for a long time (canteens, fast food outlets, collective restaurants). In the study by Lungu et al. [14], most respondents affirmed exposure to WOF defective foods, especially ready to eat meats; moreover, besides the reduced sensory quality, respondents confirmed a preference for defective foods due to their lower cost. The development of WOFs does not impair food safety, but a high oxidation rate affects the nutritional profile; therefore a huge quantity of safe food is discarded daily due to detrimental sensory defects.

In this scenario, the present research investigated the chemical and physical properties of CSS against oxidative phenomena after the cooking of chicken burgers. This study points to the nutritional advantages of including CSS as a new ingredient for chicken burger recipes, while testing some technological functionalities. Lastly, we were able to evaluate the role of CSS on the shelf life of refrigerated cooked burger, focusing on the spreading of WOF and oxidation markers along with an analysis of the volatile compounds and a sensory test with trained panellists.

2. Materials and Methods

2.1. Coffee Silver Skin

Coffee silver skin (CSS) was received from the toasting plant Marcafè Torrefazione Adriatica s.p.a. (Giulianova, Italy). After 10 cycles of roasting (10 × 240 kg of roasted coffee), 3.3 kg of CSS were recovered. CSS used for this experiment was a blend of 5 arabica varieties (*Coffea arabica*) (India Arabica, India Cherry, Vietnam, India Mysore, and India Caracolito) and 5 robusta varieties (*Coffea canephora*) (India Parchment, Santos, Uganda CRV 18, Uganda CRV 17, Togo). CSS was ground at 10,200 rpm for 1 min (Bimby®, mod. TM 31, Wuppertal, Germany) until arriving at a particle size of 125–250 μm. Then, physical, and chemical analyses were carried out. CSS was kept frozen at −20 °C until analysis. Qualitative analyses were carried out and results were reported in our previous study [2].

2.2. Preparation of Burgers

Chicken breast fillets were purchased on the market (antibiotic-free, genetically modified organism-free diet, and high welfare/partial free range system meat).

Chicken burgers were obtained from 1 kg of fresh breast fillets with the addition of 1.4% salt and 5.0% water. These ingredients were cut and mixed for 2 min at 1800 rpm with Bimby® mixer (Wuppertal, Germany), mod. TM 31, to obtain a perfectly homogenised blend.

From the whole mixture (meat, water, and salt) 3 batches were obtained: control, without any addition of coffee silver skin (CTRL), coffee silver skin addition of +1.5% (CSS 1.5%) and coffee silver skin addition of +3.0% (CSS 3.0%). CSS was added by mixing for 30 s at 500 rpm.

8 burgers (45 g each approximately) were prepared for each experimental batch (Figure 1).

(a)   (b)   (c)

Figure 1. Uncooked burgers of the three experimental batches: (**a**) control (CTRL) without CSS additions; (**b**) with addition of +1.5% of coffee silver skin (CSS 1.5%); (**c**) with +3.0% of coffee silver skin (CSS 3.0%).

Then, burgers were cooked on an electrical griddle Bosch (München, Germany), mod. TFB4431V, potency 2000 W for 4 min until reaching an internal temperature of 90–92 °C. Cooked burgers are shown in Figure 2. After cooking, some burgers were eaten during the panel test, and others were left singularly covered with a plastic oxygen permeable film at +4 °C for 72 and 120 h. The utilization of this covering film was chosen to allow the permeability of O_2 and thus the spreading of WOFs and other products of oxidation. Moreover, this condition is the closest to what can happen to consumers at home. The trial was replicated on another two independent occasions.

(a)   (b)   (c)

Figure 2. Cooked burgers of the three experimental batches: (**a**) control (CTRL); (**b**) with addition of +1.5% of coffee silver skin (CSS 1.5%); (**c**) with +3.0% of coffee silver skin (CSS 3.0%).

2.3. Physico-Chemical, Colour, and Compositional Analyses

The values of water activity (a_w) were obtained with the Aqualab 4 TE kit (Court Pullman, WA, USA). Values of pH were taken with a pH meter (model 3510, Jenway, Stone, UK). All values were measured in triplicate.

Colour was determined in different locations of burger samples by a colorimeter CR-5 (Spectrally based, Konica Minolta, Tokyo, Japan) with D_{65} light source and observer 10°. Colour was expressed as L* (lightness, intensity of white colour), a* (+a, red; −a, green) and b* (+b, yellow; −b, blue) values. Samples were measured in triplicate and at least fifteen measurements were obtained for each batch. To better define the final color observed, the saturation index (chroma, C*) was calculated according to Formula (1).

$$C^* = (a^{*2} + b^{*2})^{1/2} \tag{1}$$

Proximate analysis on moisture, proteins, and ashes was obtained following the Association of Official Analytical Chemists procedure [15]. Total lipids were measured using a modification of the chloroform to methanol procedure described by Folch et al. [16].

The determination of such micronutrients as calcium, potassium, and total dietary fibers (TDF) was performed by calculating their presence in a 45 g burger at different formulations; the estimation of Ca, K, and TDF accounts of the values found in our previous work on the characterization of CSS [2]. Quantities found refer to the presence of a defined element and not on bioavailability. We used easy proportions to find percentages.

Here, as example, we propose the calculation used for the calcium determination (2), (3) in 1.5% CSS formulation:

$$546.5 \text{ mg:}100 \text{ g} = 3.3 \text{ g:}x \tag{2}$$

$$x = \frac{(546.5 \text{ mg} * 100 \text{ g})}{3.3 \text{ g}} = 16.54 \text{ mg} \tag{3}$$

where 546.5 mg is the Ca content in 100 g of CSS, 3.3 g is the amount of CSS in a 45 g burger, and 16.54 mg is the intake of Ca in a 45 g burger formulated with 1.5% of CSS.

2.4. Cooking Yield

The cooking yield parameter is a useful and practical tool to easily calculate the quantity of meat available for consumption after the cooking process. Uncooked samples were prepared and weighted singularly, then underwent the established cooking process and weighed again. This formula was used to arrive at the result (4)

$$100 - \left(\frac{(\text{raw burger weight} - \text{cooked burger weight})}{(\text{raw burger weight})} \times 100 \right) \tag{4}$$

2.5. Thiobarbituric Acid Reacting Substances (TBARS) Assay

A thiobarbituric reactant species test was carried out following the methods of Soyer et al. [17] with some modifications. Raw meat (25 g) was ground in 125 mL of pure water for 2 min to homogenise the mixture. From this, 5 mL were filtered and transferred in falcon tubes (15 mL) with 3 mL of a solution containing trichloroacetic acid (15%, w/v) and thiobarbituric acid (80 mM) in HCl 0.25 N. Samples underwent a centrifugation step (2000 rpm for 5 min) to precipitate proteins. After centrifugation, 3 mL were transferred in tapped glass tubes and kept at 40 °C for 90 min.

Samples obtained were read at 532 nm with a spectrophotometer UV-VIS (Jenway, Stone, UK) after a further filtration with filters 0.45 μm. All samples were read in double, and data were expressed as mean ± standard deviation.

The calibration curve was obtained by using a 1,1,3,3-tetraetoxypropane (Sigma-Aldrich, St. Louis, MO, USA, ≥96%) in methanol, at a concentration range of 0.625–20 μM.

2.6. Volatile Compounds (VOCs)

The experimental plan was designed to have triplicate samples of each formulation of T_0 cooked samples, T_{72} and T_{120} samples. Cooked samples T_0 were immediately chopped and put in glassy vials of 20 mL capacity (Perkin Elmer, Waltham, MA, USA) with approximately 3 g of meat each, tightly closed and stocked at −40 °C, assuring the highest headspace, until gas chromatograph mass spectrometer (GC-MS) analysis. GC-MS analysis was performed with a gas chromatograph (Clarus 580, Perkin Elmer, Waltham, MA, USA) coupled with a mass spectrometer (SQ8S, Perkin Elmer Waltham, MA, USA). Other samples were left in refrigerated conditions for the time required to obtain T_{72} and T_{120}, then carefully chopped and stocked in 20 mL vials at −40 °C, until GC-MS analysis.

The GC-MS analysis followed the method proposed by Qi et al. [18] with some modifications. Vials were left for 1 h at room temperature, then put in a water bath at 50 °C for 20 min. Volatiles from meat were extracted with a headspace solid phase microextraction fibre (SPME 65 μm Polydimethylsiloxane/Divinylbenzene (PDMS/DVB); Supelco, Bellofonte, PA, USA) and collected for 30 min at 40 °C, then inserted into the GC injector and desorbed for 3 min at 250 °C. Volatile compounds were separated on a Capillary GC column ZB- Semi Volatiles (30 m length, 0.25 mm internal diameter, 0.25 μm film thickness: Phenomenex, Torrance, CA, USA). The oven temperature was maintained for 3 min at 40 °C, increased at 3 °C/min to 70 °C, then at 5 °C/min to 180 °C, then at 10 °C/min to 260 °C, and maintained for 5 min at 260 °C. Helium was the carrier gas with a constant flow of 1 mL/min. The mass-selective detector was operated in the electron impact mode (70 eV) and full scan mode (35–500 m/z range). The identification was performed using the National Institute of Standards and Technology mass spectral library (NIST Mass Spectral library, Search Program version 2.0, National Institute of Standards and Technology, U.S. Department of Commerce, Gaithersburg, MD, USA).

2.7. Descriptive Sensory Analysis of WOF Assessment, Rancidity and Extraneous Flavours

Chicken burger samples were evaluated for four classes of descriptors grouped as: odour, flavour, taste, and aftertaste [19]. The vocabulary used for descriptors comes from the review of recent literature about rancidity and warmed off-flavours (WOF) assessment in meat products [20–24].

A panel group of six women and two men from twenty to fifty years old was trained for evaluation of quality assessment of meat burgers. After their recruitment, panellists were screened for their ability to distinguish odours and tastes, then were trained for vocabulary development through a series of triangular tests (ISO), 8586:2012 [25]. Training duration was 80 h, including familiarization with relevant descriptive terms and ways of perceiving the selection and quantification of the sensory characteristics of cooked meat, as well as the use of intensity scales (ISO) 4121:2003 [26].

Panellists were asked to taste cooked burgers (T_0) and cooked burgers refrigerated at +4 °C for 72 h. Samples of 120 h at +4 °C were not tasted to avoid any microbial contamination. A hot bath at 72 °C was used to heat up to the core temperature of 70 °C. Panellists were provided with individual templates, where descriptors were grouped per section. They used a scale from 1 to 5 where 1 meant the absence of the attribute while 5 the maximum rate. Pie shaped pieces of warm burgers were quickly served to panellists trying to maintain the temperature between 70 and 60 °C as recommended by [27]. Samples were codified with random numbers to avoid external influences on liking rating of panellists. Meat pieces were served on white plates. All sensory tests and training sessions were carried out in the sensory laboratory of the University of Teramo that fulfils the required standards for these analyses according to (ISO) 8589:2007 [28].

2.8. Statistical Analysis

All determinations were done in triplicate. Means and relative standard deviations were calculated. Analysis of variance (ANOVA) was performed to test the significance of the effects of the factor variables (formulation, time of storage); differences among means were separated by the least significant differences (LSD) test.

Statistical analysis of data was performed using XLSTAT software version 2019.1 for Microsoft Excel (Addinsoft, New York, NY, USA). All results were considered statistically significant at $p < 0.05$.

3. Results and Discussion

Results of qualitative characteristics (proximate composition, colour, pH, a_w) of uncooked and immediately cooked burgers are shown and discussed in the first section, highlighting significant differences among treatments. Data shown and discussed in the second section (TBARS, VOCs, sensory analysis) evidence the significant differences between sampling times with respect to the oxidative phenomena occurred in the refrigerated cooked chicken burger samples.

3.1. Effect of CSS on the Qualitative Characteristics of Burger Samples

3.1.1. Compositive Characteristics

Results from proximate compositional analyses of meat are shown in Table 1 (uncooked burgers); data agree with the literature [29,30]. Generally, macro-elements checked do not change so much. Moisture, proteins, and lipids remained at around their normal values; ashes had a small significant increase reaching the highest point of 2.69% for 3% CSS sample. TDF for samples (calculated as reported in Section 2.3) are 1.70% and 3.40%, for 1.5% and 3% CSS, respectively.

Table 1. Proximate composition (mean $\pm$ standard deviation, SD) of raw burger samples (before cooking): control (CTRL), with addition of +1.5% of coffee silver skin (CSS 1.5%) and +3.0% of coffee silver skin (CSS 3.0%).

Sample	Moisture (%)	Proteins (%)	Lipids (%)	Ashes (%)
CTRL	74.56 $\pm$ 0.05	19.72 $\pm$ 0.05	2.48 $\pm$ 0.01	1.97 [a] $\pm$ 0.01
CSS 1.5%	74.14 $\pm$ 0.03	20.01 $\pm$ 0.23	2.74 $\pm$ 0.02	2.11 [b] $\pm$ 0.01
CSS 3%	73.39 $\pm$ 0.05	19.62 $\pm$ 0.05	2.69 $\pm$ 0.01	2.69 [c] $\pm$ 0.01
sign.	n.s.	n.s.	n.s.	**

Legend: data followed by different superscript letters, in the same column, are significantly different (least significant difference (LSD) test, $p < 0.05$); asterisks indicate significance at ** $p < 0.01$, n.s. not significant.

pH of breast fillets used for the burger production was 5.93 $\pm$ 0.04, not differing with reported values which indicate values around 5.89–6.00 [31]. Formulated burgers (uncooked and cooked) registered pH values, shown in Table 2, in line with other sources [32]. Cooked burgers had similar pH and a_w values.

Table 2. Results (mean $\pm$ SD) of pH, water activity (a_w) values and cooking yield (%) in burger samples (CTRL, control; CSS 1.5% and CSS 3.0%, with +1.5% and +3.0% of coffee silver skin as ingredient, respectively).

	pH $_{Uncooked}$	pH $_{Cooked}$	a_w Cooked	Cooking Yield %
CTRL	5.82 $\pm$ 0.03	6.10 $\pm$ 0.03	0.985 $\pm$ 0.0013	76.15 [a] $\pm$ 0.58
CSS 1.5%	5.75 $\pm$ 0.01	6.06 $\pm$ 0.01	0.986 $\pm$ 0.0016	89.5 [b] $\pm$ 0.36
CSS 3%	5.75 $\pm$ 0.01	6.02 $\pm$ 0.01	0.986 $\pm$ 0.0007	88.95 [b] $\pm$ 0.52
sign.	n.s.	n.s.	n.s.	**

Legend: data followed by different superscript letters, in the same column, are significantly different (LSD test, $p < 0.05$); asterisks indicate significance at ** $p < 0.01$; n.s. not significant.

In eating a newly formulated burger (45 g) with a CSS inclusion of 1.5%, 16.54 mg of calcium and 65.7 mg of potassium can be assumed. Harvard Health Publishing in 2019 reviewed the daily intake of calcium for women between 50 and 71 years old fixing this at 1200 mg [33]. The same value was established by the National Institute of Health (NIH) which also defines limits [34] for men at 1000 mg. Potassium was fixed at 3400 mg and 2600 mg for males and females from 19 to 50 years old, respectively.

For marketing within the European Union (EU), it is very important to define whether by-products, such as CSS, need to obtain an approval as ingredient for novel foods [34], with special reference to legal status within the EU and potential options for producers to obtain approval according to Novel Food [35–37].

3.1.2. Cooking Yield

The cooking yield allows calculation of how much water and fats a food item loses after a cooking process. The US Department of Agriculture (USDA) in 2014 [38] released a table of cooking yield and retention factors for many meat products. These factors can be used to calculate nutritional values where analytical data for cooked foods are unavailable. Obviously, meat represents an important class of cooked foods and in this way the cooking yield covers an important aspect. Beside this, the cooking yield parameter tells us how much in terms of weight a formulation has lost, and this is also an index of profitability.

Table 2 shows that the addition of coffee by-product allowed an increase in cooking yield (%) in respect to the control. While the control lost 23.85% of its initial weight, CSS addition limited this loss to just 10.50% (+1.5% CSS addition) and 11.06% (+3% CSS addition). This trait of CSS comes from good water holding capacity (WHC) and oil holding capacity (OHC).For CSS, values of WHC 5.11 ± 0.20 and 5.5 ± 0.2 were found; for OHC, these were 4.72 ± 0.10 and 4.8 ± 0.2 [39–43]. The increased cooking yield also marked a significant difference ($p < 0.01$) between CTRL and CSS added burgers, while burgers with CSS, at both percentages, were similar as regards cooking yield.

Losses are limited even with little addition, registering a high capacity and a possible increase in economical revenue.

3.1.3. Colour

Breast fillets used to formulate burgers had a value L* of 45.91, on average according to Ziober et al. [43], L* > 53 denoted pale soft exudative (PSE), L* < 44 is analogous to dark firm dry (DFD), and 44 ≤ L* ≤ 53 is normal meat. Colour values of burgers are in line with what Longato et al. [4] have found in their study on chicken burgers with the addition of hazelnut skin; their results on uncooked (control) samples are L* a* b* values of 53.83 ± 4.48, 0.30 ± 0.69, 9.12 ± 1.8, and 64.98 ± 2.55, 1.64 ± 0.55, 15.48 ± 1.15 for cooked burgers. Data on the colour determination (uncooked and cooked samples) are reported in Table 3.

Table 3. Colour values (mean ± SD) of L* (lightness), a* (redness), b* (yellowness), and C* (chroma) for uncooked burgers and cooked burgers, in control (CTRL) and in samples with +1.5% (CSS 1.5%) and +3.0% (CSS 3.0%) of coffee silver skin as ingredient.

Uncooked Samples	L*	a*	b*	C*
CTRL	45.91 ± 0.18 [a]	0.28 ± 0.11 [b]	10.95 ± 0.53	10.96 ± 0.53 [b]
CSS 1.5%	32.83 ± 1.33 [b]	3.81 ± 1.21 [a]	13.74 ± 1.58	15.42 ± 1.82 [a]
CSS 3%	29.40 ± 0.71 [c]	4.96 ± 0.83 [a]	14.60 ± 0.31	14.28 ± 0.52 [a]
sign.	**	**	n.s.	*
Cooked Samples				
CTRL	56.12 ± 0.76 [a]	1.18 ± 0.77	16.75 ± 0.49 [a]	16.80 ± 1.08 [a]
CSS 1.5%	47.35 ± 0.44 [b]	2.25 ± 0.60	10.94 ± 0.36 [b]	11.17 ± 0.42 [b]
CSS 3%	43.54 ± 0.27 [b]	2.83 ± 0.60	12.05 ± 0.40 [b]	12.38 ± 1.22 [b]
sign.	*	n.s.	**	**

Legend: data followed by different superscript letters, in the same column, are significantly different (LSD test, $p < 0.05$); asterisks indicate significance at * $p < 0.05$; ** $p < 0.01$; n.s. not significant.

CSS is easily reducible to a fine crumb and this determined a uniform distribution. Burgers with CSS are in fact darker, more yellow, and redder with respect to the control. Unfortunately, no studies are available to compare these data. Chroma (C*) values for uncooked and cooked chicken burgers were searched by de Oliveira et al. [44], who added chia seeds to a blend of breast and thigh chicken skinless meat and pork backfat. They reported levels of 14.0 ± 4.8 (raw samples), and 16.8 ± 2.4 (grilled samples). As in our case, burgers were darker than control. Conversely, their data do not show higher saturation. C* values of burgers here analysed are shown in Table 3. Generally, the addition of CSS increased the C* value meaning a higher saturation and, thus, a more vivid colour after cooking.

3.2. Effect of CSS on the Shelf Life of Cooked Chicken Burger Samples

3.2.1. Thiobarbituric Acid Reacting Substances (TBARS) Test

The TBARS values of cooked samples are presented in Figure 3. Generally, over time TBARS values increase in all the cases, but the CTRL set showed significantly higher values immediately after cooking ($p < 0.05$), as well as a statistically significant increase during the refrigerated storage ($p < 0.05$). In all burgers with CSS, TBARS mean values were lower than the acceptance limit of TBARS for rancidity (1.0 mg MDA(Malondialdehyde)/kg) [45] until 72 h; after 120 h CSS3% showed a TBARS mean value near to the critical content, while this limit was exceeded in all samples without CSS.

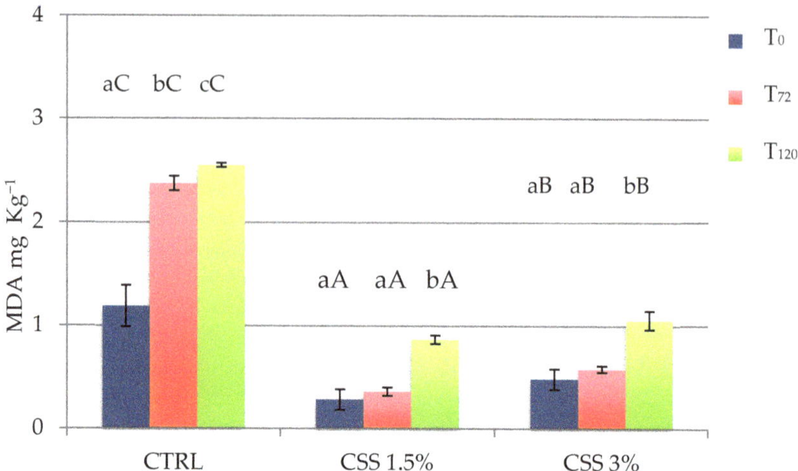

Figure 3. Results of Thiobarbituric acid reacting substances (TBARS) test, expressed as malondialdehyde (mg MDA kg^{-1}) in control (CTRL) and burgers formulated with CSS (1.5% and 3%, respectively), immediately after cooking (T$_0$) and after refrigerated storage (at 4 °C, for 72 and 120 h, T$_{72}$ and T$_{120}$ respectively). Results are expressed as means ± standard deviations. Different lowercase letters indicate significant differences ($p < 0.05$) among storage time of each batch; different uppercase letters indicate significant differences ($p < 0.05$) among different batches, at same storage time.

The TBARS test is helpful for a first screening of the oxidation rate of a food, but it does not discriminate which kind of oxidation is occurring. We cannot be sure if what is observed with this assay comes from lipid oxidation itself, or if proteins too have taken part in the process and are the principal cause starting the reactions. As is known, chicken meat is poor in lipids that are mainly unsaturated fatty acids which are the elective substrate for the oxidation. Moreover, [46] have searched for the lipidic profile of CSS and their results show a small content of lipids, which are mainly saturated fatty acids (SFAs). So, one could refer all the defective odours and tastes to the lipidic oxidation complex of reactions. Unfortunately, the oxidation process is very unstable and sometimes unpredictable. Besides lipids, proteins and iron ions boost the process within other factors (rise in temperature, oxygen and light exposure, salt addition, etc.) making TBARS not such an affordable method to establish the lipidic oxidative status of a food [47,48].

Anyway, from our results we can imagine that the contribution of CSS to the global oxidation burgers and its lipids increment are almost zero, while its protective effect seems to be promising. This may depend on the high content in phenolic and bioactive species.

3.2.2. Volatile Compounds (VOCs) and Warmed Off-Flavours (WOF) in Chicken Burgers

Among analyses used in this work, GC-MS analysis was used to trace markers of oxidation as a more reliable method than TBARS or any other faster, but less accurate, method. Results show a complex profile of compounds emerging from oxidative phenomena, Maillard reaction occurrence, and by-products addition. Table 4 contains all the volatile compounds found in CSS-containing samples. Data depicted refer to T$_0$ and T$_{72}$ samples. Chromatograms are shown as Supplementary Material (Figures S1–S3).

Table 4. Qualitative profile of main volatile compounds (area %) in cooked samples (control, addition of coffee silver skin 1.5 and 3%), immediately after cooking process (T_0) and after 72 h of storage at +4 °C (T_{72}).

VOCCs from Burgers' Samples	Burgers T_0			Burgers T_{72}			
	CTRL T_0	CSS 1.5% T_0	CSS 3% T_0	CTRL T_{72}	CSS 1.5% T_{72}	CSS 3% T_{72}	Sign.
Aldehydes							
2-Decenal, (Z)-	nd	nd	1.25	nd	nd	nd	*
2-Nonenal, (E)-	nd	nd	nd	nd	2.82	nd	*
10-Undecenal	nd	nd	4.06	nd	nd	nd	*
Benzaldehyde	nd	6.27 [c]	4.46 [b]	nd	3.62 [a]	7.42 [d]	*
Benzaldehyde, 2,5-bis[(trimethylsilyl)oxy]-	0.86 [a]	nd	nd	nd	0.72 [a]	1.73 [b]	*
Butanal, 2-methyl-	nd	nd	5.48 [a]	nd	nd	46.86 [b]	**
Butanal-3 methyl	nd	nd	5.67 [a]	nd	4.11 [a]	7.00 [b]	*
Heptanal			1.77 [a]	4.46 [b]	2.29 [a]	nd	*
Hexanal	55.1 [b]	11.7 [a]	nd	72 [c]	nd	nd	***
Octanal	41.2 [b]	nd	nd	7.94 [a]	nd	nd	*
Pentanal	2.09 [a]	nd	nd	nd	nd	nd	*
Propanal, 2-methyl-	nd	nd	3.34 [B]	nd	nd	1.98 [a]	*
Alcohols							
1,5-Pentanediol, 3-methyl-	nd	nd	8.96 [b]	nd	nd	3.90 [a]	*
1-Methylcyclopropanemethanol	nd	nd	nd	nd	nd	3.71 [a]	*
2-Hexen-1-ol, (Z)-	nd	nd	nd	nd	10.52	nd	*
3-Decyn-2-ol	nd	nd	nd	nd	5.76	nd	*
2-Nonen-1-ol, (E)-	0.58 [a]	10.76 [b]	8.38 [a]	8.80 [a]	nd	nd	*
2-Octen-1-ol, (E)-	nd	nd	2.03	nd	nd	nd	*
Ketones							
2,3-Octanedione	nd	nd	nd	5.14	nd	nd	*
7,9-Di-tert-butyl-1-oxaspiro (4,5) deca-6,9-diene-2,8-dione	nd	nd	3.78 [b]	0.20 [a]	nd	nd	*

Table 4. *Cont.*

VOCCs from Burgers' Samples	Burgers T_0			Burgers T_{72}			Sign.
	CTRL T_0	CSS 1.5% T_0	CSS 3% T_0	CTRL T_{72}	CSS 1.5% T_{72}	CSS 3% T_{72}	
Nitrogen-containing compounds							
2-(Aziridinylethyl)amine	nd	16.09 [c]	7.43 [b]	nd	15.24 [c]	4.59 [a]	*
2,6,6-Trimethyl-bicyclo [3.1.1]hept-3-ylamine	nd	nd	nd	nd	4.42	nd	*
Benzeneethanamine, 2,5-difluoro-β,3,4-trihydroxy-N-methyl-	nd	nd	4.37	nd	nd	nd	*
Oxime-, methoxy-phenyl-_	nd	nd	nd	nd	nd	1.48	*
Propanamide, 2-hydroxy-	nd	nd	8.86 [a]	nd	20.18 [b]	nd	*
Topotecan	nd	nd	nd	nd	8.81	nd	*
Hydrocarbons							
2-Trifluoroacetoxydodecane	nd	3.98	nd	nd	nd	nd	*
3-Trifluoroacetoxydodecane	nd	4.35	nd	nd	nd	nd	*
Butane, 2-nitro-	nd	3.74	nd	nd	nd	nd	*
Cyclohexene, 1-methyl-4-(1-methylethenyl)-, (S)-	nd	nd	nd	nd	nd	3.43	*
Ergosta-5,22-dien-3-ol, acetate, (3β,22E)-	nd	nd	nd	nd	10.26	nd	*
Ethylbenzene	nd	nd	1.56 [a]	nd	nd	4.17 [b]	*
Hydroperoxide, heptyl	nd	3.87	nd	nd	nd	nd	*
Propane	nd	nd	nd	nd	nd	4.05	*
p-Xylene	nd	2.57 [a]	nd	nd	2.48 [a]	3.00 [b]	*
Esters							
1,2-Benzenedicarboxylic acid, butyl octyl ester	nd	nd	nd	0.04	nd	nd	*
Other compounds							
2-Formylhistamine	nd	nd	nd	nd	5.55	nd	*
Butylated Hydroxytoluene	nd	nd	nd	0.20	nd	nd	*
Nitrous oxide	nd	nd	3.91	nd	nd	nd	*
Pregnan-18-oic acid,3,11,21-trihydroxy-20-oxo-, γ-lactone, (3β,5α,11β)-	nd	nd	1.24	nd	nd	nd	*

Legend: nd, not detectable; data followed by different superscript letters, in the same row, are significantly different (LSD test, $p < 0.05$); asterisks indicate significance at * $p < 0.05$; ** $p < 0.01$; *** $p < 0.001$.

According to Chen et al. [49], most of the typical odorants from cooked chicken meat are caused by phospholipids oxidation/degradation that led to the formation of long-chain aldehydes such as hexanal, (Z)-2-decenal and (E)-2-decenal. In any case, even if responsible for WOF development, these aldehydes are key aroma compounds of freshly cooked chicken meat. CSS samples showed these classes even if ketones and esters were not found. T_0 1.5% added samples had hydrocarbons as the first class traced, followed by aldehydes, nitrogen containing compounds and alcohols. T_0 3% added samples showed aldehydes at first place followed by Nitrogen containing compounds, alcohols, hydrocarbons, and other compounds. T_{72} containing 1.5% of CSS showed nitrogen containing compounds, hydrocarbons, other compounds, aldehydes, alcohols, and ketones, while 3% addition showed aldehydes, hydrocarbons, alcohols, other compounds, and Nitrogen containing compounds.

In general, patties tested in this study, especially CTRL, seem to have a small compounds presence if compared with other articles [50–52]. To our knowledge, this is one of the first studies where cooking conditions did not pass 92 °C and were not prolonged for more than four minutes. These conditions were selected to best simulate domestic conditions using an electric device set at medium cooking heat. Most studies on chicken meat burgers have tested grilled or oven-cooked patties. Other references on chicken meat evaluated entire boiled or roasted chicken. This is a fundamental step in explaining, for example, the absence of sulphur containing volatiles. In line with findings of other researchers [53], these compounds come from the interaction among Maillard reaction compounds and lipid oxidation products. Thus, quick cooking processes, medium/low heating, or their combination seem not to favour this interaction. These settings did not allow the development of traceable Maillard reaction products (desired and undesired). T_{72} (CTRL) samples showed an increase in concentration of hexanal, the emergence of heptanal, some alcohols such as 2-Nonen-1-ol, (E)-, ketones as 2,3-Octanedione and 7,9-Di-tert-butyl-1-oxaspiro (4,5) deca-6,9-diene-2,8-dione, and just one Sulphur containing compound: N-Methyl-taurine. These are all markers of lipid oxidation and muscle damage. The addition of by-products provoked a mitigation of some WOF species, but also gave to patties specific odorants not conducible to meat oxidation and potentially undesirable.

CSS addition reduced the occurrence of aldehydes such as hexanal that reduced from 55.1% (CTRL T_0) to 11.7% (CSS T_0 1.5%) to 0 (CSS T_0 3%). At T_{72} CTRL contained 72% and reduced to 0 in both concentrations. Heptanal was found only in CTRL T_{72}; octanal too, was just found in CTRL samples and not found in samples 1.5 and 3% at both times. Some alcohols such as 2-Nonen-1-ol, (E)- were limited in T_{72} samples, but were present in T_0 CSS 3% with other alcohols, probably from the degradation of lignocellulosic precursors. This same pathway is followed by hydrocarbons which are totally absent in CTRL samples while being present in added patties [54]. According to data here shown, Nitrogen containing compounds present in CSS formulations probably came from the degradation of CSS proteins and from the Maillard complex of reactions which takes place during the roasting process. As for CSS, the significant role of phenols' interaction with Maillard reaction products to produce specific compounds can be assumed. Unfortunately, CSS developed p-xylene and o-xylene, involved in the rise of WOF and referred to as "cardboard-like" [55,56]. No references are available to compare results obtained, especially for CSS properties.

Chromatograms, in all cases, showed a great reduction of WOF or general active odorants. As a demonstration of this, CTRL T_0 and T_{72} images had a resolution with an order of magnitude of 10^{10}, while CSS had values of 10^8. From pictures, the peak of hexanal that eluted at around 6.48 min is always visible and it is clear how much it decreases in respect of additions of by-products. In all samples at around minutes 13.44 and 16.49, long chain aldehydes were eluted (i.e., octanal, decenal). At around minutes 19.35–37 2-Nonen-1-ol, (E)- was eluted in almost all samples. After minute 19.40, the main compounds traced were siloxanes and low matched compounds.

Of course, a better characterization of the aromatic profile of these molecules is fundamental to understand if and how volatiles from these substrates can have a negative impact on the final flavour of meat products.

3.2.3. Sensory Analysis

Panellists involved in this analysis were asked to try samples T_0 and T_{72} of all formulations on separate days without knowing what they were eating, with the objective of evaluating the presence and the intensity of WOF markers and possible perceptions of extraneous flavours in cooked burgers after refrigerated storage. Descriptors were carefully explained, especially those difficult to associate with food, such as "cardboard-like" or "paint".

In Table 5 are reported all the average values for burger samples tasted immediately after cooking (T_0) and after 72 h of refrigerated storage (T_{72}).

Table 5. Average scores for cooked CSS containing samples of panelists for each descriptor in all formulations; immediately after cooking process (T_0) and after 72 h of refrigerated (+4 °C) storage (T_{72}).

Average Scores	CTRL		CSS 1.5%		CSS 3%	
	0 h	72 h	0 h	72 h	0 h	72 h
			Descriptors			
cooked meat odour	1.8 ± 0.9 [b]	4 ± 1.4 [a]	2.9 ± 1.1 [a]	2.7 ± 1.0 [ab]	2.8 ± 1.1 [a]	2.6 ± 0.7 [ab]
cardboard	1.9 ± 0.8	1.7 ± 0.9	1.5 ± 0.7	2.0 ± 0.5	1.6 ± 0.9	2.0 ± 0.1
Sulphur/rubber	1.6 ± 1.0	2.0 ± 1.0	1.3 ± 0.4	1.5 ± 1.0	1.3 ± 0.4	1.5 ± 0.7
roasted	2.1 ± 1.1	1.7 ± 0.7	2.6 ± 0.7	1.8 ± 0.8	2.8 ± 1.3	2.5 ± 0.9
painty	1.0 ± 0.0	1.3 ± 0.7	1.0 ± 0.0	1.1 ± 0.3	1.5 ± 1.0	1.2 ± 0.4
rancid	1.6 ± 1.0	1.5 ± 0.7 [a]	1.0 ± 0.0	1.1 ± 0.3	1.2 ± 0.3	1.3 ± 0.7
vegetable oil-like	1.6 ± 0.7	1.6 ± 0.7	1.6 ± 0.9	1.3 ± 0.5	1.6 ± 0.7	1.6 ± 0.7
sour	1.8 ± 0.7	1.2 ± 0.4 [a]	1.9 ± 0.8 [a]	1.6 ± 0.7	1.6 ± 0.9	1.6 ± 1.0
bitter	1.3 ⊥ 0.5 [b]	1.4 ± 0.2 [b]	1.9 ± 1.0 [ab]	1.6 ± 0.7 [ab]	2.6 ± 0.3 [a]	2.3 ± 0.9 [a]
metallic	1.3 ± 0.5	1.1 ± 0.3	1.4 ± 0.7	2.0 ± 1.2 [a]	1.3 ± 0.4	1.8 ± 1.1
astringent	1.6 ± 0.7	1.8 ± 0.7	1.8 ± 0.9	1.7 ± 0.9	2.4 ± 1.0	2.6 ± 0.8

Legend: data followed by different superscript letters, in the same row, are significantly different (LSD test, $p < 0.05$).

T_0 samples of all formulations did not show significative differences, and panellists were not able to trace significant variability from control. Statistical analysis did not show significance between samples even if, looking at the average values, some considerations can be made. The score reported for the bitter descriptor rose to 2.25 for the 3% CSS formulation, while CTRL was 1.25 and CSS 1.5% was 1.87. CSS addition did not influence the average score for all descriptors, and, in most cases, they were the same as CTRL or very near to it. CSS seemed to influence the perception of cooked meat odour rising from 1.75 in CTRL, to 2.87 in 1.5% addition, to 2.75 in 3% addition. This is one of the few situations in which 1.5% received higher scores than 3%. Astringency also registered an increase from CTRL to the 1.5% CSS addition of 3%: from 1.62 to 1.75 to 2.37, respectively.

Generally, all the descriptors for all formulations received increased scores, but some significant differences were traced via the statistical analysis or can be noted from the direct comparison among the average scores. Only two descriptors for CSS formulations were significant ($p < 0.05$): cooked meat odour and bitter. Cooked meat odour was mitigated, mainly by CSS addition. For this descriptor, the score decreased from 4 (CTRL) to 2.75 (1.5%) and 2.62 (3%). This limitation can be seen neither as negative nor positive. If the cooked meat odour can be directly linked with positive sensations, we do not know the considerations of each panellist regarding cooked meat. Nevertheless, from the explanation of each descriptor and the training, these lower scores do not directly show a positive thing. For better discrimination, a comparison with the roasted descriptor can be made; although not significant, it received lower scores than cooked meat.

The presence of this descriptor was not casual, because it helped in separating what flavours, odours, and aromas can come from a fresh grilled or oven/pan-cooked burger, instead of an already cooked and reheated burger. Panelists were not aware of the cooking process. The 'roasted' adjective generally includes those positive flavours coming from the Maillard complex of the reaction, while cooked meat is mainly linked with sensations of staling. In these terms, the mitigation obtained from by-products can be positive, while increasing the stability of the product. Bitterness perceived in CSS formulated patties can be a result of the reheating of meat. The data show an increase from CTRL at 1.37, to 1.62 (CSS 1.5%), and to 2.35 (CSS 3%). Heat can make the condensation or splitting of phenolic species easier. During the roasting process, chlorogenic acids degrade to active taste lactones which give desirable sourness and bitterness [57]. When exposed to further heating, these species undergo greater degradation which leads to the splitting of quinic acid which, in successive steps will give metallic, lingering bitter phenyl-indanes which are undesirable for coffee taste. Caffeine and, in general, methylxanthines-alkaloids give a bitter and astringent note. Probably, the double exposure to heating, even if at lower temperatures, can determine higher bitterness.

Refrigerated storage can also favour the condensation of flavonoids to tannins or bigger phenolic species, but no sources are available at this time. A general positive comment is that none of the descriptors directly linked with the development of WOF (paint, cardboard-like, vegetable oil-like and sulphur/rubber) was increased. Even if not significative, average scores of the CSS added sample were lowered in respect of CTRL. Control had 2, while CSS fell to 1.75 for 1.5% addition, and to 1.5 for 3% addition. Acidity and even metallic sensations were not increased. To better investigate the significance of cooked meat flavour and bitterness, different factors were considered. Time, Formulation, and Time x Formulation were selected as factors. For cooked meat flavour, the time factor, i.e., the effect of time, was significant for $p < 0.01$ while time x formulation factor had a $p < 0.05$ (Table 6). Formulation alone did not influence the results. Conversely, bitterness was influenced only by the formulation with a $p < 0.05$. Neither time nor time x formulation factors made an effect. Overall, T_0 samples did not show any significant difference for the cooked meat odour descriptor. After 72 h of refrigerated storage, the time effect greatly influences ($p < 0.01$) this characteristic.

Table 6. Anova matrix results for significant descriptors (cooked meat odour and bitterness).

	Cooked Meat Odour		Bitterness	
Factor	F	sign.	F	sign.
Formulation	1.8019	n.s.	3.7026	*
Time	14.6049	**	0.7326	n.s.
Formulation × Time	3.4424	*	0.4273	n.s.

Legend: asterisks indicate significance at * $p < 0.05$; ** $p < 0.01$; n.s. not significant.

4. Conclusions

The present study provides data on technological performance, nutritional aspects and effects on stability of newly formulated meat products.

Coffee silver skin could be considered as a food ingredient that can solve a complex problem in limiting the decay of meat foods (especially of poultry origin) and lowering the food waste caused by coffee production. CSS can become a cheap but valuable integrator of fibres and bioactive molecules; moreover, it is a great source of minerals such as calcium, potassium, and others.

Data obtained are the starting point of a deeper study to comprehend what are the best conditions to use this by-product and how to develop "tailor-made" formulations enjoyable to consumers. Burgers were among the easiest preparations, allowing a direct comparison with reality.

CSS has shown potential for being implemented in meat formulations to limit losses connected with the cooking process.

Considering the principal aim of this study (to understand the role of CSS on WOF occurrence), encouraging results are provided. The sensory analysis conducted gave confirmation that the spreading of WOF, or in general of oxidation markers, was arrested, even if bitterness and astringency can emerge with time. The inclusion of CSS among new ingredients for food production is hoped for, even if further analysis is needed with further consultation procedures regarding current novel food statuses.

Supplementary Materials: The following are available online at https://www.mdpi.com/article/10.3390/foods10081833/s1, Figure S1: Chromatogram of volatile compounds (VOCs) in chicken burgers immediately after cooking (a, CTRL T_0) and after 72 h of refrigerated storage (b, CTRL T_{72}); Figure S2: Chromatogram of volatile compounds (VOCs) in chicken burgers formulated with CSS 1.5%, immediately after cooking (a) and after 72 h of refrigerated storage (b); Figure S3: Chromatogram of volatile compounds (VOCs) in chicken burgers formulated with CSS 3%, immediately after cooking (a) and after 72 h of refrigerated storage (b).

Author Contributions: Conceptualization, M.M., D.M. and L.E.; methodology, M.M. and L.E.; software, M.M. and L.E.; validation, M.M.; formal analysis, L.E.; investigation, L.E., M.M.; resources, M.M. and D.M.; data curation, M.M. and L.E.; writing—original draft preparation, L.E. and M.M.; writing—review and editing, M.M., D.M. and L.E.; visualization, M.M. and D.M.; supervision, M.M., D.M. All authors have read and agreed to the published version of the manuscript.

Funding: This research received no external funding.

Institutional Review Board Statement: Not applicable.

Informed Consent Statement: Not applicable.

Acknowledgments: Authors acknowledge to the Torrefazione Adriatica spa, Giulianova (Teramo, Italy) for providing the coffee silver skin sample for this study.

Conflicts of Interest: The authors declare no conflict of interest.

References

1. Comunian, T.A.; Silva, M.P.; Souza, C.J.F. The use of food by-products as a novel for functional foods: Their use as ingredients and for the encapsulation process. *Trends Food Sci. Technol.* **2021**, *108*, 269–280. [CrossRef]
2. Martuscelli, M.; Esposito, L.; Di Mattia, C.; Ricci, A.; Mastrocola, D. Characterization of coffee silver skin as potential food safe ingredient. *Foods* **2021**, *10*, 1367. [CrossRef] [PubMed]
3. Amaral, A.B.; da Silva, M.V.; Lannes, S.C. Lipid oxidation in meat: Mechanisms and protective factors—A review. *Food Sci. Technol.* **2018**, *38*, 1–15. [CrossRef]
4. Longato, E.; Meinrei, G.; Peiretti, P.G.; Gai, F.; Viuda-Martos, M.; Pérez-Álverez, J.Á.; Amarowicz, R.; Férnandez-López, J. Effects of hazelnut skin addition on the cooking, antioxidant and sensory properties of chicken burgers. *J. Food Sci. Tech.* **2019**, *56*, 3329–333652. [CrossRef] [PubMed]
5. Oswell, N.J.; Thippareddi, H.; Pegg, R.B. Practical use of natural antioxidants in meat products in the U.S.: A review. *Meat Sci.* **2018**, *145*, 469–479. [CrossRef]
6. Faustino, M.; Veiga, M.; Sousa, P.; Costa, E.M.; Silva, S.; Pintado, M. Agro-Food Byproducts as a New Source of Natural Food Additives. *Molecules* **2019**, *24*, 1056. [CrossRef]
7. Gobena Gemechu, F. Embracing nutritional qualities, biological activities and technological properties of coffee byproducts in functional food formulation. *Trends Food Sci. Technol.* **2020**, *104*, 235–261. [CrossRef]
8. Costa, A.S.G.; Alves, R.C.; Vinha, A.F.; Costa, C.S.G.; Costa, M.A.; Nunes, A.A.; Almeida, A.; Santos-Silva Oliveira, M.B.P.P. Nutritional, chemical and antioxidant/pro-oxidant profiles of silverskin, a coffee roasting by-product. *Food Chem.* **2018**, *30*, 28–35. [CrossRef]
9. Kralik, G.; Kralik, Z.; Grčević, M.; Hanžek, D. Quality of Chicken Meat. In *Animal Husbandry and Nutrition*, 1st ed.; Yücel, B., Taşkin, T., Eds.; Intech Ope: London, UK, 2018. [CrossRef]
10. Ianni, A.; Grotta, L.; Martino, G. Feeding influences the oxidative stability of poultry meat treated with ozone. *Asian Australas. J. Anim. Sci.* **2019**, *32*, 874–880. [CrossRef]
11. Pateiro, M.; Barba, F.J.; Domínguez, R.; Sant'Ana, A.S.; Khaneghah, A.M.; Gavahian, M.; Gómez, B.; Lorenzo, J.M. Essential oils as natural additives to prevent oxidation reactions in meat and meat products: A review. *Food Res. Int.* **2018**, *113*, 156–166. [CrossRef] [PubMed]
12. Kim, J.H.; Ahn, D.U.; Eun, J.B.; Moon, S.H. Antioxidant Effect of Extracts from the Coffee Residue in Raw and Cooked Meat. *Antioxidant* **2016**, *5*, 21. [CrossRef] [PubMed]

13. Delgado-Ospina, J.; Martuscelli, M.; Grande-Tovar, C.D.; Lucas-González, R.; Molina-Hernandez, J.B.; Viuda-Martos, M.; Fernández-Lopez, J.; Pérez-Álvarez, J.A.; Chaves-Lopez, C. Cacao Pod Husk Flour as an Ingredient for Reformulating Frankfurters: Effects on Quality Properties. *Foods* **2021**, *10*, 1243. [CrossRef]

14. Lungu, N.S.; Afolayan, A.J.; Thomas, R.S.; Idamokoro, E.M. Consumer exposure to warmedover flavour and their attitudes towards the use of natural antioxidants as preservatives in meat and meat products. *Br. Food J.* **2020**, *122*, 2927–2937. [CrossRef]

15. Latimer, G.W. ; *AOAC International. Official Methods of Analysis of AOAC International, Methods 2009.01, and 2011.25*, 19th ed.; AOAC International: Rockville, MD, USA, 2012.

16. Folch, J.; Lees, M.; Sloane-Stanley, G. A simple method for the isolation and purification of total lipids from animal tissues. *J. Biol. Chem.* **1957**, *226*, 497–509. [CrossRef]

17. Soyer, A.; Özalp, B.; Dalmış, Ü.; Bilgin, V. Effects of freezing temperature and duration of frozen storage on lipid and protein oxidation in chicken meat. *Food Chem.* **2010**, *120*, 1025–1030. [CrossRef]

18. Qi, J.; Wang, H.H.; Zhou, G.H.; Xu, X.L.; Li, X.; Bai, Y.; Yu, X. Evaluation of the taste-active and volatile compounds in stewed meat from the Chinese yellow-feather chicken breed. *Int. J. Food Prop.* **2017**, *20*, S2579–S2595. [CrossRef]

19. Byrne, D.V.; Bredie, W.L.P.; Mottram, D.S.; Martens, M. Sensory and chemical investigations on the effect of oven cooking on warmed-over flavour development in chicken meat. *Meat Sci.* **2002**, *61*, 127–139. [CrossRef]

20. Byrne, D.V.; Bredie, W.L.P.; Martens, M. Development of a sensory vocabulary for warmed-over flavour: Part II, in chicken meat. *J. Sens. Stud.* **1999**, *14*, 67–78. [CrossRef]

21. O'Sullivan, M.G.; Byrne, D.V.; Jensen, M.T.; Andersen, H.J.; Vestergaard, J. A comparison of warmed-over flavour in pork by sensory analysis, GC/MS and the electronic nose. *Meat Sci.* **2003**, *65*, 1125–1138. [CrossRef]

22. Aaslyng, M.G.; Meinert, L. Meat flavour in pork and beef—From animal to meal. *Meat Sci.* **2017**, *132*, 112–117. [CrossRef]

23. Mörlein, D. Sensory evaluation of meat and meat products: Fundamentals and applications. In Proceedings of the 60th International Meat Industry Conference MEATCON2019, Kopaonik, Serbia, 22–25 September 2019; p. 012007.

24. Shahidi, F.; Abad, A. Lipid-Derived Flavours and Off-Flavours in Food. *Encycl. Food Chem.* **2019**, *2*, 182–192. [CrossRef]

25. International Organization for Standardization. *General Guidelines for the Selection, Training and Monitoring of Selected Assessors and Expert Sensory Assessors*; ISO 8586:2012 Sensory Analysis; ISO: Geneva, Switzerland, 2015.

26. International Organization for Standardization. *Guidelines for the Use of Quantitative Response Scales*; ISO 4121:2003 Sensory analysis; ISO: Geneva, Switzerland, 2003.

27. American Meat Science Association. *Research Guidelines for Cookery, Sensory Evaluation, and Instrumental Tenderness Measurements of Meat*; American Meat Science Association (AMSA): Champaign, IL, USA, 2015.

28. International Organization for Standardization. *General Guidance for the Design of Test Rooms*; ISO 8589:2007 Sensory analysis; ISO: Geneva, Switzerland, 2017.

29. Garcia, R.G.; Freitas, L.W.; de Schwingel, A.W.; Farias, R.M.; Caldara, F.R.; Gabriel, A.M.A.; Graciano, J.D.; Komiyama, C.M.; Almeida Paz, I.C.L. Incidence and Physical Properties of PSE Chicken Meat in a Commercial Processing Plant. *Rev. Bras. Cienc. Agric.* **2010**, *12*, 233–237. [CrossRef]

30. Brambila, G.S.; Chatterje, D.; Bowker, B.; Zhuang, H. Descriptive texture analyses of cooked patties made of chicken breast with the woody breast condition. *Poult. Sci.* **2017**, *96*, 3489–3494. [CrossRef] [PubMed]

31. Nardoia, M.; Ruiz-Capillas, C.; Casamassima, D.; Herrero, A.M.; Pintado, T.; Jiménez-Colmenero, F.; Chamorro, S.; Brenes, A. Effect of polyphenols dietary grape by-products on chicken patties. *Eur. Food Res. Technol.* **2017**, *244*, 367–377. [CrossRef]

32. De Marchi, M.; Riovanto, R.; Penasa, M.; Cassandro, M. At-line prediction of fatty acid profile in chicken breast using near infrared reflectance spectroscopy. *Meat Sci.* **2012**, *90*, 653–657. [CrossRef]

33. Harvard Health Publishing. Available online: https://www.health.harvard.edu/staying-healthy/how-much-calcium-do-you-really-need (accessed on 2 April 2021).

34. National Institute of Health. Available online: https://ods.od.nih.gov/factsheets/Calcium-HealthProfessional/ (accessed on 2 April 2021).

35. Klingel, T.; Kremer, J.I.I.; Vera Gottstein, V.; Rajcic de Rezende, T.; Schwarz, S.; Lachenmeier, D.W. A Review of Coffee By-Products Including Leaf, Flower, Cherry, Husk, Silver Skin, and Spent Grounds as Novel Foods within the European Union. *Foods* **2020**, *9*, 665. [CrossRef] [PubMed]

36. European Union. European Union Commission implementing regulation (EU) 2017/2468 of 20 December 2017 laying down administrative and scientific requirements concerning traditional foods from third countries in accordance with Regulation (EU) 2015/2283 of the European Parliament and of the Council on novel foods. *Off. J. Eur. Union* **2017**, *L351*, 55–63.

37. European Union. European Union Commission implementing regulation (EU) 2017/2469 of 20 December 2017 laying down administrative and scientific requirements for applications referred to in Article 10 of Regulation (EU) 2015/2283 of the European Parliament and of the Council on novel foods. *Off. J. Eur. Union* **2017**, *L351*, 64–71.

38. US Department of Agriculture. *USDA Table of Cooking Yields for Meat and Poultry, Release 2*; US Department of Agriculture: Beltsville, MD, USA, 2014; p. 20705.

39. Bresciani, L.; Calani, L.; Bruni, R.; Brighenti, F.; Del Rio, D. Phenolic composition, caffeine content and antioxidant capacity of coffee silverskin. *Food Res. Int.* **2014**, *61*, 196–201. [CrossRef]

40. Narita, Y.; Inouye, K. Review on utilization and composition of coffee silverskin. *Food Res. Int.* **2014**, *61*, 16–22. [CrossRef]

41. Ballesteros, L.F.; Teixeira, J.A.; Mussatto, S.I. Chemical, functional, and structural properties of spent coffee grounds and coffee silverskin. *Food Bioprocess. Technol.* **2014**, *7*, 3493–3503. [CrossRef]

42. Ateş, G.; Elmacı, Y. Coffee silverskin as fat replacer in cake formulations and its effect on physical, chemical and sensory attributes of cakes. *LWT Food Sci. Technol.* **2018**, *90*, 519–525. [CrossRef]

43. Ziober, I.; Paião, G.F.; Marchi, D.F.; Coutinho, L.L.; Binneck, E.; Nepomuceno, A.; Shimokomaki, M. Heat and chemical stress modulate the expression of the α-RYR gene in broiler chickens. *Genet. Mol. Res.* **2010**, *9*, 1258–1266. [CrossRef] [PubMed]

44. Paula de Oliveira, M.M.; Resende Gonçalves Silva, J.; Lamas de Oliveira, K.; Massingue, A.A.; Mendes Ramos, E.; Júnior, A.A.B.; Louzada Silva, M.H.; Olimi Silva, V.R. Technological and sensory characteristics of hamburgers added with chia seed as fat replacer. *Cienc. Rural* **2019**, *8*, 49. [CrossRef]

45. Fernández-López, J.; Lucas-González, R.; Viuda-Martos, M.; Sayas-Barberá, E.; Navarro, C.; Haros, C.M.; Pérez-Álvarez, J.A. Chia (*Salvia hispanica* L.) products as ingredients for reformulating frankfurters: Effects on quality properties and shelf-life. *Meat Sci.* **2019**, *156*, 139–145. [CrossRef] [PubMed]

46. Angeloni, S.; Scortichini, S.; Fiorini, D.; Sagratini, G.; Vittori, S.; Neiens, S.D.; Steinhaus, M.; Zheljazkov, V.D.; Maggi, F.; Caprioli, G. Characterization of Odor-Active Compounds, Polyphenols, and Fatty Acids in Coffee Silverskin. *Molecules* **2020**, *25*, 2993. [CrossRef]

47. Ito, J.; Komuro, M.; Supardi Parida, I.; Shimizu, N.; Kato, S.; Meguro, Y.; Ogura, Y.; Kuwahara, S.; Miyazawa, T.; Nakagawa, K. Evaluation of lipid oxidation mechanisms in beverages and cosmetics via analysis of lipid hydroperoxide isomers. *Sci. Rep.* **2019**, *9*, 7387. [CrossRef] [PubMed]

48. Ghani, M.A.; Barril, C.; Bedgood, D.R., Jr.; Prenzler, P.D. Measurements of antioxidant activity woth the thiobarbituric acid reactive substances assay. *Food Chem.* **2017**, *230*, 195–207. [CrossRef]

49. Chen, D.W.; Balagiannis, D.P.; Parker, J.K. Use of egg yolk phospholipids to generate chicken meat odorants. *Food Chem.* **2019**, *286*, 71–77. [CrossRef] [PubMed]

50. Ferreira, V.C.S.; Morcuende, D.; Madruga, M.S.; Hernández-López, S.H.; Silva, F.A.P.; Ventanas, S.; Estévez, M. Effect of pre-cooking methods on the chemical and sensory deterioration of ready-to-eat chicken patties during chilled storage and microwave reheating. *Food Sci. Technol.* **2016**, *53*, 2760–2769. [CrossRef]

51. Kim, S.Y.; Li, J.; Lim, N.R.; Kang, B.S.; Park, H.J. Prediction of warmed-over flavor development in cooked chicken by colorimetric sensor array. *Food Chem.* **2016**, *211*, 440–447. [CrossRef]

52. Sanchez del Pulgar, J.; Roldan, M.; Ruiz-Carrascal, J. Volatile Compounds Profile of Sous-Vide Cooked Pork Cheeks as Affected by Cooking Conditions (Vacuum Packaging, Temperature and Time). *Molecules* **2013**, *18*, 12538–12547. [CrossRef] [PubMed]

53. Apriyantonio, A.; Indrawaty, D. Comparison on Flavor Characteristics of Domestic Chicken and Broiler as Affected by Different Processing Methods. Chapter Aroma, meat. In *Food Flavors. Formation, Analysis and Packaging Influences*; Contis, E.T., Ho, C.T., Mussinas, C.J., Parliment, T.H., Shahidi, F., Spanier, A.M., Eds.; Elsevier: Amsterdam, The Netherlands, 1998; pp. 233–244.

54. Ribeiro Dias, D. Industrial uses of cocoa by-products. In *Cocoa and Coffee Fermentation*, 1st ed.; Schwan, D., Fleet, R.F., Agro, G.H., Eds.; CRC Taylor & Francis: Boca Raton, FL, USA, 2015; pp. 309–330.

55. Domínguez, R.; Gómez, M.; Fonseca, S.; Lorenzo, J.M. Influence of thermal treatment on formation of volatile compounds, cooking loss and lipid oxidation in foal meat. *LWT Food Sci. Technol.* **2014**, *58*, 439–445. [CrossRef]

56. Pegg, R.B.; Kerrihard, A.L.; Shahidi, F. Warmed Over Flavor. In *Encyclopedia of Meat Science*; Devine, C., Dikeman, M., Eds.; Elsevier: Amsterdam, The Netherlands, 2014; pp. 410–415.

57. Córdoba, N.; Moreno, F.L.; Osorio, C.; Velásquez, S.; Ruiz, Y. Chemical and sensory evaluation of cold brew coffees using different roasting profiles and brewing methods. *Food Res. Int.* **2021**, *141*, 110141. [CrossRef] [PubMed]

 *foods*

Article

Effectiveness of Oat-Hull-Based Ingredient as Fat Replacer to Produce Low Fat Burger with High Beta-Glucans Content

Carmine Summo *, Davide De Angelis, Graziana Difonzo, Francesco Caponio and Antonella Pasqualone

Department of Soil, Plant and Food Science (DISSPA), University of Bari Aldo Moro, Via Amendola, 165/a, I-70126 Bari, Italy; davide.deangelis@uniba.it (D.D.A.); graziana.difonzo@uniba.it (G.D.); francesco.caponio@uniba.it (F.C.); antonella.pasqualone@uniba.it (A.P.)
* Correspondence: carmine.summo@uniba.it

Received: 30 June 2020; Accepted: 3 August 2020; Published: 4 August 2020

Abstract: Low-fat beef burgers with high beta-glucan content was obtained using a gel made from an oat-hull-based ingredient as fat replacer. Two levels of fat substitution were considered: 50% (T1) and 100% (T2). The nutritional composition, cooking yield, textural properties, color characteristics and consumer preference were evaluated, in comparison with a burger without fat replacer (CTRL). After cooking, T2 burger showed a significant increase in the cooking yield and a very low lipid content (3.48 g 100 g^{-1}) as well as a level of beta-glucans per single portion (2.96 g 100 g^{-1}) near the recommended daily intake. In T1 burger, the decrease of lipid content was mitigated during the cooking process, because the beta-glucans added had a fat-retaining effect. Compared to CTRL, replacing fat led to a softer texture of cooked burgers evaluated by Texture Profile Analysis. The differences in color, significant in raw burgers, were smoothed with cooking. The consumer evaluation, carried out according to the duo-trio test, highlighted significant differences between CTRL and T2 burgers in terms of odor, taste, color and texture. The consumers expressed a higher preference for the T2 burger, probably due to its softer texture and greater juiciness.

Keywords: beef burgers; soluble fiber; TPA; consumer evaluation; fatty acid composition

1. Introduction

Meat and meat products play an important role in human nutrition, constituting a rich source of proteins with high biological value, vitamins (A, B1, B3 and B12), as well as iron, zinc and other micronutrients [1]. The consumption of meat and meat products dates back to antiquity, but these products are still part of the gastronomic tradition of many countries. Therefore, a high number of Protected Designation of Origin (PDO) and Protected Geographical Indication (PGI) European brands—which link the quality of food products to a specific geographical area—have been awarded to "Meat products—cooked, salted and smoked" (205 registered products, accounting for 12.95% of the total PDO and PGI products) and "Fresh meat and offal" (180 registered products, i.e., 11.37% of the total PDO and PGI products) [2]. However, the high fat content of meat products (including saturated fatty acids and cholesterol) is related to increased risk of developing coronary heart diseases [3].

In this context, researchers and private companies alike are strongly engaged in trying to improve the nutritional value of meat products by lowering the cholesterol and lipid content, as well as decreasing saturated and increasing polyunsaturated fatty acids. Fats, however, play an important role in meat products, ensuring optimal rheological and textural properties [4] and conferring pleasant sensorial characteristics in terms of flavor and juiciness [5]. Therefore, the reduction of lipid content in meat involves the use of ingredients able to mimic the properties of fat, such as polysaccharides. Several experimental trials have therefore been performed that included various mostly fiber-rich polysaccharide-based fat replacers in the formulation of meat products, such as ground poppy seeds [6], mixtures of wheat fiber and pig skin [7], legume flours [8] and other vegetable sources, as indicated in recent reviews [9]. Dietary fiber can form a compact gel due to the ability to bind water improving the structural characteristics of reduced-fat products [10].

Among dietary fibers, beta-glucans from cereal grains have been recently studied in relation to the health benefits associated with their consumption such as the reduction of cholesterol level and a chemo-preventive effect as reported by Ho et al. [11].

Moreover, beta-glucans show several technologically useful properties (gelling capacity, emulsifying activity, fat/water binding capacity), which make them suitable ingredients in health-promoting functional foods [12]. The major applications of beta-glucans in food formulation are in milk-based products, such as fermented milk products and yogurt [13] and in bakery products [14]. Several beta-glucan sources have also been considered for improving the nutritional quality of meat products, with [15–17] or without [18] fat replacement. However, the level of beta-glucan enrichment reported in previous studies on meat products does not reach the recommended daily intake for beta-glucans, which accounts for 3 g per day [19].

In this frame, the aim of this study was the production of low-fat burgers with a beta-glucan content very close to the recommended daily intake and with good textural and sensorial characteristics.

2. Materials and Methods

2.1. Preparation of the Fat Replacer

An oat-hull-based ingredient (Nutraceutica S.R.L., Monterenzio, Italy) containing, as declared by the producer, 55% beta-glucans, <10% proteins, <2% fat, was used to prepare a gel by mixing 27.27 g of flour with 72.73 mL of distilled water for 5 min at 13,500 rpm by means of a T25 Ultraturrax (IKA, Staufen, Germany). The gel was then cut into small pieces to be used, freshly prepared, as a fat replacer in burgers.

The ratio flour:water was defined in preliminary tests to obtain a gel: (i) able to mimic as much as possible the consistency and homogeneity of the beef fat conventionally used to prepare meat burgers; (ii) having a beta-glucan concentration able to achieve, when added to burgers as total fat replacer, a beta-glucan content as near as possible to the daily intake recommendation (3 g per day) [19].

2.2. Preparation of the Beef Burgers

Beef meat, purchased at a local butcher's shop, was manually sectioned with a sharp knife to separate the lean meat from the visible adipose and connective tissues. Then, the lean meat (3.5 g 100 g^{-1} fat content) and the adipose tissue (71.5 g 100 g^{-1} fat content, still containing residual proteins and moisture) were separately ground using a grinder equipped with a 4 mm plate (Kenwood MG510, Delonghi Appliances, Treviso, Italy). Adipose tissue and lean meat, both ground, were then mixed manually. During the mixing step, three batches were prepared, according to three different formulations at increasing levels of fat: control (CTRL), with 15% of beef adipose tissue added; T1, with a partial (50%) substitution of beef adipose tissue (i.e., with 7.5% beef adipose tissue and 7.5% oat-hull-based gel added); and T2, with a total substitution of beef adipose tissue (i.e., with 15% oat-hull-based gel added). With the exception of salt, no other spices or ingredients were added. The formulations of the three burgers are reported in Table 1. The burgers, weighing approximately 50 g, were finally shaped (70 mm diameter, 10 mm thickness) using a burger maker mold. The whole experiment was repeated twice.

2.3. Cooking Procedure

The burgers were cooked according to the American Meat Science Association methodology [20], i.e., were roasted in an electric oven (Delonghi EO 3275, Delonghi Appliances, Treviso, Italy) preheated at 163 °C, until their internal temperature, measured by a digital thermometer (LT−101, TFA Dostmann, Reicholzheim, Germany), reached 71 °C. Approximatively 10 min was sufficient to cook all the samples perfectly.

Cooked burgers were then submitted to the chemical and textural determinations, as well as consumer test. The colorimetric determinations, instead, were carried out on burgers both before (raw) and after cooking.

Table 1. Formulation (g kg^{-1}) of three different beef burgers without fat substitution (CTRL) and at 50% (T1) and 100% (T2) fat substitution.

Ingredient/Formulation	Samples		
	CTRL	**T1**	**T2**
Beef lean meat	835.25	835.25	835.25
Beef adipose tissue	150.00	75.00	0
Oat hull based gel *	0	75.00	150.00
Salt	14.75	14.75	14.75

* Gel as fat replacer formulated with 27.27 g of oat hull ingredient at 55% of beta-glucan concentration emulsified with 72.73 mL of distilled water.

2.4. Chemical Composition of Beef Burgers

Protein content (total nitrogen × 6.25), ash, and moisture content were determined, according to the AOAC International methods, to be 928.08, 920.153 and 950.46, respectively [21]. The lipid content was determined by Folch method [22] using chloroform and methanol (Sigma Aldrich, Milan, Italy) as extracting solvent. The carbohydrate content was determined as difference. The total beta-glucan concentration was determined by the AOAC International method 995.16 [23] by using the Megazyme mixed-linkage beta-glucan assay kit (Megazyme International, Bray, Ireland). The total energy value for each product was calculated by using the Atwater coefficients as reported in Summo et al. [24]. All determinations were carried out in triplicate.

2.5. Fatty Acid Composition of Beef Burgers

The fatty acid composition was determined by gas-chromatographic (GC) analysis of fatty acid methyl esters. The lipid fraction was cold-extracted with methanol/chloroform (1:2 *v/v*) following the method proposed by Folch et al. [22]. The methylation was carried out according to the AOCS (American Oil Chemists Society) method Ch 1–91 [25]. The GC system and conditions were the same as those reported in a previous paper [26]. The identification of each fatty acid was carried out by comparing the retention time with that of the corresponding methyl ester standard (Sigma Aldrich, Milan, Italy). All determinations were carried out in triplicate.

Atherogenic (AI) and Thrombogenic (TI) indices were calculated according to the following equations [27]:

$$AI = [C_{12:0} + (4 \times C_{14:0}) + C_{16:0}]/(\text{n-6 PUFA} + \text{n-3 PUFA} + \text{MUFA}) \tag{1}$$

$$TI = (C_{14:0} + C_{16:0} + C_{18:0})/[0.5 \times \text{MUFA} + 0.5 \times \text{n-6 PUFA} + 3 \times \text{n-3 PUFA} + (\text{n-3 PUFA/n-6 PUFA})] \tag{2}$$

where PUFA are polyunsaturated and MUFA monounsaturated fatty acids. $C_{12:0}$, $C_{14:0}$, $C_{16:0}$ and $C_{18:0}$ are lauric, myristic, palmitic and stearic acids, respectively.

2.6. Cooking Yield

The cooking yield of beef burgers was determined by measuring the weight (w) of the burgers before and after cooking according to the following equation:

$$\text{Cooking yield} = (\text{w cooked burger/w raw burger}) \times 100. \tag{3}$$

The calculation has been performed on ten burgers.

2.7. Texture Profile Analysis

Texture profile analysis (TPA) of beef burgers was performed according to Afshari et al. [17] with some modifications, using a texture analyzer model Z1.0 TN (Zwick Roell, Ulm, Germany) equipped with a 3.6 cm cylindrical probe and a 1 kN load cell. The samples were heated in an oven at 60 °C in order to simulate the serving conditions. Then, a portion of 2 cm of diameter was cut from the center of the burger. A two-compression cycle was carried out at the speed of 5 mm s^{-1}, with 5 s of pause between the two compressions, up to 70% of recorded deformation. The following parameters were assessed: hardness (N), indicating the maximum force recorded during the first compression; cohesiveness, measured as the area of work during the second compression divided by the area of work during the first compression; gumminess (N), calculated as hardness × cohesiveness; springiness, measured by the distance of the detected height during the second compression divided by the original compression distance; chewiness (N), calculated as gumminess × springiness. Ten different burgers per formulation were considered, and each burger was subjected to one measurement by TPA.

2.8. Color Determination of Burgers

Instrumental determination of the surface color of both raw and cooked burgers was carried out by using the CM-600d colorimeter (Konica Minolta, Tokyo, Japan) supported by SpectraMagic NX software (Konica Minolta, Tokyo, Japan). The CIE (International Commition on Illumination) L^*, a^*, and b^* parameters were recorded: lightness (L^*), red index (a^*) and yellow index (b^*), together with ΔE [28].

$$\Delta E = [(\Delta L^*)2 + (\Delta a^*)2 + (\Delta b^*)2]^{1/2} \tag{4}$$

Three samples per formulation were analyzed, and four readings were recorded in different areas of each sample.

2.9. Duo-Trio Consumer Test

CTRL and T2 burgers were submitted to consumer test according to the duo-trio test methodology [29] to determine if the differences between them could be recognized. Sixty people, regular consumers of meat and neither food-allergic nor intolerant, were recruited among the researchers and students of the Agricultural Faculty of the University of Bari Aldo Moro (Bari, Italy). The study protocol followed the ethical guidelines of the laboratory. Each participant was given information about study aims and individual written informed consent was obtained from each participant. The consumer test was performed at a local restaurant sited in Bari (Italy). Each participant received three samples on the same dish: one as reference (CTRL or T2 randomly, and codified with an alphanumeric code), and the other two were both CTRL and T2 randomly distributed, codified with an alphanumeric code. Each consumer was asked to indicate the sample that was different respect to the reference in terms of color, odor, taste and texture. Moreover, each panelist expressed a judgment indicating which burger preferred. The results were expressed as number of correct answers.

2.10. Statistical Analysis

Data were subjected to one-way ANOVA followed by the Tukey's HSD test. Significant differences were determined at $p < 0.05$ by the XLStat software (Addinsoft SARL, New York, NY, USA).

The results of duo–trio test were expressed as number of correct answers considering thirty-nine, forty-one and forty-four as minimum correct answers to identify statistically significant differences at $p < 0.05$, $p < 0.01$ and $p < 0.001$, respectively [30].

3. Results and Discussion

3.1. Chemical Composition

The addition of the fat replacer significantly influenced the chemical composition of cooked burgers (Table 2). An increase of moisture was observed at increasing content of fat replacer. This is principally due to the high moisture content of the fat replacer. These findings agreed with those of a previous study involving the use of oat beta-glucan as fat replacer [16]. However, in another study, the use of gelled emulsion (based on olive oil, gelatin and 9% inulin) caused an increase of moisture content only in raw patties, whereas a significantly lower moisture of cooked product was observed due to lower cooking yield and water holding capacity of the gel [31]. Therefore, our results could be due also to better moisture retention of fat-substituted burgers during cooking due to the high hydrophilicity of beta-glucans [32], able to increase the water-holding capacity of the product. The total substitution of fat (T2), indeed, caused a significantly higher moisture content than in CTRL and T1.

Table 2. Chemical composition, cooking yield and energy value of the cooked beef burger without fat substitution (CTRL) and at 50% (T1) and 100% (T2) fat substitution with an oat-hull-based gel.

	CTRL	T1	T2	*p*-Value
Moisture (% f.w.)	57.24 ± 0.22^C	58.79 ± 0.46^B	63.39 ± 0.26^A	$p < 0.001$
Protein (% f.w.)	28.41 ± 0.29^A	26.98 ± 0.15^B	25.83 ± 0.10^C	$p < 0.001$
Fat (% f.w.)	8.42 ± 0.04^A	7.25 ± 0.12^B	3.48 ± 0.03^C	$p < 0.001$
Ash (% f.w.)	2.42 ± 0.31^{AB}	2.29 ± 0.25^B	2.93 ± 0.16^A	$p = 0.045$
Total Carbohydrates (% f.w.)	3.51 ± 0.43^B	4.70 ± 0.43^A	4.38 ± 0.49^{AB}	$p = 0.044$
Beta-glucan (% f.w.)	0.01 ± 0.01^C	1.35 ± 0.13^B	2.96 ± 0.07^A	$p < 0.001$
Cooking Yield (%)	71.82 ± 1.39^B	75.14 ± 2.13^B	80.30 ± 2.50^A	$p = 0.007$
Energy Value (kcal/100 g)	203.44 ± 0.13^A	186.47 ± 3.49^B	146.24 ± 1.88^C	$p < 0.001$

Data on the chemical composition were expressed as % on fresh (f.w.) weight. Different letters in the same row indicate significant differences at $p < 0.05$.

On the contrary, the protein content of beef burgers (on fresh matter), showed a progressive and significant decrease when the fat replacement increased. Piñero et al. [15] and Afshari et al. [17] reported that the addition of a beta-glucan-based fat replacer had no significant influence on the protein content. Our findings could be related to a higher level of gel incorporation and a consequently higher moisture content. Moreover, the beef adipose tissue used in CTRL and T2 formulations contained muscular residues, which also contributed to the protein content, in accordance with other authors [33]

Compared to CTRL, the addition of the fat replacer resulted in a slight but significant fat decrease in T1 formulation, whereas the T2 burger showed a more marked decrease. Considering the lipid content of the beef adipose tissue (accounting for 71.5%) used in CTRL and T1 formulations, and the contribution of the residual intramuscular fat of the lean fraction (3.5%), the lipid content of the CTRL raw burger could be estimated at 13.6 g 100 g^{-1}. After cooking, the CTRL burger showed a lipid content of 8.42% (6.04 g of fat in 71.82 g of cooked burgers); therefore, an estimated fat loss of 56% occurred. The lipid content of the raw T1 burger could be estimated at 8.17 g 100 g^{-1}, whereas the cooked burger had a 7.25% fat content (5.45 g of fat in 75.14 g of cooked burger), with a fat loss of 34%. Therefore, even if considering estimated values, cooking induced a more limited fat loss when fat was replaced by the beta-glucan based gel than in the CTRL burger. This phenomenon could be imputable to the ability of the beta-glucans to form a tri-dimensional network which entraps fat and water within the meat protein system [15]. Therefore, it has to be considered that partial fat replacement with beta-glucans lowers fat content in the raw product, but this nutritionally positive effect is mitigated by higher fat retention during the cooking process. As a consequence, a total fat replacement has to be made to achieve a significant nutritional effect on the cooked product.

The use of the fat replacer caused, as expected, a slight but significant increase in the carbohydrate content of T1, even if no significant differences were observed comparing T1 and T2. This was imputable to the presence of carbohydrates in the oat-hull-based ingredient. The addition of vegetable fat replacer in burgers is reported to be influential on the chemical composition of the product [34]. The content of beta-glucans reached a level that made the health claim "beta-glucans contribute to the maintenance of normal blood cholesterol levels" applicable to both T1 and T2 burgers since the concentration of these compounds was always higher than 1 g per recommended portion (in meat products, this quantity corresponds to 100 g). However, the claim regulation specifies that "the beneficial effect is obtained with a daily intake of 3 g of beta-glucans" [19]. In this regard, T2 burger contained 2.96% of beta-glucans. Therefore, the recommended daily intake of beta-glucans, according to the above-mentioned regulation, could be reached by consuming a single portion (100 g) of T2 burger. This result is particularly important because it is possible to achieve a significant improvement in the nutritional characteristics of burgers. Indeed, by combining the total substitution of animal fat with the inclusion of functional macromolecules, a positive effect on cholesterol reduction could be expected. Indeed, it is known that beta-glucan has an active role on the reduction of LDL-cholesterol [11] by modulating the cholesterol metabolism and the gut microbiota [35].

The fat substitution resulted in a significant decrease in energy value, from 203.44 kcal 100 g^{-1} (CTRL) to 146.24 kcal 100 g^{-1} (T2). In particular, the T2 formulation allowed the research to obtain a product with lower fat content and, consequently, lower energy value compared to the products proposed by other studies [17–19]. An effective improvement of the nutritional value of meat products was therefore achieved, due to reduced fat content, relatively low energy value and high concentration of beta-glucans.

3.2. Cooking Yield

The fat replacement caused an increase in cooking yield. The difference, compared with the control burger, became significant in the T2 formulation. These findings agreed with previous studies [17,36] in which higher cooking yield and moisture retention with the increase of beta-glucan content was observed. This behavior can be explained with the already mentioned ability of beta-glucans to form three-dimensional structures with meat proteins, which can easily entrap water and fat, increasing the cooking yield [15].

3.3. Fatty Acid Composition

Fatty acid composition of burgers is reported in Table 3, as mg 100 g^{-1} of burger and g 100 g^{-1} of fatty acids. The nutritional value of beef burgers is also related to the composition of the lipid fraction, which usually is dominated by saturated fatty acids, palmitic and stearic acids in particular, whereas oleic acid was the most abundant unsaturated acid. The fatty acid composition of cooked burgers agreed with other studies carried out on the same category of products [17,37]. Owing to the fat substitution, a significant reduction was observed of the quantity (mg 100 g^{-1} of burger) of all fatty acids due to the general decrease of lipid content. Moreover, a different level of reduction was observed as a function of the unsaturation rate. In particular, T2 showed a content of palmitic acid 60% lower than the CTRL. The reduction was slightly lower for oleic acid (−57%), whereas linolenic, the most abundant polyunsaturated fatty acid, decreased by 45% comparing T2 with CTRL. This aspect could be better explained considering the composition of fatty acids expressed as percentage. In particular, comparing the T2 with the other formulations, we observed a significantly ($p < 0.05$) lower percentage of saturated fatty acids and a higher percentage of the polyunsaturated fatty acids, whereas the monounsaturated fatty acids remained constant across the formulations. Previous studies report significant differences in the fatty acid composition of subcutaneous and muscular beef fat, with the latter characterized by higher polyunsaturated and lower saturated fatty acids [38,39]. This could explain the differences observed in our samples, because in CTRL and T1 burgers, the fatty fraction added was mainly subcutaneous fat, while in T2 the residual fat was constituted principally by muscular fat.

Table 3. Fatty acid composition (g 100 g^{-1} of burger and g 100 g^{-1} of fatty acids) and the nutritional index of the beef burger without fat substitution (CTRL) and at 50% (T1) and 100% (T2) of fat substitution with an oat-hull-based gel.

	mg 100 g^{-1} of Burger			g 100 g^{-1} of Total Fatty Acids		
	CTRL	T1	T2	CTRL	T1	T2
Myristic C$_{14:0}$	395.29 ± 12.98^A	300.95 ± 20.07^B	111.52 ± 4.78^C	4.69± 0.15^A	4.15 ± 0.28^B	3.20 ± 0.14^C
Myristoleic C$_{14:1}$	115.74 ± 7.41^A	75.77 ± 0.59^B	37.93 ± 5.14^C	1.37 ± 0.09^A	1.05 ± 0.01^B	1.09 ± 0.15^B
Pentadecanoic C$_{15:0}$	45.57 ± 0.37^A	38.12 ± 0.94^B	17.44 ± 0.85^C	0.54 ± 0.00^A	0.53 ± 0.01^A	0.50 ± 0.02^A
Pentadecenoic C$_{15:1}$	14.58 ± 0.82^A	12.06 ± 2.50^A	8.23 ± 0.03^B	0.17 ± 0.01^B	0.17 ± 0.03^B	0.24 ± 0.00^A
Palmitic C$_{16:0}$	2300.53 ± 31.67^A	1985.39 ± 53.88^B	934.19 ± 14.64^C	27.32 ± 0.38^A	27.38 ± 0.74^A	26.84 ± 0.42^A
Palmitoleic C$_{16:1}$	460.47 ± 15.01^A	370.8 ± 14.13^B	173.9 ± 1.34^C	5.47 ± 0.18^A	5.11 ± 0.19^{AB}	5.00 ± 0.04^B
Heptadecanoic C$_{17:0}$	72.67 ± 1.48^A	71.02 ± 2.33^A	28.22 ± 0.99^B	0.86 ± 0.02^B	0.98 ± 0.03^A	0.81 ± 0.03^B
Heptadecenoic C$_{17:1}$	56.71 ± 0.17^A	57.06 ± 2.17^A	32.72 ± 0.92^B	0.67 ± 0.00^C	0.79 ± 0.03^B	0.94 ± 0.03^A
Stearic C$_{18:0}$	1146.61 ± 29.63^A	1067.91 ± 18.60^B	485.95 ± 5.80^C	13.62 ± 0.35^B	14.73 ± 0.26^A	13.96 ± 0.17^B
Oleic C$_{18:1\,n-9}$	3252.74 ± 56.21^A	2939.81 ± 113.70^B	1361.05 ± 45.98^C	38.63 ± 0.67^A	40.45 ± 1.57^A	39.11 ± 1.32^A
Linoleic C$_{18:2\,n-6}$	409.06 ± 19.51^A	266.87 ± 34.27^B	236.6 ± 22.44^B	4.86 ± 0.23^B	3.68 ± 0.47^B	6.80 ± 0.64^A
Linolenic C$_{18:3\,n-6}$	39.67 ± 4.77^A	30.33 ± 1.23^B	11.26 ± 0.42^C	0.47 ± 0.06^A	0.42 ± 0.02^A	0.32 ± 0.01^B
dihomo-γ-linolenic C$_{20:3\,n-6}$	58.61 ± 7.50^A	17.20 ± 5.95^B	23.55 ± 2.43^B	0.70 ± 0.09^A	0.24 ± 0.08^B	0.68 ± 0.07^A
Arachidonic C$_{20:4\,n-6}$	29.96 ± 11.23^A	9.61 ± 3.47^B	9.41 ± 0.95^B	0.36 ± 0.13^A	0.13 ± 0.05^B	0.27 ± 0.03^{AB}
Eicosapentaenoic C$_{20:5\,n-3}$	9.77 ± 2.08^A	3.29 ± 1.75^B	3.40 ± 2.95^B	0.12 ± 0.02^A	0.08 ± 0.02^A	0.10 ± 0.03^A
Docosapentaenoic C$_{22:5\,n-3}$	12.03 ± 0.88^A	3.82 ± 0.49^C	5.48 ± 0.41^B	0.14 ± 0.03^A	0.12 ± 0.02^A	0.16 ± 0.03^A
ΣSFA	3960.67 ± 16.87^A	3463.39 ± 53.96^B	1577.32 ± 13.74^C	47.04 ± 0.20^A	47.77 ± 0.74^A	45.33 ± 0.39^B
ΣMUFA	559.09 ± 45.97^A	331.13 ± 44.70^B	289.70 ± 27.59^B	46.32 ± 0.75^A	47.56 ± 1.36^A	46.37 ± 1.19^A
ΣPUFA	3900.24 ± 62.84^A	3455.48 ± 98.65^B	1613.83 ± 41.29^C	6.64 ± 0.55^B	4.67 ± 0.62^C	8.32 ± 0.79^A
MUFA/SFA ratio				0.98 ± 0.02^A	1.00 ± 0.04^A	1.02 ± 0.04^A
PUFA/SFA ratio				0.14 ± 0.01^B	0.10 ± 0.01^C	0.18 ± 0.02^A
AI				0.87 ± 0.02^A	0.88 ± 0.05^A	0.73 ± 0.02^B
TI				1.68 ± 0.01^A	1.71 ± 0.05^A	1.57 ± 0.02^B
n-6/n-3 PUFA				24.78 ± 1.41^A	22.35 ± 9.47^A	33.35 ± 8.89^{AB}

SFA = Saturated fatty acids; MUFA = Monounsaturated fatty acids; PUFA = Polyunsaturated fatty acids; AI = Atherogenic Index; TI = Thrombogenic Index. Different letters in the same row indicate significant differences at $p < 0.05$.

Albeit in low amounts, we detected also some polyunsaturated fatty acids important from a nutritional point of view, such as the arachidonic (C$_{20:4\,n-6}$) eicosapentaenoic (C$_{20:5\,n-3}$) and docosapentaenoic acids (C$_{22:5\,n-3}$), without significant differences among the formulations. The amount of these important fatty acids was lower than that reported in other studies carried out on the raw beef lipid fraction [40]. This difference could be related to the cooking procedure, which causes the loss of these fatty acids [37]. In studies carried out on cooked beef burgers, these fatty acids were indeed not determined [17,41].

As a consequence of the different lipid composition, the nutritional indices linked to the fatty acid composition were also influenced by the fat replacement. In particular, the PUFA/SFA ratio significantly increased in T2 compared to CTRL. Moreover, the atherogenic and thrombogenic indices related to fatty acid composition significantly decreased in T2 burger with 100% fat substitution, although the values were higher than those recommended [42]. The n-6/n-3 ratio was higher in T2 compared to CTRL and T1. It is reported that lowering the n-6/n-3 ratio to less than 4 is desirable to improve the healthiness of the product [43,44]. However, the achievement of this target in meat product is not possible solely with a fat reduction, because fat composition needs to be reformulated by the addition of oils rich in n-3 PUFA [44,45].

Similar improvements were observed by Pintado et al. [45] in fresh sausages obtained using an olive oil in water emulsion containing chia and oat as fat replacer. The authors explained the results with the high level of polyunsaturated fatty acids of chia. The oat-hull-based ingredient used in our study was characterized by a very low lipid content; therefore, its contribution to the fatty acid composition was of relevance. Several studies report that the unsaturated fatty fractions are combined with structural compounds of meat so that their loss during cooking is less influenced than saturated fatty acids [44]. The saturated fatty acids could easily be lost during cooking, and this could explain the observed results.

3.4. Texture Profile Analysis

Significant differences in the textural properties were observed among burgers with different formulation (Table 4). The incorporation of a fat replacer led to a significant decrease of hardness, cohesiveness, gumminess and chewiness in T1 and T2 burgers compared to CTRL, indicating that these burgers had a softer texture and then required less energy to be compressed. No significant differences, however, were found between T1 and T2, highlighting the fact that the level of fat substitution did not influence the textural properties of beef burgers.

Table 4. Texture profile analysis (TPA) of the beef burger without fat substitution (CTRL) and at 50% (T1) and 100% (T2) of fat substitution with an oat-hull-based gel.

	Hardness (N)	Springiness	Gumminess	Chewiness (N)	Cohesivity (N)
CTRL	159.1 ± 10.4^A	0.71 ± 0.02^A	56.2 ± 7.7^A	40.2 ± 6.4^A	0.35 ± 0.04^A
T1	116.0 ± 7.5^B	0.68 ± 0.02^B	33.9 ± 2.7^B	23.0 ± 2.3^B	0.29 ± 0.01^B
T2	113.7 ± 9.8^B	0.62 ± 0.03^C	29.8 ± 3.8^B	18.5 ± 2.7^B	0.26 ± 0.02^B
p-Value	$p < 0.001$	$p < 0.001$	$p < 0.001$	$p < 0.001$	$p < 0.001$

Different letters in the same column indicate significant differences at $p < 0.05$.

The trend of moisture and fat as an influence on texture [17] could be explained by a compensation between the differences in moisture and fat contents of T1 and T2 (Table 2), leading to similar textural properties. The effect of the fat substitution level was significant only for springiness, which showed the lowest value in T2 formulation.

Owing to the important structural functions of fat, the influence on the textural properties should be considered when the target of a new food formulation is fat substitution. The use of beta-glucans as fat replacement in beef burger or beef patties was previously studied by other authors with contrasting results, depending on whether beta-glucans were added as powder, gel or emulsion. In particular, Szpicer et al. [16] reported an increase in hardness of meat burgers after the addition of 30% beta-glucan concentrate powder. When the beta-glucans were added as gel [15] or emulsion [36], a significant reduction of hardness and other textural parameters were observed. With the increase of beta-glucans concentration, the amount of water available for proteins decreases and meat products lose springiness [46]. This behavior could be explained by a higher moisture retention of burgers and a consequently lower compactness of protein matrix [36]. Furthermore, beta-glucans have the ability to bind not only water but also fat, allowing the formation of a softer [47] and juicier product [17].

3.5. Color Indices

Color evaluations on the raw burger were made because the color characteristics of the meat products can influence the consumers' willingness to purchase, with increasing appreciation for bright red products. In raw burgers, a progressive and significant increase of lightness (L^*) and yellowness (b^*) was observed with fat replacement, while redness (a^*) was not significantly influenced (Table 5). The increase of the lightness and yellowness could be related to the presence of yellow pigments such as lutein in oat (the source of beta-glucan enriched gel), as previously reported in [48]. In contrast, a^* remained constant, indicating that the fat substitution was not significant on this index. Moreover, in a previous study, the fat substitution with a chia oil emulsion gel caused no significant variations of a^* but significant changes of L^* and b^* [49]. In the same study, L^* and b^* were slightly higher than ours, probably because of the presence of the oil in the fat replacer.

The differences observed among raw burgers were smoothed by cooking, after which no significant differences were found for all the color indices, as reported also by Gök et al. [6]. The color of burgers reformulated with fat replacers is influenced by the type of ingredients used for this purpose. In particular, Lucas-González et al. [49] reported a decrease of L^* and an increase of a^* during cooking of burgers formulated with chestnut flour and chia oil emulsion gels. By contrast, Heck et al. [43] reported an increase of L^* and a decrease of a^* in cooked burgers produced by the inclusion of linseed or chia oil microparticles. During the cooking process, meat color changes due to the heat-induced denaturation of myoglobin. Our results, assessed on the cooked burgers, were not influenced by fat substitution; however, it is reasonable to say that the primary contribution to color is given by meat.

The role of fat in influencing the color of cooked meat is not fully understood [50], but it should have a lower influence on color than other critical parameters, such as pH and storage conditions [50].

Table 5. Instrumental color determination of the beef burger without fat substitution (CTRL) and at 50% (T1) and 100% (T2) of fat substitution with an oat-hull-based gel before (Raw) and after (Cooked) cooking.

	Raw			Cooked		
	CTRL	**T1**	**T2**	**CTRL**	**T1**	**T2**
L^*	39.04 ± 0.77^C	41.07 ± 0.30^B	42.97 ± 1.28^A	48.00 ± 2.21^A	48.22 ± 2.00^A	47.69 ± 1.41^A
a^*	13.63 ± 0.40^A	13.40 ± 0.63^A	14.15 ± 1.76^A	6.09 ± 0.91^A	6.22 ± 0.73^A	6.23 ± 0.41^A
b^*	14.73 ± 0.32^C	17.84 ± 0.18^B	20.35 ± 2.17^A	13.26 ± 1.30^A	11.60 ± 0.86^{AB}	12.18 ± 0.67^B
ΔE vs. CTRL		3.89 ± 0.36	7.16 ± 2.61		3.38 ± 1.55	2.58 ± 1.62

Different letters in the same row indicate significant differences at $p < 0.05$.

The ΔE of T1 and T2 formulations, calculated by comparing them to the CTRL, was determined in order to improve evaluation of the color differences between samples. The ΔE was higher in raw than in cooked burgers, reaching the maximum of 7.16 in T2 formulation, whereas T1 showed a value of 3.89. ΔE values were between 3.5 and 5.0, meaning that the observer can clearly perceive the difference between samples; thus, T1 raw burgers could be easily distinguished from CTRL. ΔE values higher than 5 indicate the presence of two distinct colors [51]. When considering the cooked burgers, a decrease of ΔE of both T1 and T2 was observed. The changes occurring in T2 burger were particularly interesting due to the drop of ΔE at 2.58. When $2.0 < \Delta E < 3.5$, even an unexperienced observer can notice the difference in color between products [51].

3.6. Consumer Test

CTRL and T2 were submitted to a consumer test, according to the duo–trio test methodology [28], which was chosen to determine if the differences between burgers in terms of color, odor, taste and texture were recognizable by consumers. T1 burger was not considered for two main reasons. Firstly, after preliminary sensory analysis, a small group of trained panelists agreed that T1 burger was similar to CTRL. Moreover, considering the nutritional characteristics of T2 burgers, they were noticeably more interesting than T1, therefore we selected only T2 burger, which had no fat added and had a high content of beta-glucans.

As shown in Figure 1, the consumers recognized the difference between CTRL and T2 burgers for all the descriptors. In particular, forty-one people recognized CTRL and T2 for their different color ($p < 0.01$), whereas the number of correct answers increased when considering odor, texture and taste, with highly significant results ($p < 0.001$). The consumer test confirmed the results of textural and colorimetric evaluations (see for example the ΔE parameter). Szpicer et al. [16] also reported that consumers could distinguish products containing fat replacers, based on differences in color, texture, aroma and taste. Moreover, Afshari et al. [17] highlighted that fat substitution was perceived as significantly different by sensory analysis. On the whole, the substitution of fat with the beta-glucan gel changed the textural and sensorial quality of burgers, but the modification did not cause a deterioration of the general appreciation of products. In actual fact, 59.32% of panelists expressed a preference for T2 burger, and 40.68% preferred the CTRL burger. This difference was devoid of statistical significance ($p > 0.05$); therefore, the addition of beta-glucan gel did not cause a significant decrease in the sensorial acceptability of the burgers. Both texture and taste, in fact, are known to influence the acceptability of meat products, especially the juiciness and the tenderness [52]. Moreover, as reported by Desmond et al. [53], a low water binding capacity implicates a negative effect on palatability, due to the lack of juiciness and brittle texture which are both generally unacceptable to the consumers.

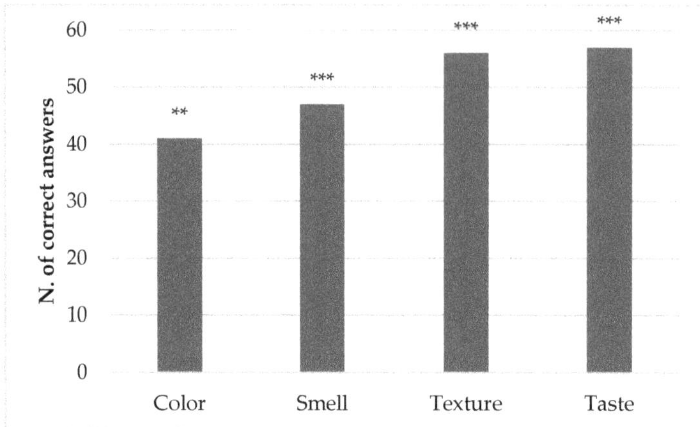

Figure 1. Number of people recognizing the difference between burger without fat substitution (CTRL) and at 100% fat substitution (T2) in a duo–trio consumer test. **: significance $p < 0.01$; ***: significance $p < 0.001$.

4. Conclusions

The use of an oat-hull-based gel as fat replacer allowed us to obtain a beef burger with a very low lipid content (3.48 g 100 g^{-1} in the formulation with a total fat substitution) and with a 2.96 g 100 g^{-1} content of beta-glucans, almost reaching the recommended daily intake per single portion of burger. With a partial substitution, the decrease of lipid content in the raw product was mitigated during the cooking process (34% and 56% of estimated fat loss in T1 and CTRL respectively). This could be related to the fat-retaining effect of beta-glucans added. Compared to CTRL, replacing fat by the oat-hull-based gel caused a significant decrease in hardness and other textural parameters of cooked burgers. Conversely, the differences in color, significant in raw burgers, were smoothed with cooking. The consumer evaluation, carried out according to the duo–trio test, highlighted significant differences between CTRL and T2 burgers in terms of odor, taste, color and texture. The consumers expressed a higher preference for the T2 burger, probably due to its softer texture and greater juiciness.

These results are a step forward for the improvement of the nutritional characteristics of meat products and indicate that the use of the oat-hull-based ingredient, rich in beta-glucans, as gel is an effective strategy for a complete fat substitution.

Author Contributions: Conceptualization, C.S.; methodology, C.S. and A.P.; formal analysis, D.D.A. and G.D.; data curation, D.D.A. and G.D.; Supervision, C.S., A.P. and F.C.; writing—original draft, C.S. and D.D.A.; writing—review and editing, C.S., A.P. and F.C. All authors have revised, read and approved the final manuscript. All authors have read and agreed to the published version of the manuscript.

Funding: This research received no external funding.

Acknowledgments: The authors would thank Raffaella Nasti and Isabella Centomani for the technical support.

Conflicts of Interest: The authors declare no conflict of interest.

References

1. Domínguez, R.; Borrajo, P.; Lorenzo, J.M. The effect of cooking methods on nutritional value of foal meat. *J. Food Compos. Anal.* **2015**, *43*, 61–67. [CrossRef]
2. eAmbrosia–the EU Geographical Indications Register. Available online: https://ec.europa.eu/info/food-farming-fisheries/food-safety-and-quality/certification/quality-labels/geographical-indications-register/# (accessed on 30 June 2020).
3. Durante, A.; Bronzato, S. A contemporary review of the relationship between red meat consumption and cardiovascular risk. *Int. J. Prev. Med.* **2017**, *8*, 40–45. [CrossRef] [PubMed]

4. Yoo, S.S.; Kook, S.H.; Park, S.Y.; Shim, J.-H.; Chin, K. Physicochemical characteristics, textural properties and volatile compounds in comminuted sausages as affected by various fat levels and fat replacers. *Int. J. Food Sci. Technol.* **2007**, *42*, 1114–1122. [CrossRef]

5. Totosaus, A.; Pérez-Chabela, M.L. Textural properties and microstructure of low-fat and sodium-reduced meat batters formulated with gellan gum and dicationic salts. *LWT Food Sci. Technol.* **2009**, *42*, 563–569. [CrossRef]

6. Gök, V.; Akkaya, L.; Obuz, E.; Bulut, S. Effect of ground poppy seed as a fat replacer on meat burgers. *Meat Sci.* **2011**, *89*, 400–404. [CrossRef]

7. Choe, J.H.; Kim, H.Y.; Lee, J.M.; Kim, Y.J.; Kim, C.J. Quality of frankfurter-type sausages with added pig skin and wheat fiber mixture as fat replacers. *Meat Sci.* **2013**, *93*, 849–854. [CrossRef] [PubMed]

8. Alakali, J.; Irtwange, S.V.; Mzer, M. Quality evaluation of beef patties formulated with bambara groundnut (Vigna subterranean L.) seed flour. *Meat Sci.* **2010**, *85*, 215–223. [CrossRef]

9. Kumar, P.; Chatli, M.K.; Verma, A.K.; Mehta, N.; Malav, O.P.; Kumar, D.; Sharma, N. Quality, functionality, and shelf life of fermented meat and meat products: A review. *Crit. Rev. Food Sci. Nutr.* **2015**, *57*, 2844–2856. [CrossRef]

10. Fernández-Ginés, J.M.; Fernández-López, J.; Sayas-Barberá, E.; Pérez-Alvarez, J. Meat Products as Functional Foods: A Review. *J. Food Sci.* **2005**, *70*, R37–R43. [CrossRef]

11. Ho, H.V.T.; Sievenpiper, J.L.; Zurbau, A.; Mejia, S.B.; Jovanovski, E.; Au-Yeung, F.; Jenkins, A.L.; Vuksan, V. The effect of oat β-glucan on LDL-cholesterol, non-HDL-cholesterol and apoB for CVD risk reduction: A systematic review and meta-analysis of randomised-controlled trials. *Br. J. Nutr.* **2016**, *116*, 1369–1382. [CrossRef]

12. Liu, N.; Nguyen, H.; Wismer, W.V.; Temelli, F.; Nguyễn, H. Development of an orange-flavoured functional beverage formulated with beta-glucan and coenzyme Q10-impregnated beta-glucan. *J. Funct. Foods* **2018**, *47*, 397–404. [CrossRef]

13. Vanegas-Azuero, A.M.; Gutiérrez, L.F. Physicochemical and sensory properties of yogurts containing sacha inchi (Plukenetia volubilis L.) seeds and β-glucans from Ganoderma lucidum. *J. Dairy Sci.* **2018**, *101*, 1020–1033. [CrossRef] [PubMed]

14. Kurek, M.A.; Małgorzata, M.; Sabina, K.; Horbańczuk, O.K.; Rodak, E. Application of rich in β-glucan flours and preparations in bread baked from frozen dough. *Food Sci. Technol. Int.* **2019**, *26*, 53–64. [CrossRef]

15. Piñero, M.; Parra, K.; Huerta-Leidenz, N.; De Moreno, L.A.; Ferrer, M.; Araujo, S.; Barboza, Y. Effect of oat's soluble fibre (β-glucan) as a fat replacer on physical, chemical, microbiological and sensory properties of low-fat beef patties. *Meat Sci.* **2008**, *80*, 675–680. [CrossRef] [PubMed]

16. Szpicer, A.; Onopiuk, A.; Półtorak, A.; Wierzbicka, A. Influence of oat β-glucan and canola oil addition on the physic-chemical properties of low-fat beef burgers. *J. Food Process Preserv.* **2018**, *42*, e13785. [CrossRef]

17. Afshari, R.; Hosseini, H.; Khaneghah, A.M.; Khaksar, R. Physico-chemical properties of functional low-fat beef burgers: Fatty acid profile modification. *LWT Food Sci. Technol.* **2017**, *78*, 325–331. [CrossRef]

18. Apostu, P.M.; Mihociu, T.E.; Nicolau, A.I. Technological and sensorial role of yeast β-glucan in meat batter reformulations. *J. Food Sci. Technol.* **2017**, *54*, 2653–2660. [CrossRef] [PubMed]

19. European Commission. Commission Regulation (EU) No 432/2012 of 16 May 2012. establishing a list of permitted health claims made on foods, other than those referring to the reduction of disease risk and to children development and health. *Off. J. Eur. Communities* **2012**, *136*, 1–40.

20. AMSA. *Research Guidelines for Cookery, Sensory Evaluation and Instrumental Tenderness Measurements of Meat*, 2nd ed.; American Meat Science Association: Champaign, IL, USA, 2015; p. 16.

21. AOAC International. *Official Methods of Analysis of AOAC International*, 17th ed.; AOAC International: Gaithersburg, MD, USA, 2006.

22. Folch, J.; Lees, M.; Stanley, G.H.S. A simple method for the isolation and purification of total lipides from animal tissues. *J. Biol. Chem.* **1957**, *226*, 497–509.

23. McCleary, B.V.; Mugford, D.C. Determination of beta-glucan in barley and oats by streamlined enzymatic method: Summary of collaborative study. *J. AOAC Int.* **1997**, *80*, 580–583. [CrossRef]

24. Summo, C.; Palasciano, M.; De Angelis, D.; Paradiso, V.M.; Caponio, F.; Pasqualone, A. Evaluation of the chemical and nutritional characteristics of almonds (Prunus dulcis (Mill). D.A. Webb) as influenced by harvest time and cultivar. *J. Sci. Food Agric.* **2018**, *98*, 5647–5655. [CrossRef] [PubMed]

25. AOCS. *Official Methods and Recommended Practices of the American Oil Chemists' Society*, 4th ed.; AOCS Press: Champaign, IL, USA, 1993.
26. Summo, C.; Centomani, I.; Paradiso, V.M.; Caponio, F.; Pasqualone, A. The effects of the type of cereal on the chemical and textural properties and on the consumer acceptance of pre-cooked, legume-based burgers. *LWT Food Sci. Technol.* **2016**, *65*, 290–296. [CrossRef]
27. Ulbricht, T.; Southgate, D. Coronary heart disease: Seven dietary factors. *Lancet* **1991**, *338*, 985–992. [CrossRef]
28. CIE. Recommendations on uniform color spaces, color-difference equations, and metric color terms. *Color. Res. Appl.* **1977**, *2*, 5–6. [CrossRef]
29. International Organization for Standardization. *Sensory Analysis–Methodology–Duo-trio Test, ISO Standard 10399*, 3rd ed.; Copyright Office: Genève, Switzerland, 2017.
30. Pagliarini, E. *Valutazione Sensoriale–Aspetti Teorici, Pratici E Metodologici*, 1st ed.; Hoepli: Milan, Italy, 2011; pp. 53–54.
31. Serdaroğlu, M.; Nacak, B.; Karabıyıkoğlu, M. Effects of beef fat replacement with gelled emulsion prepared with olive oil on quality parameters of chicken patties. *Food Sci. Anim. Resour.* **2017**, *37*, 376–384. [CrossRef]
32. Londono, D.M.; Gilissen, L.J.; Visser, R.G.F.; Smulders, M.J.M.; Hamer, R.J. Understanding the role of oat β-glucan in oat-based dough systems. *J. Cereal Sci.* **2015**, *62*, 1–7. [CrossRef]
33. Franco, D.; Martins, A.J.; López-Pedrouso, M.; Purriños, L.; Cerqueira, M.A.; Vicente, A.A.; Pastrana, L.M.; Zapata, C.; Lorenzo, J.M.; López-Pedrouso, M. Strategy towards replacing pork backfat with a linseed oleogel in frankfurter sausages and its evaluation on physicochemical, nutritional, and sensory characteristics. *Foods* **2019**, *8*, 366. [CrossRef]
34. Namir, M.; Siliha, H.; Ramadan, M.F.; Hassanien, M.F.R. Fiber pectin from tomato pomace: Characteristics, functional properties and application in low-fat beef burger. *J. Food Meas. Charact.* **2015**, *9*, 305–312. [CrossRef]
35. Joyce, S.A.; Kamil, A.; Fleige, L.; Gahan, C.G.M. The cholesterol-lowering effect of oats and oat beta glucan: Modes of action and potential role of bile acids and the microbiome. *Front. Nutr.* **2019**, *6*, 171. [CrossRef]
36. Afshari, R.; Hosseini, H.; Khaksar, R.; Mohammadifar, M.A.; Amiri, Z.; Komeili, R.; Khaneghah, A.M. Investigation of the effects of inulin and β-glucan on the physical and sensory properties of low-fat beef burgers containing vegetable oils: Optimisation of the formulation using d-optimal mixture design. *Food Technol. Biotechnol.* **2015**, *53*, 436–445. [CrossRef]
37. Alfaia, C.M.; Alves, S.P.; Lopes, A.F.; Fernandes, M.J.; Costa, A.S.; Fontes, C.M.G.A.; Castro, M.; Bessa, R.J.; Prates, J.A.M. Effect of cooking methods on fatty acids, conjugated isomers of linoleic acid and nutritional quality of beef intramuscular fat. *Meat Sci.* **2010**, *84*, 769–777. [CrossRef] [PubMed]
38. Aldai, N.; Nájera, A.I.; Dugan, M.; Celaya, R.; Osoro, K. Characterisation of intramuscular, intermuscular and subcutaneous adipose tissues in yearling bulls of different genetic groups. *Meat Sci.* **2007**, *76*, 682–691. [CrossRef] [PubMed]
39. Alam, M.K.; Rana, Z.H.; Akhtaruzzaman, M. Comparison of muscle and subcutaneous tissue fatty acid composition of bangladeshi nondescript deshi bulls finished on pasture diet. *J. Chem.* **2017**, *2017*, 1–6. [CrossRef]
40. Renna, M.; Brugiapaglia, A.; Zanardi, E.; Destefanis, G.; Prandini, A.; Moschini, M.; Sigolo, S.; Lussiana, C. Fatty acid profile, meat quality and flavour acceptability of beef from double-muscled Piemontese young bulls fed ground flaxseed. *Ital. J. Anim. Sci.* **2019**, *18*, 355–365. [CrossRef]
41. Jiang, T.; Busboom, J.; Nelson, M.; O'Fallon, J.; Ringkob, T.; Joos, D.; Piper, K. Effect of sampling fat location and cooking on fatty acid composition of beef steaks. *Meat Sci.* **2010**, *84*, 86–92. [CrossRef] [PubMed]
42. Hooper, C.L.; Martin, N.; Abdelhamid, A.; Smith, G.D. Reduction in saturated fat intake for cardiovascular disease. *Cochrane Database Syst. Rev.* **2015**, CD011737. [CrossRef] [PubMed]
43. Heck, R.T.; Vendruscolo, R.G.; de Araújo Etchepare, M.; Cichoski, A.J.; de Menezes, C.R.; Barin, J.S.; Lorenzo, J.M.; Wagner, R.; Campagnol, P.C.B. Is it possible to produce a low-fat burger with a healthy n−6/n−3 PUFA ratio without affecting the technological and sensory properties? *Meat Sci.* **2017**, *130*, 16–25. [CrossRef]
44. Selani, M.M.; Shirado, G.A.; Margiotta, G.B.; Rasera, M.L.; Marabesi, A.C.; Piedade, S.M.; Contreras-Castillo, C.J.; Canniatti-Brazaca, S.G. Pineapple by-product and canola oil as partial fat replacers in low-fat beef burger: Effects on oxidative stability, cholesterol content and fatty acid profile. *Meat Sci.* **2016**, *115*, 9–15. [CrossRef]

45. Pintado, T.; Herrero, A.; Jiménez-Colmenero, F.; Cavalheiro, C.P.; Ruiz-Capillas, C. Chia and oat emulsion gels as new animal fat replacers and healthy bioactive sources in fresh sausage formulation. *Meat Sci.* **2018**, *135*, 6–13. [CrossRef]

46. Mejia, S.M.V.; De Francisco, A.; Bohrer, B.M. Replacing starch in beef emulsion models with β-glucan, microcrystalline cellulose, or a combination of β-glucan and microcrystalline cellulose. *Meat Sci.* **2019**, *153*, 58–65. [CrossRef]

47. Warner, K.; Inglett, G.E. Flavor and texture characteristics of foods containing Z-trim corn and oat fibers as fat and flour replacers. *Cereal Foods World* **1997**, *42*, 821–825.

48. Ndolo, V.U.; Beta, T. Distribution of carotenoids in endosperm, germ, and aleurone fractions of cereal grain kernels. *Food Chem.* **2013**, *139*, 663–671. [CrossRef] [PubMed]

49. Lucas-González, R.; Roldán-Verdu, A.; Sayas-Barberá, E.; Fernández-López, J.; Pérez-Álvarez, J.A.; Viuda-Martos, M. Assessment of emulsion gels formulated with chestnut (Castanea sativa M.) flour and chia (Salvia hispanica L) oil as partial fat replacers in pork burger formulation. *J. Sci. Food Agric.* **2019**, *100*, 1265–1273. [CrossRef] [PubMed]

50. King, N.J.; Whyte, R. Does it look cooked? A Review of factors that influence cooked meat color. *J. Food Sci.* **2006**, *71*, R31–R40. [CrossRef]

51. Mokrzycki, W.S.; Tatol, M. Colour difference ΔE. *A survey. Mach Graph Vis* **2011**, *20*, 383–411.

52. Elzerman, J.E.; Hoek, A.C.; Van Boekel, M.A.; Luning, P.A. Consumer acceptance and appropriateness of meat substitutes in a meal context. *Food Qual. Prefer.* **2011**, *22*, 233–240. [CrossRef]

53. Desmond, E.; Troy, D.; Buckley, D. Comparative studies of nonmeat adjuncts used in the manufacture of low-fat beef burgers. *J. Muscle Foods* **1998**, *9*, 221–241. [CrossRef]

Article

Quality Characteristics of Healthy Dry Fermented Sausages Formulated with a Mixture of Olive and Chia Oil Structured in Oleogel or Emulsion Gel as Animal Fat Replacer

Tatiana Pintado * and **Susana Cofrades**

Institute of Food Science, Technology and Nutrition (ICTAN-CSIC), José Antonio Novais 10, 28040 Madrid, Spain; scofrades@ictan.csic.es
* Correspondence: tatianap@ictan.csic.es; Tel.: +34-91-5492-300

Received: 8 June 2020; Accepted: 23 June 2020; Published: 24 June 2020

Abstract: The present work evaluates the suitability of beeswax oleogels and emulsion gel prepared with a healthy lipid mixture (olive and chia oils) as pork fat replacers for the development of a dry fermented meat product (fuet). Because these systems offer various possibilities, this study has compared their effect on the nutritional quality and sensory acceptability of fuets and their behaviour with regard to technological properties and microbiological and oxidative stability during 30 days of chilled storage. This strategy allowed products with an improved fatty acid profile and a 12-fold decrease of the polyunsaturated fatty acids (PUFA) n-6/n-3 ratio, as compared to the control samples. Irrespective of the structuring method used as animal fat replacer, reformulated samples showed a good oxidative status during chilled storage. In general, no differences that depended on the use of oleogel or emulsion gel were observed in the technological properties and microbiological status, so the choice of one or the other would be conditioned by other factors than the characteristics that the product develops. However, further studies are needed to improve the sensory attributes of the reformulated samples.

Keywords: oleogel; emulsion gel; dry fermented sausages; healthier lipid content; chia oil; olive oil

1. Introduction

Fuet is a type of small-caliber non-acid fermented sausage from northeast Spain made with pork meat, pork fat and various seasonings. Traditionally, fermented sausages were considered safe and healthy foods, but nowadays these products have been associated with health hazards owing to the presence of some components such as saturated fats [1]. In this regard, several options have been assayed to improve lipid content in meat products based on the incorporation of vegetable oils, directly added into the meat matrix, stabilized in oil-in-water emulsion, etc. [2]. However, the interest in alternative technologies has been increasing, and therefore efforts have been made to develop healthy solid fats for foods, attaching importance to their ability to help to promote health and wellbeing [3]. Oleogels and emulsion gels are two different solid oil structured systems that offer interesting characteristics for use as animal fat replacers in the development of healthy meat products [4–8]. In oleogels, liquid oil is transformed into a 'gel-like' structure by using an organogelator, while emulsion gels may be generated from a stable liquid-like emulsion by gelling the continuous phase [3]. Regardless of the type of structured oil system, it is desirable to select an oil or a mixture of oils with a healthy fatty acid profile (reduced saturated fats, rich in unsaturated fats and good Σ polyunsaturated fatty acids (PUFA) n-6/ΣPUFA n-3 ratio, etc.), according to recommendations [9,10]. Accordingly, a mixture of olive oil, which is characterized by its high oleic fatty acid [11], and chia oil, which is the richest known botanical source of n-3 linolenic acid and does not contain any of the antinutritional compounds (total linamarin, linustatin and neolinustatin) or vitamin B6 antagonist factors that are present in other commercially-available sources of n-3 linolenic acid [12], could provide a way of obtaining new solid lipid materials with healthy fatty acid as animal fat replacers.

Some studies have been carried out to improve the fatty acid profile of cooked (frankfurter sausages) or fresh meat products (patties, longanizas, merguez) by using solid lipid material based on emulsion gels or oleogels [6,13–23]. However, there are very few studies of this kind on fermented meat products. For example, for this purpose linseed emulsion gel [24,25] or oleogel [26] was used to replace animal fat in dry fermented sausages. But we have found no studies that compare the use of emulsion gels and oleogels as animal fat replacers to improve the lipid content in meat products of any kind.

Accordingly, taking into account the particularity of this type of dry fermented meat product owing to the reactions that occur during the ripening process, the present study aimed to evaluate the quality of a functional fermented meat product (fuet) as a function of the olive-chia oil mixture structured as an oleogel or emulsion gel, used as animal fat replacer. The behaviour during one month of chilled storage was also evaluated.

2. Materials and Methods

2.1. Oleogel and Emulsion Gel Preparation

Two different animal fat replacers based on solid-structuring oil systems were made: an oleogel (OG) and an emulsion gel (EG). OG consisted mostly of oil (90%), while EG had half that oil content (45%). In both, the lipid phase consisted of a mixture of 80% olive oil (Carbonell Virgen Extra, SOS Cuétara, S.A., Madrid, Spain) and 20% chia oil (Primaria Premium Raw Materials, S.L., Valencia, Spain). The olive oil contained 13% saturated fatty acid (SFA), 75% monounsaturated fatty acid (MUFA) and 8% polyunsaturated fatty acid (PUFA), as reported by Delgado-Pando [27]. According to the information provided by the supplier, the chia oil contained 10% SFA, 5% MUFA and 80% PUFA.

Beeswax (Manuel Riesgo, S.A., Madrid, Spain), which was used as an organogelator in the OG formulation, was prepared as previously described by Gomez-Estaca [6]. Briefly, the oil mixture (90%) and beeswax (10%) were heated (65 °C) under constant stirring (500 rpm) in a food processor (Vorwerk Thermomix TM 31, Wuppertal, Germany) until complete melting and mixing. The resulting solution was then immediately poured into metal containers under pressure to compact it and prevent air bubbles, and it was stored at 3 ± 1 °C after standing for 60 min at room temperature in darkness.

EG was prepared as described by Pintado [7]. Briefly, soy protein isolate (10%) (Manuel Riesgo, S.A., Madrid, Spain) was mixed with water in a Thermomix TM 31 (Wuppertal, Germany) food processor (30 s, approx. 5600 rpm). Then, as a gelling agent, gelatin (3%) (type B, 200–220 bloom) from Manuel Riesgo, S.A. (Madrid, Spain) was added and combined (15 s, approx. 5600 rpm). The final mixture was mixed at approx. 5600 rpm with gradual addition of the appropriate amount (45%) of the oil mixture described previously. Finally, it was placed in metal containers under pressure to compact it and prevent air bubbles, and stored in a chilled room at 3 ± 1 °C for 20 h until use.

2.2. Fuet Design and Preparation

Sufficient fresh post-rigor pork meat (a mixture of biceps femoris, semimembranosus, semitendinosus, gracilis and adductor muscles) and pork backfat were obtained from a local market. Both the pork meat and the backfat were vacuum packed in batches of approximately 1000 and 500 g respectively, to be frozen and stored at −20 °C until used (less than one month).

Four different fuet-type dry fermented sausages were formulated (Table 1) in a pilot plant. Two formulations without replacement of pork backfat were prepared as references: one with normal fat content (NF/C) and the other with reduced fat content (RF/C). Additionally, two reduced-fat fuets were formulated, in which pork backfat was partially replaced by oleogel (RF/OG) or emulsion gel (RF/EG).

Although the level of fat replacement was the same, different amounts of OG and EG had to be added to obtain a similar lipid content.

Table 1. Formulation (g/100 g) of different fuets.

	Meat	Pork Back Fat	Oleogel	Emulsion Gel	Water
NF/C	74.0	20.0			0.5
RF/C	74.0	9.0			11.5
RF/OG	74.0	4.0	7.5		9.0
RF/EG	74.0	4.0		15.0	1.5

Normal fat (NF/C) and reduced-fat (RF/C) dry fermented sausages (fuet) formulated with all-animal fat. Reduced-fat fuets reformulated by partially replacing (80%) pork backfat with oleogel (RF/OG) or emulsion gel (RF/EG). All samples contain 5.5% of special commercial seasoning preparation for fuet.

Previously thawed pork meat and pork backfat (~18 h at 2 ± 2 °C) and the new lipid materials (OG in RF/OG and EG in RF/EG) were minced to a particle size of 6 mm (Van Dall S.r.l., model FTSIII, Treviglio, Italy). The ingredients for each formulation (Table 1) were placed in a mixer (MAINCA, Barcelona, Spain) and homogenized for 1 min. Half of the water and a commercial seasoning preparation for fuet (COMPLET FUETIB CU-425, Pilarica, Valencia, Spain) were added to the mixture and it was mixed for 1 min. Then the other half of the water and seasoning were added and the result was mixed again for 2 min. The mixture was stuffed (manual stuffer, MAINCA, Barcelona, Spain) into 34/36 mm-diameter natural pork casings (Julio Criado Gómez, S.A., Madrid, Spain), resulting in sausages weighing about 200 g. The sausages were dipped in a meat surface starter suspension of *Penicillium nalgiovense* and *Penicillium candidum* (TEXEL NEO 1 Danisco, DuPont™, Madrid, Spain) prepared according to the manufacturer's instructions. The sausages were placed in a ripening cabinet (BINDER model KBF 240, Tuttlingen, Germany) under the following conditions: 2 days at 19 °C and 80–85% relative humidity (RH) and 15 days at 13 °C and 75–80% RH. These conditions were set for all the products in order to have no other variables, despite the fact that the water content conditions the ripening process of fermented products [28]. The fuets were packed in plastic bags under aerobic conditions and kept in chilled storage (2 ± 2 °C) for 30 days.

Samples from each formulation were taken at 0 (the end of the ripening process and the beginning of storage), 15 and 30 days of chilled storage for analysis.

2.3. Processing Losses

Losses were calculated by weight difference during the fuet ripening period and expressed as a percentage of the initial weight.

2.4. Chemical Composition and Energy Value of Fuets

The chemical composition of the fuets was analyzed at the end of the ripening period. Each analysis was performed three times. Moisture and ash content were determined using official methods [29]. A LECO FP-2000 Nitrogen Determinator (Leco Corporation, St Joseph, MI, USA) was used to evaluate protein content and fat level was measured in accordance with Bligh and Dyer [30]. The energy value was calculated on the basis of 9 kcal/g for fat and 4 kcal/g for protein.

The fatty acid content was evaluated in triplicate by saponification and bimethylation according to Lee [31] in samples previously freeze-dried (Lyophilizer Telstar Cryodos Equipment, Tarrasa, Spain). The analysis of fatty acid methyl ester (FAME) was carried out on an Agilent gas chromatograph (Model 7820A, Santa Clara, CA, USA) fitted with a GC-7 Agilent HP-88 capillary column (60 m × 250 μm × 0.2 μm) using a flame ionization detector The temperature of the injector and the detector was 250 and 260 °C respectively. On the other hand, the temperature profile of the oven was 125 °C, increasing by 8 °C/min to 145 °C (held for 26 min) and 2 °C/min to 220 °C (held for 5 min). C13:0 was used as internal patron and for the identification of fatty acids, that was carried out by comparison of the retention times, it was used the standard 47015-U Supelco PUFA No.2 Animal Source (Sigma-Aldrich Co., St. Louis, MO, USA). Fatty acids were expressed as g of fatty acid/100 g product.

2.5. Technological Properties

Technological properties were evaluated during the chilled storage of the fuets, at 0, 15 and 30 days.

The pH was determined (in quadruplicate) at room temperature in water in a ratio of 1:10 (w/v) using a 827 Metrohm pH-meter (Metrohm AG, Zofingen, Switzerland).

Water activity (Aw) was measured (in triplicate) at 25 °C, after removing the casing, in a LabMaster-aw instrument (model 1119977, Novasina AG, Lachen SZ, Switzerland).

Colour was measured (ten times) in fuet cross-sections using a Konica Minolta CM-3500 D spectrophotometer (Konica Minolta Business Technologies, Tokyo, Japan) set to D65 illuminant/10° observer. The CIELAB colour space was used to obtain the colour coordinates L* (black (0) to white (100)), a* (green (–) to red (+)), and b* (blue (–) to yellow (+)).

Texture profile analysis (TPA), as described by Bourne [32], was carried out using a TA-XTplus Texture Analyzer (Stable Micro Systems Ltd., Godalming, UK) equipped with a 30 kg load cell. Six cores (diameter = 12 mm, height = 20 mm) per sample were axially compressed to 50% of their original height at a crosshead speed of 0.8 mm/s to calculate hardness (N). The tests were performed on the samples at room temperature immediately after refrigeration at 3 °C.

2.6. Lipid Oxidation

The fuets were assessed for oxidative stability by measuring secondary oxidation products, based on changes in concentrations of thiobarbituric acid-reactive substances (TBARs) and the main volatile aldehyde compounds formed by lipid oxidation [33].

TBARs, which were expressed as mg malonaldehyde (MDA)/kg fuet based on a standard curve prepared from 1,1,3,3-tetraethoxypropane in advance, were determined according to Delgado-Pando [34],. Volatile compounds of the fuet samples were extracted by solid phase micro-extraction and determined according to Alejandre [24]. The gas chromatograph (Agilent, model 6890N, Santa Clara, CA, USA) was equipped with a 5973 Mass Selective Detector and it used a DB-WAXetr polyethylene glycol capillary column (60 m × 320 μm × 0.25 μm). For the analysis the oven temperature was set initially at 40 °C (4 min hold), increased to 110 °C at 4 °C/min, to 180 °C at 6 °C/min, and to 240 °C at 8 °C/min (15 min hold). Helium was used as a carrier gas at 1.3 mL/min; injector and detector temperatures were held at 250 and 240 °C, respectively. Identification of the peaks was based on comparison of their mass spectra with the spectra of a commercial library (Wiley 7th edition and NIST/EPA/NIH 02 mass spectral library) and by comparison of their retention times with those of standard compounds. For semi-quantitative purposes, peak area was measured by integration of the total ion current of the spectra. Results were expressed as area/sample weight (g) × 10^3.

Determinations for each sample, volatile compounds, and TBARs were performed in triplicate at day 0 and after 30 days of chilled storage.

2.7. Microbiological Analysis

Total viable counts (TVC) and lactic acid bacteria (LAB) were evaluated as described Pintado et al. [19]. For results exposure, all microbial counts were converted to logarithms of colony-forming units per gram (Log cfu/g).

2.8. Sensory Analysis

The sensory analysis was carried out with a panel of 30 assessors selected from the Institute of Food Science, Technology and Nutrition (ICTAN-CSIC) staff. These people were chosen because they are acquainted with meat products and the terminology used for the analysis. For samples preparation, the fuets were cut into 3-mm-thick slices. Two slices per sample were presented to the panellists, who were instructed to rinse their mouth with bread and water between samples. The sensory attributes (general appearance, odour, flavour, texture and overall acceptability) were evaluated on a 10-point scale, 0 being considered as "dislike strongly" and 10 as "like strongly". The panellists were also asked to make any comments that they considered relevant about their sensory perception of the samples.

2.9. Statistical Analysis

The whole experiment was performed twice. Statistical tests were made employing the SPSS computer program (v24 SPSS Statistical Software, Inc., Chicago, IL, USA). One-way and/or two-way analyses of variance (ANOVA) were performed. Differences between pairs of means were assessed on the basis of confidence intervals using Tukey's Honestly-significant-difference (HSD) test. The level of significance was $p \leq 0.05$.

3. Results and Discussion

3.1. Processing Losses

At the end of the ripening process, the losses that products suffered were calculated to evaluate the yield for the various products as a consequence of the reformulation strategy (Figure 1). Samples with all-animal fat had the highest weight losses, 53.4% (RF/C), and the lowest, 42.2% (NF/C). Several authors [28,35,36] have observed higher losses in reduced-fat fermented sausages than in sausages with normal fat. In the present study, the strategy of reducing fat and improving the lipid profile by using oleogel (OG) and emulsion gel (EG) led to products with better binding properties than when only the fat content was reduced (RF/C) (Figure 1). No differences were observed between samples with OG or EG despite the higher quantity of water added directly during the preparation of RF/OG than in the case of RF/EG, in which water was stabilized or entrapped in an emulsion (Table 1).

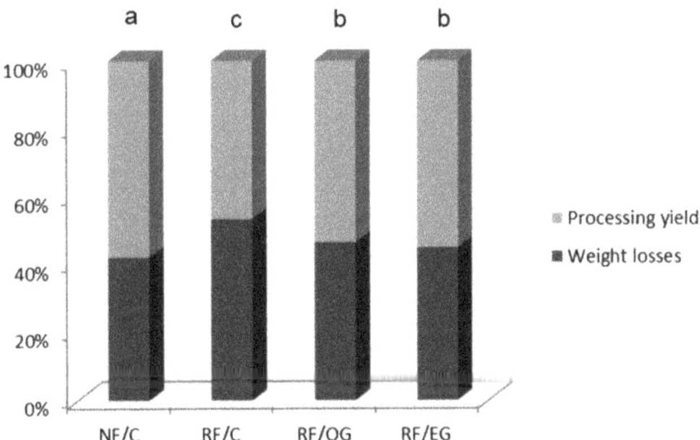

Figure 1. Weight losses and the processing yield of the fuets as a consequence of the ripening process. Normal fat (NF/C) and reduced-fat (RF/C) dry fermented sausages (fuet) formulated with all-animal fat. Reduced-fat fuets reformulated by partially replacing (80%) pork backfat with oleogel (RF/OG) or emulsion gel (RF/EG). Different letters indicate significant differences by formulation in weight losses and processing yield ($p < 0.05$).

3.2. Chemical Composition and Energy Value

The fuet composition (Table 2) was mainly influenced by the formulation (Table 1). However, for this type of meat product, the ripening process should be taken into account because during this period there is a high water loss (Figure 1), which is one of the characteristics that determine the final composition of the product because it results in concentration of the components.

Table 2. Chemical compositions and nutritional significance ratios of different fuets after the ripening process.

Parameters	Samples			
	NF/C	RF/C	RF/OG	RF/EG
Composition (%)				
Moisture	32.63 ± 0.87 [a]	36.07 ± 0.47 [b]	37.45 ± 0.70 [b]	36.84 ± 0.76 [b]
Ash	4.72 ± 0.04 [a]	6.13 ± 0.03 [c]	4.90 ± 0.01 [ab]	5.07 ± 0.15 [b]
Protein	31.17 ± 0.07 [a]	37.74 ± 0.74 [c]	31.92 ± 0.96 [a]	35.12 ± 0.24 [b]
Fat	29.73 ± 2.42 [b]	19.78 ± 1.86 [a]	22.30 ± 2.13 [a]	22.01 ± 0.61 [a]
Fatty acid profile (g/100 g of fuet)				
Myristic C14:0	0.50 ± 0.01 [c]	0.37 ± 0.04 [b]	0.17 ± 0.00 [a]	0.16 ± 0.01 [a]
Palmitic C16:0	7.85 ± 0.24 [c]	5.82 ± 0.57 [b]	3.92 ± 0.06 [a]	3.76 ± 0.12 [a]
Stearic C18:0	3.54 ± 0.17 [c]	2.63 ± 0.28 [b]	1.56 ± 0.03 [a]	1.55 ± 0.06 [a]
$\sum$ SFA	12.11 ± 0.42 [c]	8.99 ± 0.90 [b]	5.79 ± 0.09 [a]	5.61 ± 0.19 [a]
Vaccenic C18:1n7	1.19 ± 0.03 [c]	0.91 ± 0.08 [b]	0.63 ± 0.01 [a]	0.62 ± 0.02 [a]
Oleic C18:1n9	13.60 ± 0.33 [c]	10.30 ± 0.93 [a]	12.29 ± 0.20 [b]	11.94 ± 0.18 [b]
$\sum$ MUFA	16.12 ± 0.38 [c]	12.20 ± 1.10 [a]	13.51 ± 0.22 [b]	13.10 ± 0.21 [b]
Linoleic C18:2n6	2.31 ± 0.04 [b]	1.81 ± 0.14 [a]	1.81 ± 0.02 [c]	1.77 ± 0.03 [c]
Linolenic C18:3n3	0.12 ± 0.00 [a]	0.09 ± 0.01 [a]	1.68 ± 0.03 [b]	1.61 ± 0.03 [b]
$\sum$ PUFA	2.83 ± 0.05 [b]	2.25 ± 0.17 [a]	3.81 ± 0.04 [d]	3.60 ± 0.04 [c]
Nutritional significance ratios				
PUFA/SFA	0.23 ± 0.01 [a]	0.25 ± 0.01 [a]	0.66 ± 0.01 [b]	0.64 ± 0.02 [b]
n-6/n-3	14.0 ± 0.22 [b]	15.04 ± 0.47 [b]	1.12 ± 0.01 [a]	1.19 ± 0.02 [a]

SFA: saturated fatty acid; MUFA: monounsaturated fatty acid; PUFA: polyunsaturated fatty acid. For sample denominations, see Table 1. Different letters in the same row indicate significant differences ($p < 0.05$) between formulations. Means ± standard deviation.

As expected on the basis of the fuet formulations (Table 1), products with two different fat levels were obtained (Table 2). The two strategies used in this work, the replacement of animal fat by water alone or by the use of structured oils, OG (RF/OG) and EG (RF/EG), led to products with similar ($p > 0.05$) fat content (Table 2). Moisture content increased significantly as a result of the reduction in fat level, as other authors have found in dry fermented sausages [24,37]. Despite the differences in water losses (Figure 1), no significant differences were found in moisture content that depended on the type of fat used as the lipid source (all-animal-fat, OG, or EG). Reduced all-animal-fat fuet (RF/C) had the highest ($p < 0.05$) ash content, probably because it had the highest losses (Figure 1) during processing. The protein levels of the fuets were between 31.17% and 37.74% (Table 2). The use of oleogel and emulsion gel as fat replacers resulted in samples with different ($p < 0.05$) protein contents, probably because of the use of soy protein isolate as emulsifier in the preparation of the emulsion gel.

Both strategies, the pork backfat reduction and the partial pork backfat replacement by oleogel and emulsion gel systems, improved the fatty lipid profile, with decreased SFA and increased PUFA ($p < 0.05$). With regard to SFA, the use of OG and EG as animal fat replacers significantly reduced the myristic, palmitic and stearic acid contents in the fuets by more than half compared to the control (NF/C) (Table 2). The highest ($p < 0.05$) MUFA content was in the control samples (NF/C). However, MUFA represented 54% of total fat in NF/C, whereas in RF/OG and RF/EG MUFA content was approximately 60% of total fat. Oleic acid was the main fatty acid in all samples (Table 2), which is consistent with reports for fatty acid composition in pork fat [38] and in olive oil [27], which was the main oil used in the development of OG and EG. The RF/OG and RF/EG products showed the highest ($p < 0.05$) PUFA content, with a notable increase in α-linolenic fatty acid (ALA) in both samples owing to the presence of chia oil, which is the richest known botanical source of n-3 linolenic acid [12]. Consequently, owing to the technological advantages that chia seed and chia flour offer and their high lipid content (30–35%), both products have also been used (added directly or in emulsion or emulsion gel) to improve the fatty acid profile of various meat products, such as frankfurters, burgers, longanizas, etc. [19,39–41].

The PUFA/SFA ratio is one of the main parameters currently used to assess the nutritional quality of the lipid fraction of foods, and a PUFA/SFA ratio above 0.4 is recommended [38]. The PUFA/SFA ratio in the all-animal-fat samples (N/FC and R/FC) was around 0.2 (Table 2), which is consistent with reports by other authors concerning conventional meat products [20,42], whereas replacement of pork fat by the new healthy lipid materials (OG and EG) increased this ratio ($p < 0.05$) to 0.6 (Table 2), thus complying with the recommendations. The PUFA n-6/n-3 ratio is also of great interest, because diets with high PUFA n-6/n-3 ratios promote the pathogenesis of many diseases (cardiovascular diseases, cancer, etc.), whereas increased n-3 PUFA content exerts a suppressive effect [43]. The nutritional recommendation for this ratio is that it should be lower than 4, and the strategy based on the replacement of animal fat by OG or EG produced a drastic decrease to values close to 1 in the PUFA n-6/n-3 ratio in the RF/OG and RF/EG fuets, complying with the recommendations. Increasing the PUFA/SFA ratio as well as reducing the PUFA n-6 / n-3 ratio to get closer to the reference values, has been tested using new lipid materials such as EG or oleogels elaborated with oils that have a healthy profile of fatty acids (olive, flax, chia, etc.). This strategy, which has been tried on other types of meat products (fermented, cooked or fresh), has given similar results to those obtained in the present study [6,8,17,18,24].

According to the composition specified, the energy value of the normal-fat fuets (NF/C) was approximately 392 kcal/100 g. As a consequence of the reformulation strategies based on lipid content improvement, the energy value decreased to values between 328 kcal/100 g (RF/OG and RF/C samples) and 338 kcal/100 g in fuet with emulsion gel (RF/EG). These changes represent an energy reduction of around 14–16% in the reformulated products. Similar or lower energy reductions have been observed in other reduced-fat fermented sausages [24,37].

3.3. Nutritional and Health Claims

According to the composition presented in Table 2 and Regulation (European Commission) no 1924/2006 and Regulation (EU) no 432/2012 [44,45], all the fuets could be labelled with the nutritional claim "high protein content" and the corresponding health claims presented in Table 3. On the other hand, the sample with reduced all-animal-fat content (RF/C) showed a fat reduction of more than 30% with respect to the control and could therefore labelled with a "reduced fat content" claim.

Table 3. Nutrition and health claims authorised in fuets according to Regulation (EC) No 1924/2006 and Commission Regulation No 432/2012.

Claims	Conditions Applying to Them	Fuet Samples			
		NF/C	RF/C	RF/OG	RF/EG
"high protein" Proteins contribute to a growth in muscle mass and the maintenance of muscle mass and normal bones. Protein is needed for normal growth and development of bone in children.	May only be made where at least 20% of the energy value of the food is provided by protein	X	X	X	X
"reduced fat"	May only be made where the reduction in content is at least 30% compared to a similar product		X		
"high unsaturated fat" Replacing saturated fats with unsaturated fats in the diet contributes to the maintenance of normal blood cholesterol levels.	May only be made where at least 70% of fatty acids present in the product derive from unsaturated fat under the condition that unsaturated fat provides more than 20% of the energy of the product.			X	X
"high omega-3 fatty acids" Alpha-linolenic acid (ALA) contributes to the maintenance of normal blood cholesterol levels. Information shall be given to the consumer that the beneficial effect is obtained with a daily intake of 2 g of ALA.	May only be made where the product contains at least 0.6 g ALA / 100 g of product and per 100 kcal.			X	X

Furthermore, the strategy based on partial replacement of animal fat by healthy structured oil systems (OG and EG) allows other nutritional and health claims for these fuets according to European regulations [44,45]. With regard to nutritional claims, the RF/OG and RF/EG fuets could be labelled with "high unsaturated fat" and "high omega-3 fatty acids" claims (Table 3). With regard to health claims, the labelling of these samples could include the claim that "ALA contributes to the maintenance of normal blood cholesterol levels" (Information shall be given to the consumer that the beneficial effect is obtained with a daily intake of 2 g of ALA). Taking into account that it is recommended to limit the consumption of processed meat to 50 g per day [46], this amount of the RF/OG and RF/EG samples covers more than 50% of ALA needs. Accordingly, the presence of chia and olive oil oleogel or emulsion gel in the fuets reflected healthier nutritional properties when compared with the control samples.

3.4. Technological Properties

In order to know the consequences of the different composition of the fuets as well as the phenomena that occurred during the ripening process, the technological properties were evaluated during the storage period, after ripening, which is when the product is consumed. The water activity (Aw) of the fuets was affected by the formulation (Table 4), with values ranging between 0.87 and 0.90 just after the ripening period (day 0 of storage). The use of OG or EG as fat replacement in the fuets did not significantly condition the initial Aw, but their values were higher ($p < 0.05$) than those observed in the samples with all-animal fat and than those expected for this kind of product [47]. However, Triki [48] observed decreased Aw values in chorizo (a Spanish fermented sausage) fermented sausages. In the present work, what may have happened is that the use of the mixture of structured olive and chia oils in the development of the fuets conditioned the ripening process, requiring a longer time to produce a reduction in water activity levels. In general, chilled storage had hardly any effect on water activity (Table 4). Similar behavior has been observed in fermented sausages during storage [48].

Table 4. Technological properties of fuets during chilled storage: pH and water activity (Aw) values, colour parameters (L* lightness, a* redness and b* yellowness) and texture profile analysis (TPA) (Hardness, N).

	Samples	Days of Storage (5 °C)		
		0	15	30
Aw	NF/C	0.87 ± 0.02 [a1]	0.89 ± 0.01 [c1]	0.88 ± 0.00 [a1]
	RF/C	0.88 ± 0.01 [a2]	0.84 ± 0.00 [a1]	0.87 ± 0.00 [a2]
	RF/OG	0.91 ± 0.01 [b1]	0.90 ± 0.00 [c1]	0.90 ± 0.00 [b1]
	RF/EG	0.90 ± 0.01 [b2]	0.88 ± 0.01 [b1]	0.92 ± 0.00 [b3]
pH	NF/C	5.41 ± 0.01 [b1]	5.74 ± 0.11 [b2]	6.34 ± 0.04 [a3]
	RF/C	5.50 ± 0.01 [b1]	5.87 ± 0.07 [c2]	6.5 ± 0.04 [b3]
	RF/OG	5.27 ± 0.01 [a1]	5.60 ± 0.04 [a2]	6.64 ± 0.03 [c3]
	RF/EG	5.20 ± 0.01 [a1]	5.77 ± 0.12 [bc2]	6.62 ± 0.02 [c3]
Colour parameters				
L*	NF/C	41.45 ± 2.85 [a12]	42.74 ± 2.99 [a2]	38.76 ± 1.59 [ab1]
	RF/C	45.68 ± 2.79 [b2]	41.02 ± 1.63 [a1]	39.60 ± 3.99 [b1]
	RF/OG	41.69 ± 2.47 [a1]	41.69 ± 1.92 [a1]	40.76 ± 2.39 [b1]
	RF/EG	42.12 ± 1.86 [ab2]	39.85 ± 2.52 [a2]	35.31 ± 3.50 [a1]
a*	NF/C	14.11 ± 0.99 [ab2]	11.60 ± 2.99 [a1]	17.75 ± 1.73 [b3]
	RF/C	12.79 ± 1.87 [a1]	16.82 ± 1.17 [b2]	16.72 ± 1.70 [ab2]
	RF/OG	15.58 ± 1.38 [b1]	16.75 ± 0.88 [b1]	16.74 ± 2.04 [ab1]
	RF/EG	14.25 ± 1.29 [ab1]	16.52 ± 1.01 [b2]	15.33 ± 0.95 [a12]
b*	NF/C	6.07 ± 0.82 [a1]	4.66 ± 1.39 [a1]	7.77 ± 1.36 [a2]
	RF/C	8.23 ± 2.40 [b1]	7.01 ± 1.19 [b1]	7.66 ± 0.83 [a1]
	RF/OG	7.91 ± 0.98 [b1]	10.19 ± 1.08 [c2]	10.43 ± 1.10 [b2]
	RF/EG	8.95 ± 0.78 [b1]	9.70 ± 0.74 [c1]	9.32 ± 0.82 [ab1]
Texture profile analysis				
Hardness (N)	NF/C	5.81 ± 0.91 [b1]	4.54 ± 1.40 [a1]	12.30 ± 1.66 [b2]
	RF/C	9.51 ± 0.18 [c1]	12.36 ± 2.06 [c1]	16.79 ± 3.34 [c2]
	RF/OG	3.75 ± 0.49 [a1]	3.78 ± 0.65 [a1]	5.36 ± 0.52 [a2]
	RF/EG	5.84 ± 0.50 [b1]	7.71 ± 1.08 [b2]	7.97 ± 0.73 [a2]

For sample denominations, see Table 1. Different letters in the same column and different number in the same row indicate significant differences ($p < 0.05$) between formulations or chilled storage process. Means ± standard deviation.

The sausage formulations and chilled storage conditioned ($p < 0.05$) the pH values of the fuets (Table 4). However, all the pH values were within the normal range reported for similar commercial products [47] or products in which animal fat was replaced by n-3 long-chain PUFA in konjac glucomannan matrix or linseed EG [25,49]. At day 0, samples with OG or EG as fat replacer showed the lowest ($p < 0.05$) pH values. Similar behavior has been described for fuets in which animal fat was replaced by sunflower oil [28] and in higher caliber (50 mm) dry fermented sausages made with linseed oil EG as animal fat replacer [25]. On the other hand, [24] did not observe an effect on pH values as a consequence of fat replacement (26.3%, 32.8% and 39.5%) by linseed oil gelled emulsion in dry fermented sausages. During chilled storage a significant increase in pH values was observed. Similar results have been found in dry fermented sausage produced using different lactobacilli as starter culture [50]. These authors found that the pH started to increase after 28th day of ripening and the increase continued during storage at refrigeration (8° C). An increase of pH could be related to the breakdown of lactic acid following the depletion of the added sugar [50].

Table 4 shows the values obtained for lightness (L*), redness (a*) and yellowness (b*) in the control and reformulated fuets. As a result of reducing animal-fat content (comparison between NF/C and RF/C), increases ($p < 0.05$) in lightness and yellowness were observed, while no effect on redness values was found. However, as a result of the replacement of pork backfat by structured chia and olive oil systems (RF/OG and RF/EG samples), in comparison with the control (NF/C), only yellowness increased ($p < 0.05$) (Table 4). This means that the strategy of reducing and replacing animal fat with a mixture of structured olive and chia oils gives rise to products that maintain the characteristic redness of this type of product, unlike what happens when there is only a reduction in fat content (RF/C), which causes greater changes in colour. It is important to note that the healthier fuets (RF/OG and RF/EG samples) were more stable, with smaller changes in colour parameters after 30 days of chilled storage, than the products made with only animal fat (Table 4).

The hardness of the fuets varied as a result of the modifications that were assayed (Table 4). Initially it was noted that there was a significant increase in hardness in the fuets with reduced animal fat, probably owing to greater water losses in the RF/C samples (Figure 1). These results are in agreement with those found by several authors [28,51,52], who reported higher hardness in low-fat dry fermented sausages than in high-fat ones. The type of structured oil system used as the animal fat replacer conditioned the hardness of the fuets. Thus, fuets made with EG as animal-fat replacer (RF/EG) showed similar ($p > 0.05$) hardness to the control (NF/C), whereas those with oleogel (RF/OG) had the lowest ($p < 0.05$) hardness values. Hardness has a negative relation with moisture content in dry fermented meat products, as other authors have observed [51,53]. Accordingly, given that the RF/OG and RF/EG samples had similar moisture values (Table 2) and processing losses, the differences in hardness could be attributed to how the water was added during the preparation of the products, directly to the meat matrix (RF/OG) or stabilized in EG (RF/EG). In chorizo Jimenez-Colmenero [54] detected a decrease in hardness as a consequence of replacing various animal fat levels by an oil-in-konjac matrix. Similar behavior was observed by other authors when they used linseed oil EG or OG as an animal fat replacer in dry fermented sausages [25,26]. Conversely, in salchichón (a Spanish fermented sausage) and fuet, the replacement of various animal-fat levels by fish oil encapsulated in konjac gel (salchichón) or by sunflower oil added directly (fuet) resulted in harder samples [26–28]. As expected, during chilled storage all the samples experienced an increase ($p < 0.05$) in hardness (Table 4), probably because all the samples lost water during that period. However, it should be noted that, as with color, the changes in the texture of the OG and EG fuets during chilled storage were smaller than those in the control samples made with animal fat.

3.5. Lipid Oxidation

Lipid oxidation is the main non-microbial cause of quality deterioration in meat products and one of the most important reactions of fermented meat products that generates volatile compounds [33]. Accordingly, the effect of the partial replacement of pork backfat by structured chia and olive oil systems (oleogel or emulsion gel) on lipid oxidation, measured as volatile compounds and MDA levels, is shown in Table 5.

Table 5. Parameters related to lipid oxidation of fuets during chilled storage: thiobarbituric acid-reactive substances (TBARs) values (mg malonaldehyde (MDA)/kg sample) and volatile compounds (area/sample weight (g) $\times 10^3$).

Compound	Samples			
	NF/C	RF/C	RF/OG	RF/EG
TBARs				
day 0	0.103 ± 0.025 [a1]	0.085 ± 0.010 [a1]	0.404 ± 0.028 [b1]	0.404 ± 0.028 [b1]
day 30	0.107 ± 0.016 [a1]	0.091 ± 0.010 [a1]	0.384 ± 0.040 [b1]	0.365 ± 0.041[b1]
Hexanal				
day 0	156.0 ± 0.4 [a1]	152.3 ± 20.8 [a1]	239.8 ± 26.2 [b2]	367.0 ± 15.2 [c2]
day 30	119.4 ± 18.4 [a1]	121.3 ± 9.9 [a1]	153.1 ± 0.5 [a1]	279.4 ± 26.2 [b1]
Heptanal				
day 0	17.5 ± 3.1 [a2]	30.8 ± 4.8 [a2]	65.3 ± 36.8 [b2]	60.1 ± 3.8 [b2]
day 30	11.9 ± 1.3 [a1]	12.3 ± 5.6 [a1]	35.6 ± 1.1 [b1]	48.0 ± 1.2 [c1]
Octanal				
day 0	74.2 ± 4.0 [a1]	55.4 ± 1.9 [a1]	473.3 ± 188.5 [b2]	462.2 ± 139.3 [b2]
day 30	51.4 ± 3.4 [a1]	51.0 ± 6.1 [a1]	96.5 ± 0.5 [b1]	137.9 ± 11.5 [c1]
Nonanal				
day 0	366.9 ± 20.7 [a1]	352.6 ± 39.4 [a1]	971.8 ± 164.4 [b1]	809.2 ± 1.1 [b1]
day 30	324.9 ± 18.1 [a1]	387.8 ± 66.2 [a1]	762.2 ± 22.3 [b1]	882.5 ± 20.9 [b1]

For sample denominations, see Table 1. Different letters in the same row indicate significant differences by formulation and different number in the same column indicate differences by chilled storage ($p < 0.05$). Means ± standard deviation.

TBARs values were significantly higher in the RF/OG and RF/EG samples, reflecting increased lipid oxidation in the fuets owing to the higher level of unsaturated fat, although their oxidation levels remained well below the rancidity threshold which is usually when the MDA concentration is above 1 mg per kg of sample [33]. Chilled storage did not have a significant effect on TBARs values, probably because of the stability provided by the structured systems in which the oil mixture was located, unlike what occurs when the oil is incorporated directly [55]. Similar results have been found in various meat products with an improved lipid profile based on plant and marine oils stabilized in different ways [48].

Aldehydes are the most abundant volatile compounds produced by lipid oxidation, and hexanal is the aldehyde that has been considered to be the best indicator [33]. As expected, higher ($p < 0.05$) levels of all volatile compounds were observed after the ripening process (day 0) in samples with OG (RF/OG) or EG (RF/EG) used as animal fat replacer (Table 5). These results are in agreement with those obtained in the determination of TBARs and those found by some other authors. Thus, Alejandre [24] and Glisic [25] observed higher levels for aldehydes in dry fermented sausages in which the lipid content was improved by using linseed emulsion gel as an animal fat replacer. On the other hand, although RF/EG showed higher ($p < 0.05$) hexanal levels than RF/OG, non-significant differences were observed in heptanal, octanal and nonanal levels depending on the structured oil system used as healthier lipid material (Table 5). Josquin [56] assayed the replacement of pork backfat with pure, pre-emulsified or encapsulated fish oil in fermented sausages and observed differences in volatile levels, depending on the strategy used to incorporate the oil. The sausages in which encapsulated oil was incorporated had lower volatile compound levels than the others.

After chilled storage, a significant decrease was observed in the volatiles studied, except for nonanal in the samples made with OG or EG, whereas the samples with all-animal fat generally showed values (Table 5) similar to those at the beginning of storage.

3.6. Microbiological Analysis

Microbiological factors during chilled storage are known to affect the stability and shelf life of meat products. Figure 2 shows changes in total viable count (TVC) and lactic acid bacteria (LAB). All samples presented high initial microbial counts (>8 log cfu/g) of TVC and LAB, which in general were maintained during chilled storage. However, fuet formulated with emulsion gel (RF/EG) experienced a significant increase in TVC and LAB counts after 30 days in refrigeration, reaching levels close to log 9 cfu/g (Figure 2). These results are in accordance with others observed in dry fermented sausages in which various animal-fat levels were replaced [35,48].

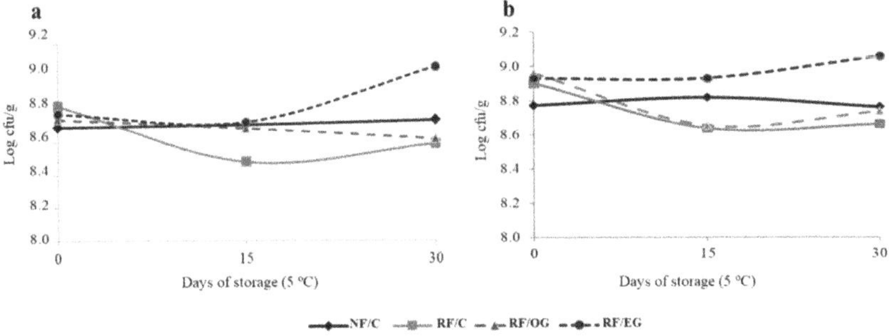

Figure 2. Microorganism (**a**: total viable count; **b**: lactic acid bacteria) counts (log cfu/g) of fuets during 30 days of chilled storage. For sample denominations see Table 1.

3.7. Sensory Analysis

The external appearance of the fuets was similar regardless of the formulation strategy used (Figure 3). However, some differences were observed in their cross-sectional appearance, depending on the lipid source that was used. Thus, while the animal fat was perfectly differentiated in the meat matrix, the oleogel or EG in R/OG and R/EG, respectively, could not be seen (Figure 3).

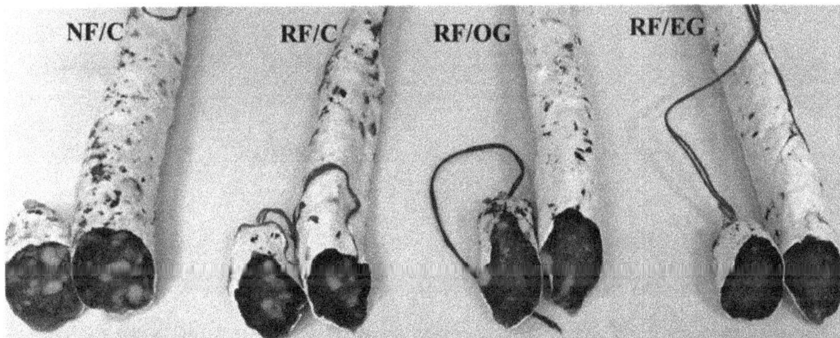

Figure 3. Effect of formulation strategies on the external and cross-sectional appearance of the fuets after the ripening process. For sample denominations see Table 1.

The results of the hedonic analysis for the attributes evaluated are shown in Figure 4. In general, for all of them, the samples made with all-animal fat received higher scores than the others. With regard to RF/OG and RF/EG, the panelists evaluated them with similar scores for all attributes. The lower scores that the reformulated samples received could be attributed to the high aldehyde content as compared to the control (Table 5), as other authors have reported for this type of meat product [57]. On the other hand, the differences observed between their appearances (Figure 3) may have conditioned how the panelists evaluated other sensory attributes [58]. Furthermore, after 30 days of storage, when they showed lower aldehyde contents (Table 4), RF/OG and RF/EG received higher scores for flavor or general acceptability. Alejandre et al. [24] did not observe differences in taste and juiciness but found differences in odor between control dry fermented sausages and others made with linseed emulsion gel as animal-fat replacer. However, the sensory attributes could be further improved by slight modifications to the product, including modifications to the conditions associated with the ripening process.

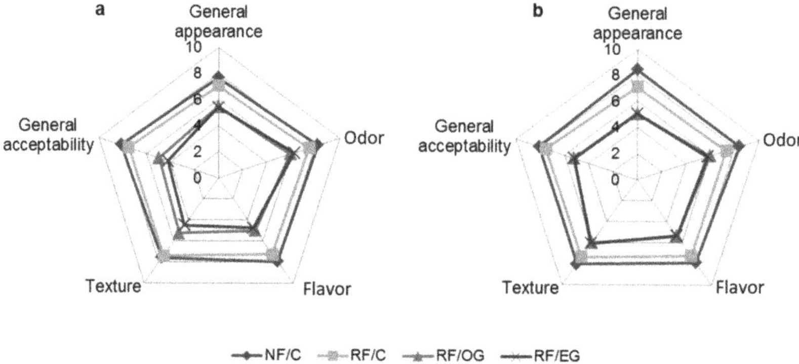

Figure 4. Sensory analysis scores for general appearance, odor, flavor, texture and general acceptability of the fuets: **a)** after ripening process; **b)** after 30 days of chilled storage. For sample denominations see Table 1.

4. Conclusions

The healthy oil mixture based on chia and olive oil, structured into an oleogel or emulsion gel, was proved to be an interesting option for the development of functional dry fermented sausages. These products could be labelled with certain nutritional and health claims according to European legislation, mainly because of the high α-linolenic fatty acid content. The strategy of reducing and replacing animal fat with a mixture of structured olive and chia oils gives rise to products that maintain the color characteristic of this type of product and a good oxidative and microbiological status during chilled storage. Fuets made with EG as animal-fat replacer had similar hardness to the control whereas those with oleogel were softer. Nevertheless, further studies are necessary to improve sensory attributes of the reformulated fuets with this type of lipid material but no great differences resulting from the use of one or the other were observed. Moreover, the strategy based on reduction and improvement of the lipid fraction yielded products that were stable during chilled storage.

Author Contributions: T.P. and S.C. contributed equally to this work. Both, designed this study, performed the experiments and collaborated in the statistical analysis and drafted the main manuscript. In addition, Paloma González and María Solano participated in the experimental phase of this study during their academic practices. All authors have read and agreed to the published version of the manuscript.

Funding: This research was funded by the Spanish Ministry of Economy and Competitiveness, AGL2014-53207-C2-1-R and PID2019-103872RB-I00, and the Intramural 202070E177 was funded by The Spanish National Research Council (CSIC)

Acknowledgments: We are grateful to the Analysis Service Unit facilities of Institute of Food Science, Technology and Nutrition (ICTAN) for the analysis of chromatography and mass spectrometry. Furthermore, the authors wish to thank Paloma González and María Solano for their support during the experimental work.

Conflicts of Interest: The authors declare no conflict of interest.

References

1. Bou, R.; Cofrades, S.; Jiménez-Colmenero, F. Fermented meat sausages. In *Fermented foods in health and disease prevention*; Frias, J., Martínez-Villaluenga, C., Penas, E., Eds.; Elsevier Science LTD: London, UK, 2017; pp. 203–235.
2. Jiménez-Colmenero, F. Healthier lipid formulation approaches in meat-based functional foods. Technological options for replacement of meat fats by non-meat fats. *Trends Food Sci. Tech.* **2007**, *18*, 567–578. [CrossRef]
3. Jiménez-Colmenero, F.; Salcedo-Sandoval, L.; Bou, R.; Cofrades, S.; Herrero, A.M.; Ruiz-Capillas, C. Novel applications of oil-structuring methods as a strategy to improve the fat content of meat products. *Trends Food Sci. Tech.* **2015**, *44*, 177–188. [CrossRef]
4. Barbut, S.; Wood, J.; Marangoni, A. Potential use of organogels to replace animal fat in comminuted meat products. *Meat Sci.* **2016**, *122*, 155–162. [CrossRef]
5. Fayaz, G.; Goli, S.A.H.; Kadivar, M. A novel propolis wax-based organogel: Effect of oil type on its formation, crystal structure and thermal properties. *J. Am. Oil Chem. Soc.* **2017**, *94*, 47–55. [CrossRef]
6. Gómez-Estaca, J.; Herrero, A.M.; Herranz, B.; Alvarez, M.D.; Jiménez-Colmenero, F.; Cofrades, S. Characterization of ethyl cellulose and beeswax oleogels and their suitability as fat replacers in healthier lipid pates development. *Food Hydrocolloid* **2019**, *87*, 960–969. [CrossRef]
7. Pintado, T.; Ruiz-Capillas, C.; Jiménez-Colmenero, F.; Carmona, P.; Herrero, A.M. Oil-in-water emulsion gels stabilized with chia (*Salvia hispanica* L.) and cold gelling agents: Technological and infrared spectroscopic characterization. *Food Chem.* **2015**, *185*, 470–478. [CrossRef] [PubMed]
8. Pintado, T.; Herrero, A.M.; Jiménez-Colmenero, F.; Ruiz-Capillas, C. Emulsion gels as potential fat replacers delivering beta-glucan and healthy lipid content for food applications. *J. Food Sci. Tech. Mys.* **2016**, *53*, 4336–4347. [CrossRef]
9. EFSA (European Food Safety Agency). Scientific opinion on dietary reference values for fats, including saturated fatty acids, polyunsaturated fatty acids, monounsaturated fatty acids, trans fatty acids, and cholesterol. *EFSA J.* **2010**, *8*, 1461–1568.
10. WHO. *Diet, Nutrition and the Prevention of chronic diseases*; World Health Organization Technical Report Series; WHO: Geneve, Switzerland, 2003.
11. Borges, T.H.; Pereira, J.A.; Cabrera-Vique, C.; Lara, L.; Oliveira, A.F.; Seiquer, I. Characterization of Arbequina virgin olive oils produced in different regions of Brazil and Spain: Physicochemical properties, oxidative stability and fatty acid profile. *Food Chem.* **2017**, *215*, 454–462. [CrossRef]
12. Ayerza, R. Oil content and fatty-acid composition of chia (*Salvia hispanica* L.) from 5 Northwestern locations in Argentina. *J. Am. Oil Chem. Soc.* **1995**, *72*, 1079–1081. [CrossRef]
13. Agregan, R.; Barba, F.J.; Gavahian, M.; Franco, D.; Khaneghah, A.M.; Carballo, J.; Ferreira, I.C.F.R.; Barretto, A.C.D.; Lorenzo, J.M. Fucus vesiculosus extracts as natural antioxidants for improvement of physicochemical properties and shelf life of pork patties formulated with oleogels. *J. Sci. Foof. Agr.* **2019**, *99*, 4561–4570. [CrossRef] [PubMed]
14. Alejandre, M.; Passarini, D.; Astiasaran, I.; Ansorena, D. The effect of low-fat beef patties formulated with a low-energy fat analogue enriched in long-chain polyunsaturated fatty acids on lipid oxidation and sensory attributes. *Meat Sci.* **2017**, *134*, 7–13. [CrossRef] [PubMed]
15. Barbut, S.; Marangoni, A.G.; Thode, U.; Tiensa, B.E. Using canola oil organogels as fat replacement in liver pate. *J. Food Sci.* **2019**, *84*, 2646–2651. [CrossRef] [PubMed]
16. Franco, D.; Martins, A.J.; López-Pedrouso, M.; Purrinos, L.; Cerqueira, M.A.; Vicente, A.A.; Pastrana, L.M.; Zapata, C.; Lorenzo, J.M. Strategy towards replacingp backfat with a linseed oleogel in frankfurter sausages and its evaluation on physicochemical, nutritional, and sensory characteristics. *Foods* **2019**, *8*, 366. [CrossRef] [PubMed]
17. Freire, M.; Cofrades, S.; Serrano-Casas, V.; Pintado, T.; Jiménez, M.J.; Jiménez-Colmenero, F. Gelled double emulsions as delivery systems for hydroxytyrosol and n-3 fatty acids in healthy pork patties. *J. Food Sci. Tech. Mys.* **2017**, *54*, 3959–3968. [CrossRef]
18. Gómez-Estaca, J.; Pintado, T.; Jiménez-Colmenero, F.; Cofrades, S. Assessment of a healthy oil combination structured in ethyl cellulose and beeswax oleogels as animal fat replacers in low-fat, PUFA-enriched pork burgers. *Food Bioprocess Tech.* **2019**, *12*, 1068–1081. [CrossRef]

19. Pintado, T.; Herrero, A.M.; Jiménez-Colmenero, F.; Pasqualin-Cavalheiro, C.; Ruiz-Capillas, C. Chia and oat emulsion gels as new animal fat replacers and healthy bioactive sources in fresh sausage formulation. *Meat Sci.* **2018**, *135*, 6–13. [CrossRef]

20. Pintado, T.; Herrero, A.M.; Ruiz-Capillas, C.; Triki, M.; Carmona, P.; Jiménez-Colmenero, F. Effects of emulsion gels containing bioactive compounds on sensorial, technological, and structural properties of frankfurters. *Food Sci. Technol. Int.* **2015**, *22*, 132–145. [CrossRef]

21. Poyato, C.; Astiasaran, I.; Barriuso, B.; Ansorena, D. A new polyunsaturated gelled emulsion as replacer of pork back-fat in burger patties: Effect on lipid composition, oxidative stability and sensory acceptability. *LWT* **2015**, *62*, 1069–1075. [CrossRef]

22. Serdaroglu, M.; Nacak, B.; Karabiyikoglu, M. Effects of beef fat replacement with gelled emulsion prepared with olive oil on quality parameters of chicken patties. *Korean J. Food Sci. An.* **2017**, *37*, 376–384.

23. Wolfer, T.L.; Acevedo, N.C.; Prusa, K.J.; Sebranek, J.G.; Tarte, R. Replacement of pork fat in frankfurter-type sausages by soybean oil oleogels structured with rice bran wax. *Meat Sci.* **2018**, *145*, 352–362. [CrossRef] [PubMed]

24. Alejandre, M.; Poyato, C.; Ansorena, D.; Astiasarán, I. Linseed oil gelled emulsion: A successful fat replacer in dry fermented sausages. *Meat Sci.* **2016**, *121*, 107–113. [CrossRef] [PubMed]

25. Glisic, M.; Baltic, M.; Glisic, M.; Trbovic, D.; Jokanovic, M.; Parunovic, N.; Dimitrijevic , M.; Suvajdzic, S.; Boskovic , M.; Vasilev, D.; et al. Inulin-based emulsion-filled gel as a fat replacer in prebiotic- and PUFA-enriched dry fermented sausages. *Int. J. Food Sci. Tech.* **2019**, *54*, 787–797. [CrossRef]

26. Franco, D.; Martins, A.J.; López-Pedrouso, M.; Cerqueira, M.A.; Purrinos, L.; Pastrana, L.M.; Vicente , A.A.; Zapata, C.; Lorenzo, J.M. Evaluation of linseed oil oleogels to partially replace pork backfat in fermented sausages. *J. Sci. Food Agr.* **2020**, *100*, 218–224. [CrossRef] [PubMed]

27. Delgado-Pando, G.; Cofrades, S.; Ruiz-Capillas, C.; Solas, M.T.; Jiménez-Colmenero, F. Healthier lipid combination oil-in-water emulsions prepared with various protein systems: an approach for development of functional meat products. *Eur. J. Lipid Sci. Tech.* **2010**, *112*, 791–801. [CrossRef]

28. Mora-Gallego, H.; Guardia, M.D.; Serra, X.; Gou, P.; Arnau, J. Sensory characterisation and consumer acceptability of potassium chloride and sunflower oil addition in small-caliber non-acid fermented sausages with a reduced content of sodium chloride and fat. *Meat Sci.* **2016**, *112*, 9–15. [CrossRef] [PubMed]

29. AOAC. *Official Methods of Analysis*, 18th ed.; AOAC International: Gaithersburg, MD, USA, 2005.

30. Bligh, E.G.; Dyer, W.J. A rapid method of total lipid extraction and purification. *Can J. Biochem. Phys.* **1959**, *37*, 911–917. [CrossRef] [PubMed]

31. Lee, M.R.F.; Tweed, J.K.S.; Kim, E.J.; Scollan, N.D. Beef, chicken and lamb fatty acid analysis -a symplified direct bimethylation procedureusing freeze-dried material. *Meat Sci.* **2012**, *92*, 863–866. [CrossRef]

32. Bourne, M.C. Texture profile analysis. *Food Tech.* **1978**, *32*, 62.

33. Domínguez, R.; Pateiro, M.; Gagaoua, M.; Barba, F.J.; Zhang, W.; Lorenzo, J.M. A comprehensive review on lipid oxidation in meat and meat products. *Antioxidants* **2019**, *8*, 429. [CrossRef]

34. Delgado-Pando, G.; Cofrades, S.; Ruiz-Capillas, C.; Triki, M.; Jiménez-Colmenero, F. Enriched n-3 PUFA/konjac gel low-fat pork liver pate: Lipid oxidation, microbiological properties and biogenic amine formation during chilling storage. *Meat Sci.* **2012**, *92*, 762–767. [CrossRef] [PubMed]

35. Campagnol, P.C.B.; dos Santos, B.A.; Wagner, R.; Terra, N.N.; Pollonio, M.A.R. Amorphous cellulose gel as a fat substitute in fermented sausages. *Meat Sci.* **2012**, *90*, 36–42. [CrossRef] [PubMed]

36. Muguerza, E.; Fista, G.; Ansorena, D.; Astiasaran, I.; Bloukas, J.G. Effect of fat level and partial replacement of pork backfat with olive oil on processing and quality characteristics of fermented sausages. *Meat Sci.* **2002**, *61*, 397–404. [CrossRef]

37. Ruiz-Capillas, C.; Triki, M.; Herrero, A.M.; Rodríguez-Salas, L.; Jiménez-Colmenero, F. Konjac gel as pork backfat replacer in dry fermented sausages: Processing and quality characteristics. *Meat Sci.* **2012**, *92*, 144–150. [CrossRef] [PubMed]

38. Wood, J.D.; Enser, M.; Fisher, A.V.; Nute, G.R.; Sheard, P.R.; Richardson, R.I.; Hughes, S.I.; Whittington, F.M. Fat deposition, fatty acid composition and meat quality: A review. *Meat Sci.* **2008**, *78*, 343–358. [CrossRef]

39. Angiolillo, L.; Conte, A.; Del Nobile, M.A. Technological strategies to produce functional meat burgers. *Lwt* **2015**, *62*, 697–703. [CrossRef]

40. Barros, J.C.; Munekata, P.E.; Pires, M.A.; Rodrigues, I.; Andaloussi, O.S.; Rodrigues, C.E.C.; Trindade, M.A. Omega-3-and fibre-enriched chicken nuggets by replacement of chicken skin with chia (*Salvia hispanica* L.) flour. *Lwt* **2018**, *90*, 283–289. [CrossRef]

41. Pintado, T.; Herrero, A.M.; Jiménez-Colmenero, F.; Ruiz-Capillas, C. Strategies for incorporation of chia (*Salvia hispanica* L.) in frankfurters as a health-promoting ingredient. *Meat Sci.* **2016**, *114*, 75–84. [CrossRef]

42. Salcedo-Sandoval, L.; Cofrades, S.; Ruiz-Capillas, C.; Jiménez-Colmenero, F. Effect of cooking method on the fatty acid content of reduced-fat and PUFA-enriched pork patties formulated with a konjac-based oil bulking system. *Meat Sci.* **2014**, *98*, 795–803. [CrossRef]

43. Simopoulos, A.P. The importance of the omega-6/omega-3 fatty acid ratio in cardiovascular disease and other chronic diseases. *Exp. Biol. Med.* **2008**, *233*, 674–688. [CrossRef]

44. European Commission. Regulation (EC) No 1924/2006 of the European Parliament and of the Council of 20 December 2006 on nutrition and health claims made on foods. *OJEU* **2006**, *L 404*, 3–18.

45. European Commission. Regulation (EU) No 432/2012 of the European Parliament and of the Council of 16 may 2012 establishing a list of permited health claims made on foods other than those referring to the reduction of disease risk and to children's development and health. *OJEU* **2012**, *L 136*, 1–408.

46. IARC (International Agency for Research on Cancer). IARC monographs evaluate consumption of red meat and processed meat. Available online: https://www.iarc.fr/wp-content/uploads/2018/07/pr240_E.pdf (accessed on 5 May 2020).

47. Herrero, A.M.; Ordonez, J.A.; de Avila, R.; Herranz, B.; de la Hoz, L.; Cambero, M.I. Breaking strength of dry fermented sausages and their correlation with texture profile analysis (TPA) and physico-chemical characteristics. *Meat Sci.* **2007**, *77*, 331–338. [CrossRef] [PubMed]

48. Triki, M.; Herrero, A.M.; Rodríguez-Salas, L.; Jiménez-Colmenero, F.; Ruiz-Capillas, C. Chilled storage characteristics of low-fat, n-3 PUFA-enriched dry fermented sausage reformulated with a healthy oil combination stabilized in a konjac matrix. *Food Control* **2013**, *31*, 158–165. [CrossRef]

49. Munekata, P.E.S.; Domínguez, R.; Franco, D.; Bermúdez, R.; Trindade, M.A.; Lorenzo, J.M. Effect of natural antioxidants in Spanish salchichón elaborated with encapsulated n-3 long chain fatty acids in konjac glucomannan matrix. *Meat Sci.* **2017**, *124*, 54–60. [CrossRef] [PubMed]

50. Komprda, T.; Neznalova, J.; Standara, S.; Bover-Cid, S. Effect of starter culture and storage temperature on the content of biogenic amines in dry fermented sausage polican. *Meat Sci.* **2001**, *59*, 267–276. [CrossRef]

51. Fonseca, S.; Gómez, M.; Domínguez, R.; Lorenzo, J.M. Physicochemical and sensory properties of Celta dry-ripened "salchichón" as affected by fat content. *Grasas Aceites* **2015**, *66*, 59.

52. Lorenzo, J.M.; Franco, D. Fat effect on physico-chemical, microbial and textural changes through the manufactured of dry-cured foal sausage: Lipolysis, proteolysis and sensory properties. *Meat Sci.* **2012**, *92*, 704–714. [CrossRef]

53. Gómez, M.; Lorenzo, J.M. Effect of fat level on physicochemical, volatile compounds and sensory characteristics of dry-ripened "chorizo" from Celta pig breed. *Meat Sci.* **2013**, *95*, 658–666.

54. Jiménez-Colmenero, F.; Triki, M.; Herrero, A.M.; Rodríguez-Salas, L.; Ruiz-Capillas, C. Healthy oil combination stabilized in a konjac matrix as pork fat replacement in low-fat, PUFA-enriched, dry fermented sausages. *LWT* **2013**, *51*, 158–163. [CrossRef]

55. Bloukas, J.G.; Paneras, E.D.; Fournitzis, G.C. Effect of replacing pork backfat with olive oil on processing and quality characteristics of fermented sausages. *Meat Sci.* **1997**, *45*, 133–144. [CrossRef]

56. Josquin, N.M.; Linssen, J.P.H.; Houben, J.H. Quality characteristics of Dutch-style fermented sausages manufactured with partial replacement of pork back-fat with pure, pre-emulsified or encapsulated fish oil. *Meat Sci.* **2012**, *90*, 81–86. [CrossRef] [PubMed]

57. Ansorena, D.; Astiasarán, I. The use of linseed oil improves nutritional quality of the lipid fraction of dry-fermented sausages. *Food Chem.* **2004**, *87*, 69–74. [CrossRef]

58. Stone, H.; Sidel, J.L. *Sensory evaluation practices*, 3rd ed.; Elsevier: Redwood City, CA, USA, 2004.

Review

Clean Label Alternatives in Meat Products

Gonzalo Delgado-Pando [1], Sotirios I. Ekonomou [2], Alexandros C. Stratakos [2] and Tatiana Pintado [1,*]

1   Institute of Food Science, Technology and Nutrition (CSIC), José Antonio Novais 10, 28040 Madrid, Spain; g.delgado@ictan.csic.es
2   Centre for Research in Biosciences, Coldharbour Lane, Faculty of Health and Applied Sciences, University of the West of England, Bristol BS16 1QY, UK; Sotirios.Oikonomou@uwe.ac.uk (S.I.E.); Alexandros.Stratakos@uwe.ac.uk (A.C.S.)
*   Correspondence: tatianap@ictan.csic.es

Abstract: Food authorities have not yet provided a definition for the term "clean label". However, food producers and consumers frequently use this terminology for food products with few and recognisable ingredients. The meat industry faces important challenges in the development of clean-label meat products, as these contain an important number of functional additives. Nitrites are an essential additive that acts as an antimicrobial and antioxidant in several meat products, making it difficult to find a clean-label alternative with all functionalities. Another important additive not complying with the clean-label requirements are phosphates. Phosphates are essential for the correct development of texture and sensory properties in several meat products. In this review, we address the potential clean-label alternatives to the most common additives in meat products, including antimicrobials, antioxidants, texturisers and colours. Some novel technologies applied for the development of clean label meat products are also covered.

Keywords: clean label; meat products; nitrites alternatives; phosphates alternatives

Citation: Delgado-Pando, G.; Ekonomou, S.I.; Stratakos, A.C.; Pintado, T. Clean Label Alternatives in Meat Products. *Foods* **2021**, *10*, 1615. https://doi.org/10.3390/foods10071615

Academic Editor: Rubén Domínguez

Received: 23 June 2021
Accepted: 9 July 2021
Published: 13 July 2021

Publisher's Note: MDPI stays neutral with regard to jurisdictional claims in published maps and institutional affiliations.

1. Introduction

Over the last few years, food producers have identified the term "clean label" as an important market trend. Nevertheless, what does "clean label" mean? So far there, is no official nor clear definition of the term [1,2]. Asioli et al. [3] proposed two ways the consumers can interpret a product as being clean label. In a broad sense, by looking at the front of pack, consumers might assume a product is clean label if related visual claims appear, such as "free from … ", "organic", "no additives", etc. In a strict sense, the authors conclude that, on the back of the pack, consumers associate clean-label products with those that have a short list of ingredients, are non-synthetic, are common for the consumers, etc. Therefore, a definition of clean label should relate to the number and type of additives (synthetic or not) a product has as well as its wholesomeness. An attempt of a definition was released in the official blog of the Institute of Food Technologists: "clean label means making a product using as few ingredients as possible, and making sure those ingredients are items that consumers recognize and think of as wholesome" [4]. We believe that this is a very accurate definition of the term. It relates to all the three important aspects of the clean-label trend: short list of ingredients, trust in the ingredients and perceived healthiness. In line with this, Aschemann-Witzel et al. [5] found that consumers perceived ingredients as belonging to one of these two opposing categories: known-"natural"-good or unknown-synthetic-bad. The former being the one related with the clean-label option. It is important to remark the following finding: there is a correlation for an additive of being perceived as potentially unsafe, unhealthy or of low quality if the name is not common or difficult to pronounce [6,7]. A survey in the USA showed that, depending on the ingredient name, the perceived naturalness differs. When asked about added salt, 65.6% of the respondents considered it natural. However, when they were asked about added sodium chloride, only 32% considered it natural [8]. As with the term "clean label", the term "natural" does not have a proper definition given by the regulators. Although consumer might understand it as a synonym of non-chemical, good and healthy, this is far from the reality where sodium chloride is the same as common salt or nitrites from synthetic origin are the same as the ones extracted from the Swiss chard. Nonetheless, consumer perception must be taken into account for product success and we do not need to forget that safety plays an essential role for the consumer, along with health, being a top-ten consumer trend in 2021 [9].

The meat industry faces important challenges, and as part of the food industry conglomerate, it needs to address changes towards clean-label options. Meat products, per definition, need to utilise an important amount of additives during their processing, so that the typical technological and organoleptic characteristics are met. In addition, many of the additives also employed during meat processing are essential to preserve the safety and shelf life of the products. Many synthetic-sounding ingredients offer functionalities that are paramount for meat quality. For this reason, nowhere else are these challenges greater than in meat production.

Additives are one of the most researched substances in the world, as they are constantly monitored by food-safety agencies, such as The European Food Safety Authority (EFSA) in the EU and (Food and Drug Administration) FDA in the USA. Within the EU, there is a list of permitted additives and their maximum level of use depending on the type of product [10]. For meat products, the list is long, including antimicrobials, antioxidants and texturisers as the most used ones, but also some other additives (such as colours, stabilisers and acidity regulators) are allowed to be used in some of the European meat products (Table 1). Consumers might perceive these additives as unhealthy or unnecessary due to their chemical-sounding name. However, all the additives used in meat processing are considered safe within the established limits by the food safety authorities.

In this article, we present a thorough review of the clean-label options in the form of ingredients or novel technologies that can offer a real clean-label alternative to the most common additives used in meat processing.

Table 1. Additives permitted in the EU for meat product according to Reference [10].

E-Number	Additive Names	Max Dosage (mg/kg)	Permitted Products
E120	Cochineal, carminic acid, carmines	100	Sausages, pates, terrines, breakfast sausages (min 6% cereal) and burger meat (4% vegetables or cereal)
		200	Chorizo, salchichón
		quantum satis	pasturmas
E129	Allura Red AG	25	Luncheon meat, breakfast sausages (min 6% cereal) and burger meat (4% vegetables or cereal)
E124	Ponceau 4R, Cochineal Red A	250	Chorizo, salchichón
		200	Sobrasada
E150a–d	Caramels	*quantum satis*	Sausages, pates, terrines, breakfast sausages (min 6% cereal) and burger meat (4% vegetables or cereal)
E160a	Carotenes	20	Sausages, pates, terrines
E160c	Paprika extract, capsanthin, capsorubin	10	Sausages, pates, terrines
E162	Beetroot Red, betanin	*quantum satis*	Sausages, pates, terrines
E200–203	Sorbic acid-sorbates	1000	Pates, aspic
E210–213	Benzoic acid-benzoates	500	aspic
E214–219	p-hydroxybenzoates	1000	pates

Table 1. *Cont.*

E-Number	Additive Names	Max Dosage (mg/kg)	Permitted Products
E220–228	Sulphur dioxide-sulphites	450	breakfast sausages, burger meat (4% vegetables or cereal), salsicha fresca, longaniza fresca, butifarra fresca
E249–250	Nitrites	150	Non-sterilised meat products
		100	Sterilised meat products ($F_0 > 3.00$)
E251–E252	Nitrates	150	Non heat treated meat products
E300–301	Ascorbic acid, sodium ascorbate	*quantum satis*	Foie gras, foie gras entier, blocs de foie gras/Libamáj, libamáj egészben, libamáj tombben
E310–320	Gallates, TBHQ and BHA	200	Dehydrated meat
E315–316	Erythorbic acid, sodium erythorbate	500	Cured meat products and preserved meat products
E338–452	Phosphoric acid-phosphates-di-, tri- and polyphosphates	5000	Except foie gras, foie gras entier, blocs de foie gras, Libamáj, libamáj egészben, libamáj tömbben
E385	Calcium disodium ethylene diamine tetra-acetate (Calcium disodium EDTA)	250	Libamáj, libamáj egészben, libamáj tömbben
E392	Extracts of rosemary	150	Dehydrated meat, heat treated and non-heat treated meat products excluding dried sausage
		100	Dried sausage
E427	Cassia gum	1500	Heat treated meat products
E473–474	Sucrose esters of fatty acids-sucroglycerides	5000	Heat treated meat products except foie gras, foie gras entier, blocs de foie gras, Libamáj, libamáj egészben, libamáj tömbben
E481–482	Stearoyl-2-lactylates	4000	Minced and diced canned meat products
E959	Neohesperidine DC	5	As flavour enhancer only, except for foie gras, foie gras entier, blocs de foie gras, Libamáj, libamáj egészben, libamáj tömbben

2. Clean-Label Ingredients in Meat Products

2.1. Antimicrobial

Consumers' demand for safe and high-quality meat and meat products is more dynamic and diversified nowadays than in the past. They want minimally processed, easily prepared, all-natural ready-to-eat (RTE) meat products [11]. To date, the trend in consumers' food demands, clean labelling has rapidly increased, particularly for meat products containing many food additives [12]. Researchers in parallel with producers and manufacturers have been challenged to develop healthy meat products with high quality and safety criteria. The microorganisms associated with the spoilage of meat and meat products are bacteria such as *Pseudomonas*, *Acinetobacter*, *Brochothrix thermosphacta*, *lactobacillus* spp., *Enterobacter*, as well as yeasts and moulds that can affect the organoleptic characteristics of food [13].

The extended use of nitrites led to growing awareness and concern about the healthiness of meat products. Numerous safety issues about nitrite have been raised because it can be converted into N-nitroso with amines in meat products, known as carcinogenic compounds to humans [14,15]. Therefore, several studies counter this challenge and help produce meat products with low or no-nitrite salts using potential alternatives with similar antimicrobial effects without causing any health hazards [16,17]. Additionally, nitrite play a major role in inhibiting the growth of foodborne pathogens such as *Listeria monocytogenes*, *salmonella* spp., *Campylobacter jejuni*, *Escherichia coli O157:H7*, *Flavobacterium*, *micrococcus* spp. and *clostridium* spp. that can cause important public health problems with million cases of foodborne diseases occurring each year [15,18].

Another additive used as preservative in meat products is sulphites. Sulphites or SO_2 are antibacterial agents more powerful against gram-negative bacteria [19]. These additives are considered allergens as certain people have adverse reactions to their consumption, especially those sensitive to asthma, including triggering of anaphylactic reactions, hypotension, abdominal pain, dermatitis, etc. [20]. In addition to be declared as allergen content, sulphites and sulphiting agents are controlled and, in the EU, sulphites and SO_2 are the only ones permitted at a maximum dose of 450 mg/kg and only for the following meat products: breakfast sausage, longaniza fresca, butifarra fresca and burger meat when it has 4% of cereal or vegetable.

In the meat processing industry, several traditional thermal and novel non-thermal preservation techniques are being used to increase the products' shelf life and enhance the sensory properties. To achieve this, meat curing is a well-developed processing stage that includes the addition of salt, nitrite and nitrate even on fresh-cut meat imparting several distinctive properties to the meat products [21,22]. The main synthetic nitrites used in the meat industry are sodium nitrite ($NaNO_2$) and potassium nitrite (KNO_2) because they are cost-effective, stable, and easy to prepare and use [23]. Before using compounds of natural origin as a replacement for nitrite, their antimicrobial efficacy should be examined, and this review provides a comparison of the published data. Foodborne pathogens can easily contaminate raw meat or meat products, and during prolonged periods of storage, spoilage microorganisms may produce an unwanted visual appearance and diminish their organoleptic properties. Research for additives of natural origin with antimicrobial activities, especially of plant origin, has notably increased in recent years [23]. Numerous natural extracts have been applied to meat and meat products, with herbs and spices being the most used as clean-label alternatives to nitrites and sulphites [24]. Among these, some plant extracts can serve as natural nitrate sources, as nitrate naturally occurs in the environment (plants, soils, water, etc.) [25]. However, nitrites of natural origin do not offer any healthier advantage towards synthetic nitrites, and they only provide a clean-label option for the consumer. Table 2 presents some potential antimicrobial alternatives from natural origin for nitrite and sulphites that can be used effectively in clean-label meat products.

Table 2. Studies on the application of clean-label antimicrobial compounds on meat products.

Antimicrobial	Dosage	Product	Target	Main Effects	References
Clove (*Syzygium aromaticum*) EO	5 and 10%	Ground beef	*L. monocytogenes*	*L. monocytogenes* population completely inactivated after 3 days of storage at 0, 8 and −18 °C (10% clove oil) and inhibited with 5% clove oil	[26]
Cinnamon (*Cinnamomum cassia*) EO	2.5 and 5.0%			*L. monocytogenes* counts reduced by 3.5–4.0 Log CFU/g after 7 days at 0 and 8 °C and after 60 days at −18 °C (5% cinnamon oil)	
Oregano oil and Sodium nitrite	400 pm and 50–100 ppm	Minced pork	*C. botulinum*	The synergistic effect of oregano oil and $NaNO_2$ inhibited the growth of *C. botulinum*	[27]
Cinnamon EO and Grape seed extract	0.02–0.04% and 0.08–0.16% individually and in combination	Lyoner-type sausages	Lactic acid bacteria (LAB), Total viable count (TVC), Psychrotrophic count, mould and yeast count, and *C. perfringens*	Combination of cinnamon oil with grape extract 0.04 and 0.08%, respectively, reduced the final population of all counted microorganisms after 40 days, at 4 °C The combined effect of cinnamon oil with grape extract 0.04 and 0.16% reduced *C. perfringens* by 1.72 Log CFU/g at the end of storage	[24]

Table 2. *Cont.*

Antimicrobial	Dosage	Product	Target	Main Effects	References
Grape seed extract, Pine bark extract, and Rosemary extract	1% for each extract applied separately	Ground beef	*E. coli* O157:H7, *L. monocytogenes* and *S. Typhimurium*	After 9 days of storage at 4 °C, *E. coli* O157:H7 reduced by 0.62, 0.66 and 0.18 Log CFU/g; *L. monocytogenes* by 1.01, 1.34 and 0.89 Log CFU/g; and *S. Typhimurium* by 1.11, 1.33 and 1.06 Log CFU/g, respectively, by 1% grape seed, 1% pine bark and 1% rosemary extract, compared with the control samples	[28]
Ziziphora clinopodioides EO and Nisin	0.1–0.2% and 250–500 IU/g individually and in combination	Raw beef patty	TVC, psychrotrophic and Enterobacteriaceae count and *Staphylococcus aureus* and *E. coli* O157:H7	All treatments affected the growth of TVC, psychrotrophic and Enterobacteriaceae count, as well as *S. aureus* and *E. coli* O157:H7 Treatment with 0.2% EO+ 500 IU/g nisin presented the highest effect on microorganisms during storage for 9 days, at 4 °C *E. coli* O157:H7 and *S. aureus* counts were under the detection limit after 7 days, at 4 °C	[29]
Nisin and Lactoferrin	0, 100 and 200 µg/g individually and in combination	Turkish style meatball	TVC, LAB, Total psychrophilic bacteria, *Pseudomonas* spp., sulfite-reducing anaerobic bacteria, yeast and mould, and coliforms, *E. coli*, Total staphylococcae count, and *S. aureus*	All groups of microorganisms significantly reduced after treatment with nisin and lactoferrin alone or in combination after 12 days of storage at 4 °C Nisin (100 µg/g) and lactoferrin (200 µg/g) reduced the coliform (> 5-Log CFU/g) and *E. coli* population to undetectable level after 3 days, at 4 °C Nisin (200 µg/g) and lactoferrin (100 µg/g) effectively reduced *S. aureus* by 3.50 Log CFU/g	[30]
Lysozyme Nisin and Disodium ethylenediaminetetraacetic acid (EDTA)	250 ppm, 250 ppm and 20 mM in combination	Ostrich Meat Patties	TVC, LAB, *Pseudomonas* spp., Enterobacteriaceae and *L. monocytogenes*	*L. monocytogenes* population decreased below the detection limit of 2.00 Log CFU/g and LAB counts reduced about 2.00 Log CFU/g after treatment on patties packaged in air and vacuum and stored at 3 °C for 8 days	[31]
Tomato, red grape, olive and pomegranate by-product extracts	1000 mg/kg	Lamb meat patties	Mesophilic bacteria, Psychrotrophic counts, LAB, Enterobacteriaceae, and *L. monocytogenes* and *Salmonella* spp.	Microbial counts on lamb patties packed in MAP and stored at 2 °C (7-day storage) after treatment with by-product extracts were significantly lower than control samples Results showed the absence of *L. monocytogenes* and *Salmonella* spp.	[32]

Removing nitrite from meat products could be problematic because of its high antimicrobial efficacy. Hence, McDonnell et al. [33] evaluated several compounds for their antimicrobial efficacy against *L. monocytogenes* to uncured and alternative cured RTE processed meat and poultry products. The addition of vinegar, lemon and cherry powder blend (1.5%) delayed the growth of *L. monocytogenes* inoculated on the surface of cured ham and deli-style turkey breast. They suggested using the three antimicrobials on uncured roast beef as no growth of *L. monocytogenes* was observed after 12 weeks of storage at 4 °C. Moreover, *L. monocytogenes* effectively inhibited and decreased by 4 and 3 Log on RTE bologna type turkey meat coated with Nisaplin and Guardian (antimicrobial gelatin) films, respectively, after 56 days of refrigeration (4 °C) storage [34]. The efficacy of chitosan coating as an alternative to chemical protective additives demonstrated by Bostan and Mahan [35] on sausages. All sausages were dipped into 0.25, 0.50 and 1.00% chitosan solutions prepared with 1.00% acetic acid. The authors observed that the shelf life of the products increased and that 0.25% chitosan concentration was enough to inhibit the growth of aerobic bacteria, whereas higher concentrations were needed to inhibit the lactic acid bacteria (LAB). Soultos et al. [36] observed a positive effect of chitosan (0.50 and 1.00%) against the total viable count, LAB, *pseudomonas* spp., *B. thermosphacta*, *Enterobacteriaceae*, yeasts and moulds on Greek-style fresh pork sausages. Golden et al. [37] evaluated the efficacy of antimicrobial blends containing dried vinegar (DV), together with fruit and spice extracts with salt, against *C. perfringens* in uncured ham compared to traditionally cured ham. They manifested that combining the clean-label antimicrobials used had similar inhibition effects against *C. perfringens* in uncured compared to traditionally cured ham.

Additionally, a broad range of essential oils (EOs) with antimicrobial effects is widely used on meat products to prevent the growth of foodborne pathogens and spoilage microorganisms and extend the shelf life. EOs are secondary metabolites obtained from plants [38], are composed of a complex mixture of volatile compounds of low molecular weight and are characterised by being mainly liquid at room temperature [39]. Oregano oil has been extensively used on meat with positive results against common spoilage microbiota [40–42] and pathogens such as S. Enteritidis [43], S. typhimurium [28,41], *S. aureus* and *L. monocytogenes* [44]. Interestingly, Hernández-Hernández et al. [45] used a novel method to encapsulate Mexican oregano (*Lippia graveolens Kunth*) EO and found that it was efficient to control the naturally occurring microbiota of fresh pork meat during cold storage. Although it is challenging to replace nitrite with a single antimicrobial compound owing to its broad-spectrum activity [46], especially against inactivation of *C. botulinum* spores in cured meat products [21], a combination of nitrite and different antimicrobial agents may be successful. In this way, De Oliveira et al. [47] reported that different levels of winter savoury with 100 ppm of sodium nitrite allowed them to control the growth of *C. perfringens* on mortadella sausages. The authors attributed the antimicrobial activity of the EOs to the presence of carvacrol, ρ-cymene, linalool and thymol. The study by Bellés et al. [48] showed that the use of carvacrol in lamb burgers could be an option as an alternative to sulphites, as it showed a delay on microbial growth. Cui et al. [49] evaluated the antimicrobial efficacy of nutmeg, sage and clove plant extracts in a model meat food. They observed a synergistic effect of the natural extracts with 10 ppm NaNO$_2$ against *C. botulinum*, showing a potential combination in the control of botulism in minimally processed meat. Furthermore, Xi et al. [50] reported that lemon and lime powders and grape seed extract are less effective against *L. monocytogenes*. Still, cranberry powder together with nitrite (150 ppm) reduced the growth of *L. monocytogenes* by 2–4 Log CFU/g in cured cooked meat. Cranberry powder, long recognised as a source of natural antimicrobials, combined with nitrite (150 ppm) and grape seed extract, also offers a potential combination to inhibit *L. monocytogenes* growth in natural and organic processed meats [50]. The antimicrobial activity of the EOs is commonly attributed to the presence of the phenolic compounds [12,44,51] that can disturb the phospholipid bilayer of the cytoplasmic membrane and damage the membrane proteins leading to increased permeability of the cell membrane. However, there are several other mechanisms leading to the inactivation of the target microorganism, such as the disruption of a variety of enzyme systems [52] and destruction of genetic material [53].

The application of EOs is partially limited due to their intense aroma, which may cause adverse organoleptic effects and limited consumer's acceptance. To overcome this problem, novel thermal and non-thermal techniques [53,54] and the use of EOs as part of the hurdle technology together with other compounds and other processing technologies, such as the encapsulation of EOs in nanostructures, are essential to improve the shelf life and the sensory attributes of meat products.

2.2. Antioxidants

Antioxidants are added to meat and meat products to extend their shelf life through the deactivation of free radicals, and thus slowing down the rancidity. Various factors can promote lipid oxidation in meat products. Based on their mode of action, primary antioxidants prevent lipid peroxidation by preventing a chain reaction, reacting directly with lipid radicals and converting them into relatively stable products; and secondary antioxidants act by donating a hydrogen atom (H·) and binding to catalysts such as metal ions [55,56]. The list of approved antioxidants is small within the EU but larger for the USA. The only synthetic "pure" antioxidants approved in the EU list are gallates, tert-Butylhydroquinone (TBHQ) and butylated hydroxyanisole (BHA), which are allowed for only one specific meat product: dried meat. Other additives that provide antioxidant capacity but also have other functions are nitrites, ascorbates, erythorbates and citrates. Even though the safety of synthetic antioxidants has been questioned, the safety of antioxidants of natural origin is not much different [57], as the chemical compounds are the same irrespective of their origin. However, consumers relate the word "natural" to "good", as we mentioned before. For this reason, there has been an increase of the research and use of antioxidants of natural origin.

Antioxidants of natural origin have been identified in spices, herbs, fruits or vegetables and applied on meat and meat products primarily for their flavours and aroma. However, several natural extracts have been proven to offer the same functionality as their synthetic alternatives, with the advantage of being label-friendly and process compatible. Phenolic compounds are well known as a major group of natural antioxidants [28,58,59]. A growing list of clean-label natural extracts with antioxidant activity Generally Recognised as Safe (GRAS) by the FDA in the last years (USFDA, 2018) can be used in the meat industry. To name some of the commercially available antioxidants used throughout the meat industry, these are coffee, grape seed, green tea, oregano, sage (Greek and Spanish), lavender, lime, dill, parsley and rosemary extract between them being the most used in the meat industry [60,61]. Conversely, the EU has only approved rosemary extract as antioxidant additives for meat products [10], but the spices can be used as ingredients in the formulation following all the safety controls.

One of the most important natural antioxidants is 3,4-dihydroxyphenylethanol or hydroxytyrosol (HXT), showing interesting antioxidant characteristics and having beneficial effects on health [62]. Martinez-Zamora et al. [63] tested both natural (HXTo) and synthetic (HXTs) antioxidants on lamb meat burgers. Natural HXTo consisted of organic hydroxytyrosol (HXTo, sample 7% purity from olive tree leaves, 200 ppm) showed higher preservative activity in maintaining the nutritional value than the control synthetic HTX (HXTs, 99% purity, 200 ppm) made with sulphites. Rosemary, orange and lemon extracts were investigated in cooked Swedish-style meatballs, with the citrus extracts showing a 50% control of rancidity. The rosemary (water and oil soluble) extracts presented a complete elimination of rancidity after 12 days of storage at 8 °C [64]. In the same way, Kim et al. [65] also observed that rosemary extract had high antioxidant properties that could delay the onset of rancidity in meat fats. In this context, to explore for alternatives to synthetic additives, numerous industrial by-products of chestnuts (wood, flowers, leaves, shells, etc.) [66–69] and various fruits [32,70–74] have been used for their antioxidant activity on meat and meat products. The use of industrial by-products agrees with the circular economy concept [67]. It reduces the environmental impact of food processing and waste production while bringing benefits for the meat industry that avoids significant losses by protecting the meat products from oxidation, increasing their quality and shelf life.

As we mentioned earlier, many natural extracts can negatively affect the aroma of meat products. However, there are several plants, such as spinach, radishes and celery, that contain more than 2500 mg nitrate/kg [25,75], and their extracts can be used as natural sources of nitrate in meat products. Celery has been extensively studied and used commercially because it does not affect the sensory attributes of meat products [76]. The addition of celery powder in cooked sausages significantly inhibited the quality deterioration during cold storage for four weeks [77]. Sausages containing celery powder (0.8%) showed comparable pH, thiobarbituric acid reactive substances (TBARS) and volatile basic nitrogen (VBN) values to the control samples containing sodium nitrite (0.01%). These results manifested that celery powder effectively protected sausages from quality deterioration and can be used as nitrite source from natural origin. Similarly, added celery juice powder and starter culture in emulsified sausages presented good quality characteristics without significant differences with the control samples containing sodium nitrite [78]. Nitrate obtained from plant sources can be used directly in the brine solution or the product together with a starter culture (to form nitrate into nitrite) or as a "cultured", "prefermented" or "pre-converted" nitrate-containing plant source. The meat industry mainly applies the second method because they can control the specific natural pre-converted nitrites they use and their concentrations [76,79].

When evaluating natural antioxidant compounds that may prevent or retard protein and lipid oxidation, it is essential to consider the compound's fat solubility, effective dose, optimum temperature, pH and thermal stability, as well as cost, availability and regulatory status. The meat industry has an excellent opportunity to utilise antioxidants of natural origin in their products, following the consumers' demands for clean-label meat products.

2.3. Texturisers

Phosphates are the most widely used additive in processed-meat products because of their functional effects. Phosphates possess a certain antimicrobial effect and inhibit lipid oxidation, which condition the colour and the flavour of the products; but the main reason for their use is that they increase the water-holding capacity (WHC) affecting texture and sensory qualities [80]. Based on this, their replacement can lead to several technological limitations; therefore, it is essential to find alternatives that will not compromise the functions phosphates provide. Fibres, seaweeds and vegetable powders are ingredients with similar capacities to phosphates and could offer an opportunity towards clean-label meat products [80]. Phosphates are of concern for people with chronic kidney disease, as their excess in blood is associated with cardiovascular risk [81]. For the healthy individuals, even though phosphates present no concern with respect to genotoxicity or carcinogenicity and their acute oral toxicity is low, the EFSA found that the exposure was higher than the acceptable daily intake for some population groups in their re-evaluation of these additives in 2019 [82]. This is another reason for trying to find alternatives to phosphates in meat products.

In general, strategies based on the reduction or elimination of phosphates have been studied in emulsion-type sausages (Table 3); however, they have been used in others, such as ham, bacon, delicatessen meats, breaded chicken products or injected poultry pieces [80].

Fibres present potential as functional alternatives to phosphate due to their technological advantages (high water- and fat-holding capacity, improved emulsion stability, and texture enhancement) and their positive effect on health [95]. In that sense, several rich-fibres components (whole seeds, fibre extracts, etc.) have been used to improve the texture and sensory attributes of meat products, mainly in those with reduced fat or reduced salt content [95]. However, in the development of free-phosphates meat products, the use of fibres as replacers is not so widespread.

Chia seed presents several functional advantages but can also affect consumers' health positively due to its high content of soluble dietary fibre [96]. In that sense, chia mucilage (formed after soaking chia seeds in water) has been used in powder and gelled form in two concentrations (2 and 4%) as sodium tripolyphosphate replacer in the development of bologna sausages [87]. New healthier products showed similar yield than controls, with both concentrations of mucilage, and in the two forms (powder and gel). Other alternative could be the use of mushrooms due to their high levels of nutrients (protein, polysaccharides, fibre and vitamins) and several biological benefits. Lyophilized and pulverized winter mushrooms were used in different concentrations (0, 0.5, 1.0, 1.5 and 2.0%) as sodium pyrophosphate (0.3%) replacer in emulsion-type sausages to evaluate their technological properties [89]. Over 1% of mushrooms powder, the exudation of fat from sausages was inhibited and an increase of pH was noted. Moreover, lipid oxidation of sausages was inhibited. However, it was observed that free-phosphates samples were softer [89] (Table 3).

Table 3. Ingredients used as phosphates alternatives in the development of clean-label meat products.

Ingredient	Meat Product	Effects in Meat Products	Reference
Inulin (powder or gelled)	restructured chicken steaks	Maintain sensory scores.Better juiciness scores with gelled form. Oxidative and microbiological stability during frozen storage.	[83]
Citrus fibre	Cured bologna sausages	Similar emulsion stability and yield. Good behaviour during chilled storage.	[84]
Bamboo fibre	Bologna sausages	Sensorially accepted	[85]
Mango peel	Chicken marinade breast	Similar cooking/thawing yield	[86]
Chia mucilage (powder and gelled)	Bologna sausages	Reduced chewy and firm. With 2% of mucilage better emulsion stability and sensory acceptability	[87]
Sea tangle	emulsion type sausage	Similar cooking loss, overall acceptability	[88]
Winter mushroom powder	emulsion type sausage	No negative effects in colour and sensory parameters with <2%	[89]
Dried Plum Products	Chicken marinade fillets	similar sensory characteristics and yield	
SavorPhosp (commercial blend)	Rotisserie chickens and chicken breasts	Yield improved. No negative effects on technological and sensory properties	[90]
Porcine blood plasma	Frankfurter sausages	Similar water holding capacity, cooking loss and texture. Modified flavour.	[91]
Dehydrated beef protein	Beef strip loin steaks	Similar sensory characteristics, colour and microbial stability. Lower oxidation stability and tenderness.	[92]
Fructo-oligosaccharides (FOS)	Cooked hams	Higher cooking loss, satisfactory technological quality.	[93]
Calcium powders from egg and oyster	Cooked meat products	Similar yield and texture properties lighter colour.	[94]

Fructo-oligosaccharides (FOSs) are soluble prebiotic fibres that have been used as an alternative clean-label ingredient to phosphates in the production of restructured chicken steaks and cooked hams [83,93]. For phosphates-free restructured steaks' development, inulin was added in gel and powder form (4.5%). In the case of hams, FOSs were employed in different concentrations as substitutes for phosphates and dextrose, using response surface methodology. In general, the behaviour of these healthier products was similar when comparing with samples with phosphates. However, authors indicated the need to tolerate some processing compromises, such as a reduction in yield [83,93]. Other type of fibres used to avoid the use of phosphates was bamboo fibre. Its use in Bologna sausages (2.5 and 5%) resulted in being similar to others cited. Although some technological properties were conditioned with bamboo fibres, sausages maintained emulsion stability and yields [85].

By-products of the food industry that have a high fibre content could be a phosphate replacement that would allow for the industry to obtain healthier meat products while improving sustainability (many of them would otherwise go unutilised) (Table 3). Citrus fibre, a by-product of the fruit-juice industry, has been used in different concentrations (0.50, 0.75 and 1.00%) instead of tripolyphosphate with optimal results for some functional properties, such as adequate emulsion stability and yield [84]. However, authors considered that citrus-fibre levels must be assayed more critically depending on the content and type of protein present in the products. Aside from applying phosphates replacement strategies directly in the reformulation of the product, others have tried it in marinades for chicken products. Plum ingredients, dried plum powder and dried plum fibre (0.06%), and a blend of them (0.06%) were used to replace sodium tripolyphosphate in chicken breast fillets marinade [97]. A hedonic analysis and a 5-point just-about-right (JAR) demonstrated that the marinade of the blend of plum fibre and powder was not distinguishable from the control. Moreover, no differences were observed in cooking and thawing losses. Mango peel is another by-product that has been evaluated as a phosphate substitute to marinade chicken breast. Samples treated with mango peel showed similar cooking and thawing yield than those with marinate solution containing tripolyphosphate [86].

By-products obtained from the meat industry, such as porcine blood plasma or dehydrated beef proteins, could be used as phosphates alternatives and have been studied added directly to meat products (frankfurters) or through brines (for beef strip loins) [91,92]. The use of both meat industry by-products as phosphate replacers resulted in being positive regarding their yield; however, sensory quality was affected, as it increased animal taste and odour in frankfurters [91] and decreased tenderness in beef steaks [92].

Sea tangle (*Lamina japonica*) is a type of brown algae with water retention and binding ability that has been added to totally replace the sodium pyrophosphate (0.2%) in an emulsion-type sausage. Both 1.5 and 3% of sea tangle offered similar cooking loss to sausages without negative effects on sensory acceptability [88]. Natural calcium powders obtained from eggs and oyster shells were used individually or in combination as phosphate alternatives to formulate pork meat products [94]. It was observed that the combination of oyster (0.2%) and egg (0.3%) shell powder would enable the replacement of synthetic phosphate with desirable qualities in the reformulated products.

Based on some of the ingredients mentioned, commercial alternatives to phosphates have been patented. An example that has been evaluated in marinade chicken-meat products is SavorPhos (Formtech Solutions Inc., College Station, TX, USA), a proprietary blend labelled as citrus flour, all natural flavourings and less than 2% of sodium carbonate [90]. The use of SavorPhos blend as replacer of a commercial phosphate blend, both in water and oil-based marinades, resulted in an optimal option in rotisserie chickens and chicken breasts. Similar yields were obtained with water-based marinades; however, the use of SavorPhos improved the yield with oil-based marinades. Moreover, texture values of breast were improved with the use of SavorPhos and without negatively affecting colour or sensory acceptability [90].

2.4. Colours

Food colours are used to help improve the appearance of food products that could be affected by exposure to light, moisture, air and temperature variations, as well as to enhance the naturally occurring colours or give colour to otherwise colourless products. This type of additives comes from natural and synthetic origin and according to EU legislation [10] only a few are accepted and most of them limited to some dosage and specific products. From the additives of synthetic origin, only two are permitted for meat products within the EU: Allura Red AG and Ponceau 4R. The former can be applied for luncheon meat, breakfast sausages and burger meat, whereas Ponceau 4R can only be applied in three specific products: chorizo, salchichón and sobrasada. The clean-label alternatives for these colours are the food colours from natural origin, such as cochineal and carminic acids, as well as caramels, carotenes, paprika extracts or beetroot red. However, not all of these colours are permitted in the aforementioned products (Table 1). In addition, some of the food colours might present poor stability to light and time (such as beetroot red or paprika extracts), are not soluble in fat (such as cochineal) or are not soluble in water (such as carotenes) [19]. A problematic with food colours is the consumer perception of their use. Some might have a negative perception as food colours can mask other colours in the food product [98] and also for the relationship of some of them with attention-deficit/hyperactivity disorder in children [99]. Consumers might perceive a meat product as clean label if it has food colours of natural origin in it, but even these food colours can dissuade the consumer if the food colour is not a recognizable ingredient in that product, e.g., caramel in sausages. For this reason, the use of food colours in clean-label meat products should be limited to the few already accepted in the traditional recipes.

3. Novel Technologies for the Development of Clean-Label Meat Products

Thermal processing in addition to the use of additives, have been the only generally recognized methods for reducing food spoilage. However, the high temperatures used during these processes induce changes in the structure of food and losses of consistency and, in addition, lipid oxidation, which is the main cause of rancidity. These negative effects on the nutritional and sensory properties and the probable health risks have given rise to new technologies called non-thermal processing/mild processing/hurdle techniques [100]. High-pressure processing (HPP), ultrasound and packaging—mainly modified atmospheric packaging (MAP)—are non-thermal techniques that currently are gaining interest in the development of minimally processed food products. However, these techniques also need of an optimisation step to maintain the product quality while also extending or maintaining its shelf life.

High-pressure processing (HPP) is a treatment based on the application of high pressure (100–800 MPa), at mild temperatures (<45 °C), that is uniformly distributed through the product by a liquid transmitter. The utilization of HPP allows us to inactivate microorganisms and enzymes for a longer period without the need of chemical additives. Nonetheless, to assure food safety and to extend shelf life, the applied pressure and the temperature must be chosen according to the characteristics of the product [101]. In general, the treatment involves a minimal impact on sensory quality and nutritional value, but the noticeable differences in thermal and aggregative behaviour of proteins can condition the products' colour and texture [102]. In beef patties, the texture and cooking loss increased with higher pressure levels [103], but a contrary effect was observed in beef gels in which HPP treatment improved the yield and texture parameters [104]. Furthermore, Maksimenko et al. [104] observed a decrease in colour values of beef gels under HHP treatment. On the other hand, as a consequence of the aggregation that HPP caused on proteins, the digestion of the meat can be improved [105]. However, high-pressure treatment may also induce lipid oxidation depending on the processing time and the pressure level applied [101]. However, this negative effect could be solved by using antioxidants of natural origin, thus maintaining the condition of clean label. For example, the use of sage powder on beef burgers pressurized at 600 MPa retarded the lipid oxidation of products over 60 days of chilled storage [106].

The introduction of the ultrasonic treatment promotes the production of pro-health, minimally processed food, which is currently very popular among consumers. Power ultrasound is a non-thermal processing technology that uses sound energy at frequencies higher than human audible range (>20 kHz) and lower than microwave frequencies (10 MHz) with many applications on muscle products, included meat tenderization, acceleration of maturation and mass transfer, and shelf-life extension [107]. Moreover, is a treatment characterized with a low impact on the organoleptic properties and the nutritional value of meat products. The use of ultrasound reduces microbial contamination due to its capacity to cause damage on biological cells, especially microbial cell membranes [108]. In addition, the use of ultrasound may allow us to reduce the use of additives, such as phosphates, due to its ability to improve the emulsification and gelling properties of proteins [109,110]. The characteristics of this technology make it attractive to reduce or even eliminate the use of additives and obtain clean-label meat products [108,111].

In addition, PEF (Pulsating Electric Field) or Pulse Light are non-thermal technologies that are receiving increased attention. Both technologies, in comparison with conventional thermal sterilization make it possible to achieve effective inactivation of microorganisms in a much shorter processing time and using less energy [108]. Moreover, the impact on nutritional and sensory characteristics of the final products is, in general, minimal.

Food packaging is an indispensable element that serves as the protection from contamination, external environment and mechanical damage. Currently, a new generation of packaging is emerging with several functions that, among others, extend the shelf life of meat products. For example, it has been observed that the combination of vacuum-packaging technology and shrinking largely extends shelf-life in comparison with traditional packaging [108]. In addition, this packaging is growing as an eco-friendly technology due to the use of biodegradable films. The new packaging materials are developed by considering not only the sustainability of their materials but also to extend shelf life, in a healthier and convenient way. The packaging that not only acts as a barrier from the outside environment but also has some active functions towards improving the shelf life is called active packaging. There are four classes of active packaging depending on the function: scavenging or absorbing, emitting, creating barriers and regulating [112]. The first class comprises mainly gas or liquid absorbers and is barely used in meat products; however, in fresh meat, they are more popular (e.g., sachets that absorb losses from fresh meat). Within the active packaging, emitting antioxidants and creating antimicrobial barriers are the most popular functions for meat products in order to prevent oxidation and microbial spoilage, and thus improving shelf life. The use of edible coatings with antioxidants and/or bioactive compounds (as the ones mentioned in Section 2.1) are being tested in different meat products. Zhao et al. [113] found that chitosan and carvacrol starch packaging films delayed microbial spoilage by up to 25 days in ham. A novel edible film made up of calcium alginate was developed by Noor et al. [114] that included *Asparagus racemosus* as bioactive ingredient. The use of this film prevented the lipid oxidation and improved the storage quality of a model meat product. A recent and thorough review of edible coatings as active packaging in meat products can be found in [115]. Consumers might perceive some risks associated with this new active packaging (technology acceptance, toxicity of new materials, economic risk, malfunction, etc.) and, thus, reject it. Although most of the attitudes towards active packaging are neutral to mildly positive, there is low familiarity with it, and if educational communication is not provided of the information of its value (i.e., extending shelf life), consumers might reject this technology [116].

4. Conclusions and Future Trends

The use of some additive is so extended in the manufacturing of meat products that the meat industry did not worry about finding alternatives until very recently. Consumers are demanding safe, nutritious and healthier meat products and have put the focus on the additives they contain. A clean-label meat product should only contain the ingredients from the traditional recipes easily recognised by the consumers. Some additives, such as texturisers or colours, are being replaced with alternative options. However, avoiding the use of some additives can create situations where food safety is at risk. Some alternatives rely on the origin of the additive: natural vs. synthetic (e.g., nitrites from green vegetables vs. synthetic nitrites), as natural is perceived as a good trait for most of the consumers. This would be enough for the industry, as products with "natural" alternatives will be perceived as being clean label. Nonetheless, the health problems associated with some additives do not distinguish if the substance is extracted from the nature or synthesised in a laboratory, the chemical component remains the same. We believe that future research should focus on the application of synergistic alternatives, such as a combination of novel technologies and the use of preservatives with no health implications. There is a surge in different antioxidants and antimicrobials from natural sources, but these would need to be thoroughly evaluated before being utilised as alternatives just for being "natural". Innovations in the packaging industry are yet to be widely applied in the meat industry. Once they are fully developed, they will make an important impact on the products' shelf life in a sustainable manner. The meat industry and meat scientists should explore further the clean-label alternatives to develop safer, nutritious and healthier meat products.

Author Contributions: Conceptualization, G.D.-P. and T.P.; investigation, G.D.-P., S.I.E., A.C.S. and T.P.; writing—original draft preparation, G.D.-P., S.I.E., A.C.S. and T.P.; writing—review and editing, G.D.-P., S.I.E., A.C.S. and T.P. All authors have read and agreed to the published version of the manuscript.

Funding: This research received no external funding.

Conflicts of Interest: The authors declare no conflict of interest.

References

1. Cegiełka, A. "Clean Label" as one of the leading trends in the meat industry in the world and in poland—A review. *Rocz. Panstw. Zakl. Hig.* **2020**, *71*, 43–55. [CrossRef]
2. Yong, H.; Kim, T.; Choi, H.; Jung, S.; Choi, Y. Technological strategy of clean label meat products. *Food Life* **2020**, *1*, 13–20. [CrossRef]
3. Asioli, D.; Aschemann-Witzel, J.; Caputo, V.; Vecchio, R.; Annunziata, A.; Næs, T.; Varela, P. Making sense of the "Clean Label" trends: A review of consumer food choice behavior and discussion of industry implications. *Food Res. Int.* **2017**, *99*, 58–71. [CrossRef]
4. Velissariou, M. What Is Clean Label? Available online: http://blog.ift.org/what-is-clean-label (accessed on 3 June 2021).
5. Aschemann-Witzel, J.; Varela, P.; Peschel, A.O. Consumers' categorization of food ingredients: Do consumers perceive them as 'Clean Label' producers expect? An exploration with projective mapping. *Food Qual. Prefer.* **2019**, *71*, 117–128. [CrossRef]
6. Song, H.; Schwarz, N. If it's difficult to pronounce, it must be risky: Fluency, familiarity, and risk perception. *Psychol. Sci.* **2009**, *20*, 135–138. [CrossRef]
7. Román, S.; Sánchez-Siles, L.M.; Siegrist, M. The importance of food naturalness for consumers: Results of a systematic review. *Trends Food Sci. Technol.* **2017**, *67*, 44–57. [CrossRef]
8. Lusk, J. What Is Natural Food Anyway? Available online: http://jaysonlusk.com/blog/2013/6/11/what-is-natural-food-anyway (accessed on 13 June 2021).
9. Westbrook, G.; Angus, A. *Top 10 Global Consumer Trends 2021*; Euromonitor International: London, UK, 2021. Available online: https://go.euromonitor.com/white-paper-EC-2021-Top-10-Global-Consumer-Trends.html (accessed on 3 June 2021).
10. Commission Regulation (EU) No 1129/2011 of 11 November 2011 Amending Annex II to Regulation (EC) No 1333/2008 of the European Parliament and of the Council by Establishing a Union List of Food AdditivesText with EEA Relevance. *Off. J. Eur. Union* **2011**, *L295*, 1–177.
11. Font-i-Furnols, M.; Guerrero, L. Consumer Preference, Behavior and Perception about Meat and Meat Products: An Overview. *Meat Sci.* **2014**, *98*, 361–371. [CrossRef]
12. Jayasena, D.D.; Jo, C. Essential Oils as Potential Antimicrobial Agents in Meat and Meat Products: A Review. *Trends Food Sci. Technol.* **2013**, *34*, 96–108. [CrossRef]
13. Adams, M.R.; Moss, M.O. The Microbiology of Food Preservation. In *Food Microbiology*; Royal Society of Chemistry: Cambridge, UK, 2008.
14. Sen, N.P.; Baddoo, P.A. Trends in the levels of residual nitrite in canadian cured meat products over the past 25 years. *J. Agric. Food Chem.* **1997**, *45*, 4714–4718. [CrossRef]
15. Krause, B.L.; Sebranek, J.G.; Rust, R.E.; Mendonca, A. Incubation of curing brines for the production of ready-to-eat, uncured, no-nitrite-or-nitrate-added, ground, cooked and sliced ham. *Meat Sci.* **2011**, *89*, 507–513. [CrossRef]
16. Alahakoon, A.U.; Jayasena, D.D.; Ramachandra, S.; Jo, C. Alternatives to Nitrite in Processed Meat: Up to Date. *Trends Food Sci. Technol.* **2015**, *45*, 37–49. [CrossRef]
17. Kim, T.-K.; Hwang, K.-E.; Song, D.-H.; Ham, Y.-K.; Kim, Y.-B.; Paik, H.-D.; Choi, Y.-S. Effects of natural nitrite source from swiss chard on quality characteristics of cured pork loin. *Asian Australas. J. Anim. Sci.* **2019**, *32*, 1933–1941. [CrossRef]
18. Tauxe, R.V.; Doyle, M.P.; Kuchenmüller, T.; Schlundt, J.; Stein, C.E. Evolving public health approaches to the global challenge of foodborne infections. *Int. J. Food Microbiol.* **2010**, *139*, S16–S28. [CrossRef]
19. Feiner, G. Meat Products Handbook. In *Practical Science and Technology*, 1st ed.; Woodhead Publishing: Sawston, UK, 2006.
20. Vally, H.; Misso, N.L. Adverse reactions to the sulphite additives. *Gastroenterol. Hepatol. Bed Bench* **2012**, *5*, 16–23. [PubMed]
21. Sindelar, J.J.; Milkowski, A.L. *Sodium Nitrite in Processed Meat and Poultry Meats: A Review of Curing and Examining the Risk/Benefit of Its Use*; White Paper Series; American Meat Science Association: Savoy, IL, USA, 2011; Volume 3.
22. Parthasarathy, D.K.; Bryan, N.S. Sodium nitrite: The "cure" for nitric oxide insufficiency. *Meat Sci.* **2012**, *92*, 274–279. [CrossRef] [PubMed]
23. Yong, H.I.; Kim, T.-K.; Choi, H.-D.; Jang, H.W.; Jung, S.; Choi, Y.-S. Clean label meat technology: Pre-converted nitrite as a natural curing. *Food Sci. Anim. Resour.* **2021**, *41*, 173–184. [CrossRef] [PubMed]
24. Aminzare, M.; Hashemi, M.; Hassanzadazar, H.; Hejazi, J. The Use of Herbal Extracts and Essential Oils as a Potential Antimicrobial in Meat and Meat Products; a Review. *J. Hum. Environ. Health Promot.* **2016**, *1*, 63–74. [CrossRef]
25. Gassara, F.; Kouassi, A.P.; Brar, S.K.; Belkacemi, K. Green Alternatives to Nitrates and Nitrites in Meat-Based Products—A Review. *Crit. Rev. Food Sci. Nutr.* **2016**, *56*, 2133–2148. [CrossRef] [PubMed]
26. Khaleque, M.A.; Keya, C.A.; Hasan, K.N.; Hoque, M.M.; Inatsu, Y.; Bari, M.L. Use of Cloves and Cinnamon Essential Oil to Inactivate Listeria Monocytogenes in Ground Beef at Freezing and Refrigeration Temperatures. *LWT* **2016**, *74*, 219–223. [CrossRef]
27. Ismaiel, A.A.; Pierson, M.D. Effect of Sodium Nitrite and Origanum Oil on Growth and Toxin Production of Clostridium Botulinum in TYG Broth and Ground Pork. *J. Food Prot.* **1990**, *53*, 958–960. [CrossRef] [PubMed]
28. Ahn, J.; Grün, I.U.; Mustapha, A. Antimicrobial and Antioxidant Activities of Natural Extracts in Vitro and in Ground Beef. *J. Food Prot.* **2004**, *67*, 148–155. [CrossRef]
29. Shahbazi, Y.; Shavisi, N.; Mohebi, E. Effects of Z iziphora clinopodioides essential oil and nisin, both separately and in combination, to extend shelf life and control Escherichia Coli O157:H7 and *Staphylococcus aureus* in raw beef patty during refrigerated storage. *J. Food Saf.* **2016**, *36*, 227–236. [CrossRef]

30. Colak, H.; Hampikyan, H.; Bingol, E.B.; Aksu, H. The Effect of Nisin and Bovine Lactoferrin on the microbiological quality of turkish-style meatball (TEKIRDAĞ KÖFTE). *J. Food Saf.* **2008**, *28*, 355–375. [CrossRef]
31. Mastromatteo, M.; Lucera, A.; Sinigaglia, M.; Corbo, M.R. Synergic Antimicrobial Activity of Lysozyme, Nisin, and EDTA against Listeria Monocytogenes in Ostrich Meat Patties. *J. Food Sci.* **2010**, *75*, M422–M429. [CrossRef] [PubMed]
32. Andrés, A.I.; Petrón, M.J.; Adámez, J.D.; López, M.; Timón, M.L. Food By-Products as Potential Antioxidant and Antimicrobial Additives in Chill Stored Raw Lamb Patties. *Meat Sci.* **2017**, *129*, 62–70. [CrossRef] [PubMed]
33. McDonnell, L.M.; Glass, K.A.; Sindelar, J.J. Identifying Ingredients That Delay Outgrowth of Listeria Monocytogenes in Natural, Organic, and Clean-Label Ready-to-Eat Meat and Poultry Products. *J. Food Prot.* **2013**, *76*, 1366–1376. [CrossRef]
34. Min, B.J.; Han, I.Y.; Dawson, P.L. Antimicrobial Gelatin Films Reduce Listeria Monocytogenes on Turkey Bologna. *Poult. Sci.* **2010**, *89*, 1307–1314. [CrossRef]
35. Bostan, K.; Mahan, F. Microbiological Quality and Shelf-Life of Sausage Treated with Chitosan. *J. Fac. Vet. Med. Istanb. Univ.* **2011**, *37*, 117–126.
36. Soultos, N.; Tzikas, Z.; Abraham, A.; Georgantelis, D.; Ambrosiadis, I. Chitosan Effects on Quality Properties of Greek Style Fresh Pork Sausages. *Meat Sci.* **2008**, *80*, 1150–1156. [CrossRef]
37. Golden, M.; Wanless, B.; Glass, K. Comparison of Clean Label Antimicrobials with Nitrite on the Inhibition of Clostridium Perfringens during Extended Cooling of a Model Deli-Style Ham Product. In Proceedings of the IAFP 2019 Annual Meeting, IAFP, Louisville, KY, USA, 21–24 July 2019.
38. Wińska, K.; Mączka, W.; Łyczko, J.; Grabarczyk, M.; Czubaszek, A.; Szumny, A. Essential Oils as Antimicrobial Agents—Myth or Real Alternative? *Molecules* **2019**, *24*, 2130. [CrossRef]
39. Pateiro, M.; Munekata, P.E.; Sant'Ana, A.S.; Domínguez, R.; Rodríguez-Lázaro, D.; Lorenzo, J.M. Application of Essential Oils as Antimicrobial Agents against Spoilage and Pathogenic Microorganisms in Meat Products. *Int. J. Food Microbiol.* **2021**, *337*, 108966. [CrossRef] [PubMed]
40. Tsigarida, E.; Skandamis, P.; Nychas, G.J. Behaviour of Listeria Monocytogenes and Autochthonous Flora on Meat Stored under Aerobic, Vacuum and Modified Atmosphere Packaging Conditions with or without the Presence of Oregano Essential Oil at 5 °C. *J. Appl. Microbiol.* **2000**, *89*, 901–909. [CrossRef] [PubMed]
41. Skandamis, P.; Tsigarida, E.; Nychas, G.E. The Effect of Oregano Essential Oil on Survival/Death of Salmonella Typhimurium in Meat Stored at 5 C under Aerobic, VP/MAP Conditions. *Food Microbiol.* **2002**, *19*, 97–103. [CrossRef]
42. Chouliara, E.; Karatapanis, A.; Savvaidis, I.N.; Kontominas, M.G. Combined Effect of Oregano Essential Oil and Modified Atmosphere Packaging on Shelf-Life Extension of Fresh Chicken Breast Meat, Stored at 4 °C. *Food Microbiol.* **2007**, *24*, 607–617. [CrossRef] [PubMed]
43. Govaris, A.; Solomakos, N.; Pexara, A.; Chatzopoulou, P.S. The Antimicrobial Effect of Oregano Essential Oil, Nisin and Their Combination against Salmonella Enteritidis in Minced Sheep Meat during Refrigerated Storage. *Int. J. Food Microbiol.* **2010**, *137*, 175–180. [CrossRef]
44. Barbosa, L.N.; Rall, V.L.M.; Fernandes, A.A.H.; Ushimaru, P.I.; da Silva Probst, I.; Fernandes, A. Essential Oils Against Foodborne Pathogens and Spoilage Bacteria in Minced Meat. *Foodborne Pathog. Dis.* **2009**, *6*, 725–728. [CrossRef]
45. Hernández-Hernández, E.; Lira-Moreno, C.Y.; Guerrero-Legarreta, I.; Wild-Padua, G.; Di Pierro, P.; García-Almendárez, B.E.; Regalado-González, C. Effect of Nanoemulsified and Microencapsulated Mexican Oregano (*Lippia graveolens* Kunth) Essential Oil Coatings on Quality of Fresh Pork Meat. *J. Food Sci.* **2017**, *82*, 1423–1432. [CrossRef]
46. Pegg, R.B.; Shahidi, F. *Nitrite Curing of Meat: The Nnitrosamine Problem and Nitrite Alternatives*; Food and Nutrition Press. Inc.: Trumbull, CT, USA, 2000.
47. de Oliveira, T.L.C.; de Araújo Soares, R.; Ramos, E.M.; das Graças Cardoso, M.; Alves, E.; Piccoli, R.H. Antimicrobial Activity of Satureja Montana, L. Essential Oil against Clostridium Perfringens Type A Inoculated in Mortadella-Type Sausages Formulated with Different Levels of Sodium Nitrite. *Int. J. Food Microbiol.* **2011**, *144*, 546–555. [CrossRef]
48. Bellés, M.; Alonso, V.; Roncalés, P.; Beltrán, J.A. Sulfite-Free Lamb Burger Meat: Antimicrobial and Antioxidant Properties of Green Tea and Carvacrol. *J. Sci. Food Agric.* **2019**, *99*, 464–472. [CrossRef]
49. Cui, H.; Gabriel, A.A.; Nakano, H. Antimicrobial Efficacies of Plant Extracts and Sodium Nitrite against Clostridium Botulinum. *Food Control* **2010**, *21*, 1030–1036. [CrossRef]
50. Xi, Y.; Sullivan, G.A.; Jackson, A.L.; Zhou, G.H.; Sebranek, J.G. Use of Natural Antimicrobials to Improve the Control of Listeria Monocytogenes in a Cured Cooked Meat Model System. *Meat Sci.* **2011**, *88*, 503–511. [CrossRef] [PubMed]
51. Hyldgaard, M.; Mygind, T.; Meyer, R.L. Essential Oils in Food Preservation: Mode of Action, Synergies, and Interactions with Food Matrix Components. *Front. Microbiol.* **2012**, *3*, 12. [CrossRef] [PubMed]
52. Burt, S. Essential Oils: Their Antibacterial Properties and Potential Applications in Foods—A Review. *Int. J. Food Microbiol.* **2004**, *94*, 223–253. [CrossRef]
53. Solomakos, N.; Govaris, A.; Koidis, P.; Botsoglou, N. The antimicrobial effect of thyme essential oil, nisin and their combination against Escherichia Coli O157:H7 in minced beef during refrigerated storage. *Meat Sci.* **2008**, *80*, 159–166. [CrossRef]
54. Stratakos, A.C.; Koidis, A. Suitability, Efficiency and Microbiological Safety of Novel Physical Technologies for the Processing of Ready-to-eat Meals, Meats and Pumpable Products. *Int. J. Food Sci. Technol.* **2015**, *50*, 1283–1302. [CrossRef]
55. Rice-Evans, C.; Miller, N.; Paganga, G. Antioxidant Properties of Phenolic Compounds. *Trends Plant Sci.* **1997**, *2*, 152–159. [CrossRef]

56. Kumar, Y.; Yadav, D.N.; Ahmad, T.; Narsaiah, K. Recent Trends in the Use of Natural Antioxidants for Meat and Meat Products. *Compr. Rev. Food Sci. Food Saf.* **2015**, *14*, 796–812. [CrossRef]
57. Pokorný, J. Are Natural Antioxidants Better—and Safer—than Synthetic Antioxidants? *Eur. J. Lipid Sci. Technol.* **2007**, *109*, 629–642. [CrossRef]
58. Zheng, W.; Wang, S.Y. Antioxidant Activity and Phenolic Compounds in Selected Herbs. *J. Agric. Food Chem.* **2001**, *49*, 5165–5170. [CrossRef]
59. Bozin, B.; Mimica-Dukic, N.; Simin, N.; Anackov, G. Characterisation of the Volatile Composition of Essential Oils of Some Lamiaceae Spices and the Antimicrobial and Antioxidant Activities of the Entire Oils. *J. Agric. Food Chem.* **2006**, *54*, 1822–1828. [CrossRef]
60. Shah, M.A.; Bosco, S.J.D.; Mir, S.A. Plant extracts as natural antioxidants in meat and meat products. *Meat Sci.* **2014**, *98*, 21–33. [CrossRef]
61. Oswell, N.J.; Thippareddi, H.; Pegg, R.B. Practical Use of Natural Antioxidants in Meat Products in the U.S.: A Review. *Meat Sci.* **2018**, *145*, 469–479. [CrossRef]
62. Martínez, L.; Ros, G.; Nieto, G. Hydroxytyrosol: Health Benefits and Use as Functional Ingredient in Meat. *Medicines* **2018**, *5*, 13. [CrossRef]
63. Martínez-Zamora, L.; Ros, G.; Nieto, G. Synthetic vs. Natural Hydroxytyrosol for Clean Label Lamb Burgers. *Antioxidants* **2020**, *9*, 851. [CrossRef] [PubMed]
64. Fernandez-Lopez, J.; Zhi, N.; Aleson-Carbonell, L.; Pérez-Alvarez, J.A.; Kuri, V. Antioxidant and Antibacterial Activities of Natural Extracts: Application in Beef Meatballs. *Meat Sci.* **2005**, *69*, 371–380. [CrossRef] [PubMed]
65. Kim, I.S.; Yang, M.R.; Lee, O.H.; Kang, S.N. Antioxidant Activities of Hot Water Extracts from Various Spices. *Int. J. Mol. Sci.* **2011**, *12*, 4120–4131. [CrossRef] [PubMed]
66. Munekata, P.E.S.; Domínguez, R.; Franco, D.; Bermúdez, R.; Trindade, M.A.; Lorenzo, J.M. Effect of Natural Antioxidants in Spanish Salchichón Elaborated with Encapsulated N-3 Long Chain Fatty Acids in Konjac Glucomannan Matrix. *Meat Sci.* **2017**, *124*, 54–60. [CrossRef] [PubMed]
67. Echegaray, N.; Gómez, B.; Barba, F.J.; Franco, D.; Estévez, M.; Carballo, J.; Marszałek, K.; Lorenzo, J.M. Chestnuts and By-Products as Source of Natural Antioxidants in Meat and Meat Products: A Review. *Trends Food Sci. Technol.* **2018**, *82*, 110–121. [CrossRef]
68. Zamuz, S.; López-Pedrouso, M.; Barba, F.J.; Lorenzo, J.M.; Domínguez, H.; Franco, D. Application of Hull, Bur and Leaf Chestnut Extracts on the Shelf-Life of Beef Patties Stored under MAP: Evaluation of Their Impact on Physicochemical Properties, Lipid Oxidation, Antioxidant, and Antimicrobial Potential. *Food Res. Int.* **2018**, *112*, 263–273. [CrossRef]
69. Lorenzo, J.M.; Franco, D.; Carballo, J. Effect of the Inclusion of Chestnut in the Finishing Diet on Volatile Compounds during the Manufacture of Dry-Cured "Lacón" from Celta Pig Breed. *Meat Sci.* **2014**, *96*, 211–223. [CrossRef] [PubMed]
70. Fernández-López, J.; Fernández-Ginés, J.M.; Aleson-Carbonell, L.; Sendra, E.; Sayas-Barberá, E.; Pérez-Alvarez, J.A. Application of Functional Citrus By-Products to Meat Products. *Trends Food Sci. Technol.* **2004**, *15*, 176–185. [CrossRef]
71. Devatkal, S.K.; Naveena, B.M. Effect of Salt, Kinnow and Pomegranate Fruit by-Product Powders on Color and Oxidative Stability of Raw Ground Goat Meat during Refrigerated Storage. *Meat Sci.* **2010**, *85*, 306–311. [CrossRef] [PubMed]
72. Ahmad, S.R.; Gokulakrishnan, P.; Giriprasad, R.; Yatoo, M.A. Fruit-Based Natural Antioxidants in Meat and Meat Products: A Review. *Crit. Rev. Food Sci. Nutr.* **2015**, *55*, 1503–1513. [CrossRef] [PubMed]
73. Qi, S.; Huang, H.; Huang, J.; Wang, Q.; Wei, Q. Lychee (*Litchi chinensis* Sonn.) Seed Water Extract as Potential Antioxidant and Anti-Obese Natural Additive in Meat Products. *Food Control* **2015**, *50*, 195–201. [CrossRef]
74. Lorenzo, J.M.; Pateiro, M.; Domínguez, R.; Barba, F.J.; Putnik, P.; Kovačević, D.B.; Shpigelman, A.; Granato, D.; Franco, D. Berries Extracts as Natural Antioxidants in Meat Products: A Review. *Food Res. Int.* **2018**, *106*, 1095–1104. [CrossRef] [PubMed]
75. Schullehner, J.; Hansen, B.; Thygesen, M.; Pedersen, C.B.; Sigsgaard, T. Nitrate in Drinking Water and Colorectal Cancer Risk: A Nationwide Population-based Cohort Study. *Int. J. Cancer* **2018**, *143*, 73–79. [CrossRef] [PubMed]
76. Sebranek, J.G.; Jackson-Davis, A.L.; Myers, K.L.; Lavieri, N.A. Beyond Celery and Starter Culture: Advances in Natural/Organic Curing Processes in the United States. *Meat Sci.* **2012**, *92*, 267–273. [CrossRef]
77. Jin, S.-K.; Choi, J.S.; Yang, H.-S.; Park, T.-S.; Yim, D.-G. Natural Curing Agents as Nitrite Alternatives and Their Effects on the Physicochemical, Microbiological Properties and Sensory Evaluation of Sausages during Storage. *Meat Sci.* **2018**, *146*, 34–40. [CrossRef]
78. Sindelar, J.J.; Cordray, J.C.; Sebranek, J.G.; Love, J.A.; Ahn, D.U. Effects of Varying Levels of Vegetable Juice Powder and Incubation Time on Color, Residual Nitrate and Nitrite, Pigment, PH, and Trained Sensory Attributes of Ready-to-Eat Uncured Ham. *J. Food Sci.* **2007**, *72*, 388–395. [CrossRef]
79. Flores, M.; Toldrá, F. Chemistry, Safety, and Regulatory Considerations in the Use of Nitrite and Nitrate from Natural Origin in Meat Products—Invited Review. *Meat Sci.* **2021**, *171*, 108272. [CrossRef] [PubMed]
80. Thangavelu, K.P.; Kerry, J.P.; Tiwari, B.K.; McDonnell, C.K. Novel Processing Technologies and Ingredient Strategies for the Reduction of Phosphate Additives in Processed Meat. *Trends Food Sci. Technol.* **2019**, *94*, 43–53. [CrossRef]
81. Savica, V.; Maiolino, G.; Calò, L.A. To Reconsider (Limit) the Use of Phosphate Based Food and Beverages Additives. A Real Need for Health Preservation. *Clin. Nutr.* **2016**, *35*, 240. [CrossRef]

82. Younes, M.; Aquilina, G.; Castle, L.; Engel, K.-H.; Fowler, P.; Fernandez, M.J.F.; Fürst, P.; Gürtler, R.; Husøy, T.; Mennes, W.; et al. Re-Evaluation of Phosphoric Acid–Phosphates—Di-, Tri- and Polyphosphates (E 338–341, E 343, E 450–452) as Food Additives and the Safety of Proposed Extension of Use. *EFSA J.* **2019**, *17*, e05674. [CrossRef] [PubMed]
83. Öztürk-Kerimoğlu, B.; Serdaroğlu, M. Powder/Gelled Inulin and Sodium Carbonate as Novel Phosphate Replacers in Restructured Chicken Steaks. *J. Food Process. Preserv.* **2019**, *43*, e14243. [CrossRef]
84. Powell, M.J.; Sebranek, J.G.; Prusa, K.J.; Tarté, R. Evaluation of Citrus Fiber as a Natural Replacer of Sodium Phosphate in Alternatively-Cured All-Pork Bologna Sausage. *Meat Sci.* **2019**, *157*, 107883. [CrossRef]
85. Magalhaes, I.M.C.; de Souza Paglarini, C.; Vidal, V.A.S.; Pollonio, M.A.R. Bamboo Fiber Improves the Functional Properties of Reduced Salt and Phosphate-Free Bologna Sausage. *J. Food Process. Preserv.* **2020**, *44*. [CrossRef]
86. Roidoung, S.; Ponta, N.; Intisan, R. Mango Peel Ingredient as Salt and Phosphate Replacement in Chicken Breast Marinade. *Int. J. Food Stud.* **2020**, *9*, 193–202. [CrossRef]
87. Câmara, A.K.F.I.; Vidal, V.A.S.; Santos, M.; Bernardinelli, O.D.; Sabadini, E.; Pollonio, M.A.R. Reducing Phosphate in Emulsified Meat Products by Adding Chia (*Salvia hispanica* L.) Mucilage in Powder or Gel Format: A Clean Label Technological Strategy. *Meat Sci.* **2020**, *163*, 108085. [CrossRef]
88. Lee, H.; Choe, J.; Yong, H.I.; Lee, H.J.; Kim, H.-J.; Jo, C. Combination of Sea Tangle Powder and High-Pressure Treatment as an Alternative to Phosphate in Emulsion-Type Sausage. *J. Food Process. Preserv.* **2018**, *42*, e13712. [CrossRef]
89. Choe, J.; Lee, J.; Jo, K.; Jo, C.; Song, M.; Jung, S. Application of Winter Mushroom Powder as an Alternative to Phosphates in Emulsion-Type Sausages. *Meat Sci.* **2018**, *143*, 114–118. [CrossRef]
90. Casco, G.; Veluz, G.A.; Alvarado, C.Z. SavorPhos as an All-Natural Phosphate Replacer in Water- and Oil-Based Marinades for Rotisserie Birds and Boneless-Skinless Breast. *Poult. Sci.* **2013**, *92*, 3236–3243. [CrossRef] [PubMed]
91. Hurtado, S.; Saguer, E.; Toldrà, M.; Parés, D.; Carretero, C. Porcine Plasma as Polyphosphate and Caseinate Replacer in Frankfurters. *Meat Sci.* **2012**, *90*, 624–628. [CrossRef] [PubMed]
92. Lowder, A.C.; Goad, C.L.; Lou, X.; Morgan, J.B.; DeWitt, C.A.M. Evaluation of a Dehydrated Beef Protein to Replace Sodium-Based Phosphates in Injected Beef Strip Loins. *Meat Sci.* **2011**, *89*, 491–499. [CrossRef]
93. Resconi, V.C.; Keenan, D.F.; Gough, S.; Doran, L.; Allen, P.; Kerry, J.P.; Hamill, R.M. Response Surface Methodology Analysis of Rice Starch and Fructo-Oligosaccharides as Substitutes for Phosphate and Dextrose in Whole Muscle Cooked Hams. *LWT Food Sci. Technol.* **2015**, *64*, 946–958. [CrossRef]
94. Cho, M.G.; Bae, S.M.; Jeong, J.Y. Egg Shell and Oyster Shell Powder as Alternatives for Synthetic Phosphate: Effects on the Quality of Cooked Ground Pork Products. *Korean J. Food Sci. Anim. Resour.* **2017**, *37*, 571–578. [CrossRef]
95. Jimenez-Colmenero, F.; Delgado-Pando, G. Fibre-enriched meat products. In *Fibre-Rich and Wholegrain Foods: Improving Quality*; Delcour, J.A., Poutanen, K., Eds.; Woodhead Publishing: New Delhi, India, 2013; Volume 237, pp. 329–347. ISBN 978-0-85709-578-7.
96. Munoz, L.A.; Cobos, A.; Diaz, O.; Aguilera, J.M. Chia Seed (*Salvia hispanica*): An Ancient Grain and a New Functional Food. *Food Rev. Int.* **2013**, *29*, 394–408. [CrossRef]
97. Jarvis, N.; Clement, A.R.; O'Bryan, C.A.; Babu, D.; Crandall, P.G.; Owens, C.M.; Meullenet, J.-F.; Ricke, S.C. Dried Plum Products as a Substitute for Phosphate in Chicken Marinade. *J. Food Sci.* **2012**, *77*, S253–S257. [CrossRef]
98. Bearth, A.; Cousin, M.-E.; Siegrist, M. The Consumer's Perception of Artificial Food Additives: Influences on Acceptance, Risk and Benefit Perceptions. *Food Qual. Prefer.* **2014**, *38*, 14–23. [CrossRef]
99. Nigg, J.T.; Lewis, K.; Edinger, T.; Falk, M. Meta-Analysis of Attention-Deficit/Hyperactivity Disorder or Attention-Deficit/Hyperactivity Disorder Symptoms, Restriction Diet, and Synthetic Food Color Additives. *J. Am. Acad. Child Adolesc. Psychiatry* **2012**, *51*, 86–97.e8. [CrossRef]
100. Hygreeva, D.; Pandey, M.C. Novel Approaches in Improving the Quality and Safety Aspects of Processed Meat Products through High Pressure Processing Technology—A Review. *Trends Food Sci. Technol.* **2016**, *54*, 175–185. [CrossRef]
101. Bolumar, T.; Orlien, V.; Sikes, A.; Aganovic, K.; Bak, K.H.; Guyon, C.; Stübler, A.-S.; de Lamballerie, M.; Hertel, C.; Brüggemann, D.A. High-Pressure Processing of Meat: Molecular Impacts and Industrial Applications. *Compr. Rev. Food Sci. Food Saf.* **2021**, *20*, 332–368. [CrossRef]
102. Speroni, F.; Szerman, N.; Vaudagna, S.R. High Hydrostatic Pressure Processing of Beef Patties: Effects of Pressure Level and Sodium Tripolyphosphate and Sodium Chloride Concentrations on Thermal and Aggregative Properties of Proteins. *Innov. Food Sci. Emerg. Technol.* **2014**, *23*, 10–17. [CrossRef]
103. Szerman, N.; Ferrari, R.; Sancho, A.M.; Vaudagna, S. Response Surface Methodology Study on the Effects of Sodium Chloride and Sodium Tripolyphosphate Concentrations, Pressure Level and Holding Time on Beef Patties Properties. *LWT* **2019**, *109*, 93–100. [CrossRef]
104. Maksimenko, A.; Kikuchi, R.; Tsutsuura, S.; Nishiumi, T. Effect of High Hydrostatic Pressure and Reducing Sodium Chloride and Phosphate on Physicochemical Properties of Beef Gels. *High Press. Res.* **2019**, *39*, 385–397. [CrossRef]
105. Rakotondramavo, A.; Rabesona, H.; Brou, C.; de Lamballerie, M.; Pottier, L. Ham Processing: Effects of Tumbling, Cooking and High Pressure on Proteins. *Eur. Food Res. Technol.* **2019**, *245*, 273–284. [CrossRef]
106. Mizi, L.; Cofrades, S.; Bou, R.; Pintado, T.; Lopez-Caballero, M.E.; Zaidi, F.; Jimenez-Colmenero, F. Antimicrobial and Antioxidant Effects of Combined High Pressure Processing and Sage in Beef Burgers during Prolonged Chilled Storage. *Innov. Food Sci. Emerg. Technol.* **2019**, *51*, 32–40. [CrossRef]

107. Alarcon-Rojo, A.D.; Carrillo-Lopez, L.M.; Reyes-Villagrana, R.; Huerta-Jiménez, M.; Garcia-Galicia, I.A. Ultrasound and Meat Quality: A Review. *Ultrason. Sonochem.* **2019**, *55*, 369–382. [CrossRef]
108. Rudy, M.; Kucharyk, S.; Duma-Kocan, P.; Stanisławczyk, R.; Gil, M. Unconventional Methods of Preserving Meat Products and Their Impact on Health and the Environment. *Sustainability* **2020**, *12*, 5948. [CrossRef]
109. Pinton, M.B.; Correa, L.P.; Facchi, M.M.X.; Heck, R.T.; Leães, Y.S.V.; Cichoski, A.J.; Lorenzo, J.M.; dos Santos, M.; Pollonio, M.A.R.; Campagnol, P.C.B. Ultrasound: A New Approach to Reduce Phosphate Content of Meat Emulsions. *Meat Sci.* **2019**, *152*, 88–95. [CrossRef] [PubMed]
110. Zhang, F.; Zhao, H.; Cao, C.; Kong, B.; Xia, X.; Liu, Q. Application of Temperature-Controlled Ultrasound Treatment and Its Potential to Reduce Phosphate Content in Frankfurter-Type Sausages by 50%. *Ultrason. Sonochem.* **2021**, *71*, 105379. [CrossRef] [PubMed]
111. Al-Hilphy, A.R.; Al-Temimi, A.B.; Al Rubaiy, H.H.M.; Anand, U.; Delgado-Pando, G.; Lakhssassi, N. Ultrasound Applications in Poultry Meat Processing: A Systematic Review. *J. Food Sci.* **2020**, *85*, 1386–1396. [CrossRef]
112. Kuswandi, B. Jumina 12—Active and intelligent packaging, safety, and quality controls. In *Fresh-Cut Fruits and Vegetables*; Siddiqui, M.W., Ed.; Academic Press: Cambridge, MA, USA, 2020; pp. 243–294. ISBN 978-0-12-816184-5.
113. Zhao, Y.; Teixeira, J.S.; Saldaña, M.D.A.; Gänzle, M.G. Antimicrobial Activity of Bioactive Starch Packaging Films against Listeria Monocytogenes and Reconstituted Meat Microbiota on Ham. *Int. J. Food Microbiol.* **2019**, *305*, 108253. [CrossRef]
114. Noor, S.; Bhat, Z.F.; Kumar, S.; Mudiyanselage, R.J. Preservative Effect of Asparagus Racemosus: A Novel Additive for Bioactive Edible Films for Improved Lipid Oxidative Stability and Storage Quality of Meat Products. *Meat Sci.* **2018**, *139*, 207–212. [CrossRef] [PubMed]
115. Umaraw, P.; Munekata, P.E.S.; Verma, A.K.; Barba, F.J.; Singh, V.P.; Kumar, P.; Lorenzo, J.M. Edible Films/Coating with Tailored Properties for Active Packaging of Meat, Fish and Derived Products. *Trends Food Sci. Technol.* **2020**, *98*, 10–24. [CrossRef]
116. Young, E.; Mirosa, M.; Bremer, P. A Systematic Review of Consumer Perceptions of Smart Packaging Technologies for Food. *Front. Sustain. Food Syst.* **2020**, *4*. [CrossRef]

Article

New Formulation towards Healthier Meat Products: *Juniperus communis* L. Essential Oil as Alternative for Sodium Nitrite in Dry Fermented Sausages

Vladimir Tomović [1], Branislav Šojić [1,*], Jovo Savanović [1,2], Sunčica Kocić-Tanackov [1], Branimir Pavlić [1], Marija Jokanović [1], Vesna Đorđević [3], Nenad Parunović [3], Aleksandra Martinović [4] and Dragan Vujadinović [5]

[1] Faculty of Technology Novi Sad, Bulevar cara Lazara 1, University of Novi Sad, 21000 Novi Sad, Serbia; tomovic@uns.ac.rs (V.T.); jovosavanovicdimdim@gmail.com (J.S.); suncicat@uns.ac.rs (S.K.-T.); bpavlic@uns.ac.rs (B.P.); marijaj@tf.uns.ac.rs (M.J.)
[2] "DIM-DIM" M.I. d.o.o, Trn-Laktaši Svetosavska bb, 78252 Trn Laktaši, Bosnia and Herzegovina
[3] Institute of Meat Hygiene and Technology (INMES), Kaćanskog 13, 11040 Belgrade, Serbia; vesna.djordjevic@inmes.rs (V.Đ.); nenad.parunovic@inmes.rs (N.P.)
[4] Faculty for Food Technology, Food Safety and Ecology, Donja Gorica, University of Donja Gorica, 81000 Podgorica, Montenegro; aleksandra.martinovic@udg.edu.me
[5] Faculty of Technology Zvornik, Karakaj 1, University of East Sarajevo, 75400 Zvornik, Bosnia and Herzegovina; dragan.vujadinovic@tfzv.ues.rs.ba
* Correspondence: sojic@tf.uns.ac.rs; Tel.: +381-214-853-714

Received: 8 July 2020; Accepted: 3 August 2020; Published: 6 August 2020

Abstract: The effect of *Juniperus communis* L. essential oil (JEO) addition at concentrations of 0.01, 0.05 and 0.10 µL/g on pH, instrumental parameters of color, lipid oxidation (2-Thiobarbituric acid reactive substances (TBARS)), microbial growth, texture and sensory attributes of dry fermented sausages produced with different levels of fat (15 and 25%) and sodium nitrite (0, 75 and 150 mg/kg) was assessed. Reduced level of sodium nitrite (75 mg/kg) in combination with all three concentrations of JEO (0.01–0.10 µL/g) resulted in satisfying physico-chemical (color and texture) properties and improved oxidative stability (TBARS < 0.3 mg MDA/kg) of dry fermented sausages produced with 25% of fat. However, sausages produced with 0.10 µL/g of JEO had untypical flavor. No foodborne pathogens (*Escherichia coli*, *Listeria monocytogenes*, *Salmonella* spp. and sulfite-reducing clostridia) were detected in any sample throughout the storage period (225 days). The results of this study revealed significant antioxidative activity of JEO and consequently its high potential as effective partial replacement for sodium nitrite in dry fermented sausages.

Keywords: *Juniperus communis* L.; essential oil; sodium nitrite; dry fermented sausage

---

1. Introduction

Fermented sausages have been manufactured in many countries worldwide. Currently, customers are becoming progressively aware of these meat products for their unique sensory characteristics and important health benefits [1]. Dry fermented sausages are produced using fresh or frozen meat (70–80%) and back fat (20–30%), salt, starter cultures, spices and food additives [2,3]. Owing to the relatively high level of fat and distinctive processing technology (e.g., using diverse raw materials, absence of thermal treatment), fermented sausages are highly susceptible to quality deterioration, including lipid oxidation and bacterial growth [2,3].

Lipid oxidation is one of the chief non-microbial factors in quality deterioration in meat and meat-derived products [4]. It is well known that meat products become very susceptible to oxidative deterioration due to high levels of unsaturated lipids (e.g., polyunsaturated fatty acids, phospholipids and cholesterol), a variety of oxidizing agents in the muscle tissue, the presence of metal catalysts, heme pigments, etc. Lipids (triacyl-glycerides, phospholipids and sterols) are largely spread in both the intra- and extracellular space of muscle tissue. Oxidation of lipids is a three-step radical chain reaction which involves: initiation, propagation and termination with the free radical's formation [5]. It should be highlighted that lipid oxidation leads to loss of nutritional quality, reduced shelf life, intensified toxicity and reduction of the market value of meat and meat-derived products [4].

Spoilage (*Acinetobacter*, *Lactobacillus* spp., *Pseudomonas*, *Proteus* spp., *Enterobacter*, *Leuconostoc* spp., *Moraxella*, etc., yeasts and molds) and pathogenic (e.g., *Salmonella* spp., *Campylobacter jejuni*, *Listeria monocytogenes*, *Escherichia coli* O157:H7, *Clostridium* spp.) microorganisms can diminish the quality of meat and meat products and consequently induce numerous foodborne contaminations [6]. The growth of spoilage microorganisms causes the degradation of lipids and proteins present in meat and meat products and affects the development of unpleasant quality characteristics (e.g., discoloration, slime and gas production, off-odors and off-flavors). On the other hand, pathogenic bacteria are primarily responsible for foodborne diseases and food poisoning of meat and meat-derived products. Furthermore, in past decades, foodborne diseases have been marked as essential factors of growing public health and economic problems all around the world.

Therefore, lipid oxidation and microbiological deterioration of meat and meat products can be marked as major limitations in the modern meat industry [6].

The use of synthetic additives is one of the main approaches for preventing microbial growth and oxidative reactions in meat products [7]. Nitrites (sodium and potassium nitrite) are well known food additives and curing agents in meat processing [8]. They are officially registered as preservatives by European Union directives [9]. During the process of curing, nitrites are applied in order to improve the product's shelf life because they efficiently suppress the growth of many harmful microorganisms and impart significant antioxidant potential to meat products [10,11]. Besides the strong preservative effect, the use of nitrites contributes to the of development of the typical reddish-pink color and flavor of cured meat products [11,12]. However, these preservatives were recently marked as unhealthy to humans because they promote the formation of carcinogenic N-nitroso-compounds [13,14].

Hence, consumers are increasingly demanding fresh, natural, and negligibly processed products with lower content of artificial additives [12,15]. Essential oils are defined as volatile oils with peculiar scents isolated from aromatic and medicinal plants by hydro-distillation or by cold pressing from citrus fruit peel. It is well known that essential oils obtained from different aromatic and medicinal plants possess a significant antioxidant and antimicrobial potential and therefore they are progressively used as natural additives in the modern food industries [16,17]. They represent the complex mixture of terpenoid compounds which can be present in different parts of herbs, particularly in their waxy channels, glands and trichomes. From a chemical point of view, essential oils are usually multipart mixtures of different organic compounds (e.g., terpenoids), aldehydes, ketones, esters, acids and alcohols, where the main constituents commonly constitute up to 85% of the essential oils, while minor compounds and trace elements constitute up to 15% [18]. Predominantly, essential oils are attracting attention as natural food additives (antioxidants and/or antimicrobials), as they are "generally recognized as safe" (GRAS) and have a wide customer acceptance [19]. Hence, several authors have investigated the application of essential oils as natural additives in dry fermented sausages [20–22], as well as potential replacements for nitrites in processing of cooked [12,16,23] and dry cured meat products [24].

Juniperus communis L. is an evergreen coniferous plant widespread throughout Europe, North America and North Asia [25]. The berries obtained from the medicinal herb *Juniperus communis* L. are conventionally well known as a strong immune system booster and powerful detoxifier [26]. *Juniperus communis* L. is most frequently used in natural remedies for respiratory infections, sore throat, arthritis, muscle aches and fatigue. It has been found that plant stems have also been used in order to prevent both short- and long-term illnesses. *Juniperus communis* L. essential oil has been assessed and established for its in vitro antiradical and antioxidant activities which are mostly dependent on its chemical shape [25,26].

Due to its strong antioxidant, antibacterial, antifungal, and anti-inflammatory properties, *Juniperus communis* L. and its essential oil are widely used in food processing, and in the pharmaceutical and cosmetic industries. Terpenoids (e.g., α-pinene, limonene and myrcene) determine the strong and distinctive aroma of juniper essential oil [27].

Recently, the application of *Juniperus communis* L. essential oil as natural additive was investigated in several studies [28–30]. Selim et al. [29] found that *Juniperus communis* L. essential oil added at concentrations of 0.1, 0.5, and 1% possesses a weak inhibitory effect towards *Enterococci* and *Escherichia coli* O157:H7 that were inoculated in ground beef meat, stored at a temperature of 7 °C for 14 days. However, in an earlier study, Schelz et al. [28] determined the strong antimicrobial potential of *Juniperus communis* L. essential oil against *Saccharomyces cerevisiae*. In our previous research, we found that *Juniperus communis* L. essential oil efficiently suppressed lipid oxidation and microbial growth and enhanced the color of cooked pork sausages [30].

A literature review has exposed only a few published research papers that discuss the application of essential oil as natural additive in dry fermented sausage processing. There is also a lack of data regarding the application of essential oils as sodium nitrite replacements in this type of dry cured meat product. Regarding its strong antioxidant and antimicrobial potential, we hypothesized that *Juniperus communis* L. essential oil could be used as an alternative for sodium nitrite in meat processing. Thus, the aim of this study was to assess the effect of *Juniperus communis* L. essential oil as an alternative for sodium nitrite in dry fermented sausages. For these purposes, several physicochemical (pH, color and texture), microbiological (total plate count, lactic acid bacteria) and sensory (color, odor and flavor) parameters of dry fermented sausages were determined.

2. Materials and Methods

2.1. Juniperus communis L. Essential Oil

GC-MS Profile of Terpenoid Compounds

Juniperus communis L. essential oil (JEO) was purchased from the manufacturer Herba doo (Belgrade, Serbia). JEO was kept in dark glass bottles at 4 °C prior to the experiments.

For identification of volatile terpenoids from JEO, GC-MS analysis was used according to the method described by Pavlić et al. [31]. Agilent GC890N system coupled to mass spectrometer Agilent MS 5759, with HP-5MS column (0.25 mm inner diameter and 0.25 μm film thickness, 30 m length), was applied for the characterization of terpenoid profile. Flow rate of helium was 2 mL/min. JEO was dissolved in dichloromethane (approx. 1 mg/mL) and 5 μL of solution was injected in the device with split ratio 30:1. Temperature conditions were: injector temperature 250 °C, detector temperature 300 °C, initial 60 °C with linear increase of 4 °C/min up to 150 °C. The NIST 05 and Wiley 7n data base were used for compound identification. Retention equations, which describe dependence of peak area on different concentration ($R^2 > 0.99$), were obtained using standard compounds dissolved in dichloromethane at different concentrations (1–500 μg/mL). Results were expressed as relative percentage (%).

2.2. Samples

Dry fermented sausages were created with two levels (15 and 25%) of pork back fat (FC). In both obtained batters, sodium nitrite (NC) was added at three concentrations (0, 75 and 150 mg/kg). Next, each batter was divided into four parts, and into each part the corresponding concentrations of JEO (0.00, 0.01, 0.05 and 0.10 μL/g) were added. The total number of batches (B) was: FC (2) × NC (3) × JEO (4) = 24 (Figure 1). Samples were collected at different storage periods (SD) involving three randomly selected dry fermented sausages from each batch at the end of drying (0) and after 75, 150 and 225 days of storage. The total number of samples was: B (24) × SD (4) × 3 = 288.

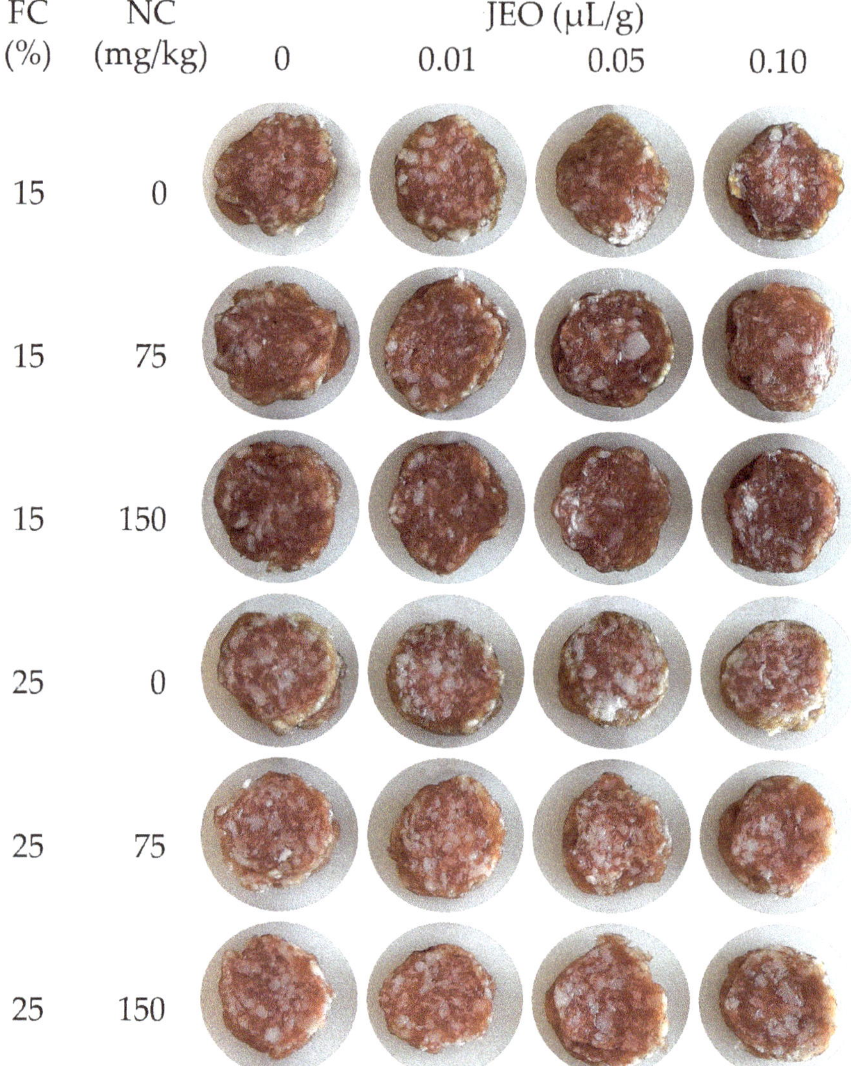

Figure 1. Photograph of the inside surfaces of the sausages at the end of storage.

2.3. Preparation of Dry Fermented Pork Sausages

Dry fermented sausages were produced in a local industrial plant (A.D. Dim-Dim, Laktaši, Bosnia and Herzegovina). Batters were produced using lean pork shoulder and pork back fat in the ratio 75:25 and 85:15%. The amounts of other ingredients were calculated in relation to raw material weight, and were as follows: NaCl (2.50%), gluconic delta-lactone (0.70%), spice mix (0.50%), dextrose (0.10%), sodium nitrite (0, 75 and 150 mg/kg) and JEO (0.00, 0.01, 0.05 and 0.10 µL/g.) The meat and back fat were minced using a cutter (Krämer & Grebe, Germany), and then the other ingredients were added and mixed with them until the required temperature (1 °C) was achieved. The sausages were stuffed in 37 mm diameter collagen casings and were placed in a climate chamber (Frigovent, Serbia) for 21 days. The processes of fermentation, smoking, drying and ripening were performed at a temperature of 14–16 °C and a relative humidity of 80–95%. Produced sausages were vacuum packed (Multivac C500, Wolfertschwenden, Germany) and stored at 15 ± 1 °C for 225 days. The proximate chemical compositions of the sausages produced with 15 and 25% of back fat at the end of drying process are presented in Table S1 (Supplementary material).

2.4. Physico-Chemical Analysis

The proximate chemical composition (moisture, protein, fat and ash) was determined according to International Organization for Standardization (ISO) procedures [32–35].

The pH was evaluated using a digital pH meter Testo 205 (Testo AG, USA). Before measurement it was calibrated using standard buffers (pH = 4.00 ± 0.05 and pH = 7.00 ± 0.01 at 20 ± 2 °C). pH values were determined for three samples, from each group of dry fermented sausages, in duplicate.

Color (CIE-*LAB* values: L^*—lightness; a^*—redness; b^*—yellowness) of each sample of the dry fermented sausages was measured on fresh cross cut immediately after slicing. The L^*, a^* and b^* color coordinates were determined using a MINOLTA Chroma Meter CR-400 (Minolta Co., Ltd., Osaka, Japan) using D-65 lighting, a 2° standard observer angle and an 8-mm aperture in the measuring head [16]. Prior to measurement it was calibrated using a Minolta calibration plate (No. 11333090; Y = 92.9, x = 0.3159; y = 0.3322). Color was measured for three samples (2 cm thick) from each group of dry fermented sausages in triplicate.

The TPA (Texture profile analysis) test was conducted at room temperature using TA.XT2 Texture Analyzer (Texture Technologies Corp., Scarsdale, NY/Stable MicroSystems, Godalming, UK) equipped with a standard ∅ 75 mm cylindrical plate. TPA parameters hardness (g), springiness, cohesiveness, and chewiness (g) were determined as described by Ikonić et al. [36]. The cylindrical shape samples (2.54 cm in diameter, 2 cm thick) were taken from the central part of the sausage, and were analyzed in two cycle compressions to 50% of their original thickness at a constant test speed of 1 mm/s. Peak force during the first compression cycle was marked as hardness. The rate at which a deformed sample goes back to its undeformed condition after the deforming force is removed was defined as springiness. The ratio of the area under the second and first curve was defined as cohesiveness. Lastly, by multiplying hardness, cohesiveness and springiness, chewiness was obtained. TPA was performed for three samples from each group of dry fermented sausages in duplicate.

Lipid oxidation of dry fermented sausages was assessed using the 2-Thiobarbituric acid reactive substances (TBARS) test according to the method of Botsoglou et al. [37], with some modifications. The final step of the extraction procedure was carried out with total volume (10 mL) of TCA (trichloroacetic acid) in ultrasonic bath XUB 12 (Grant Instruments, Cambridge, UK). Spectrophotometer Jenway 6300 (Jenway, Felsted, UK) was used for absorbance measurement at 532 nm. The results of the TBARS test were expressed as milligrams of malondialdehyde per kilogram of sample (mg MDA/kg). TBARS was determined on three samples from each group of dry fermented sausages in duplicate.

2.5. Microbiological Analysis

Microbiological analyses were performed on three samples from each group of dry fermented sausages in duplicate. Samples (20 g) were homogenized in 180 mL 1 g/L buffered peptone water (Merk, Darmstadt) for 10 min at 200 rpm (Unimax 1010, Heidolph, Germany) and the serial of decimal dilutions were prepared (up to 7–10). From each dilution 1 mL was placed in a sterile Petri plate and poured with appropriate media depending on the type of tested microorganisms. The following microorganisms were determined: total plate count (TPC), lactic acid bacteria (LAB), *Escherichia coli*, *Salmonella* spp., *Listeria monocytogenes* and sulfite-reducing clostridia count [38–43]. TPC was enumerated in Plate Count Agar (PCA) (Merk, Darmstadt, Germany) and incubated at 30 °C for 72 h; LAB was enumerated in de Man, Rogosa and Sharpe (MRS) Agar (Merk, Darmstadt, Germany) and incubated at 30 °C for 72 h; *Escherichia coli* was determined on Tryptone Bile Glucuronic Agar (TBX agar) (Merk, Darmstadt, Germany) after an incubation at 44 °C for 24 h; *Salmonella* spp. was determined on Xylose Lysine Deoxycholate (XLD) agar (Merk, Darmstadt, Germany) after an incubation at 37 °C for 24 h; *Listeria monocytogenes* was determined on Listeria agar acc. Ottaviani and Agosti (ALOA) (Merk, Darmstadt, Germany) after an incubation at 37 °C for 24 h; sulfite-reducing clostridia count was determined on Tryptone Sulfite Cycloserine (TSC) Agar (Merk, Darmstadt, Germany) after an incubation at 37 °C for 24–48 h under anaerobic conditions. After incubation, microscopic observation of cell morphology and biochemical tests were used for typical and atypical grown colonies identification. Results were expressed as a log number of colony forming units per gram (log CFU/g).

2.6. Sensory Analyses

Sensory analysis was carried out by a trained panel consisting of ten members, aged 25 to 50 years, per two sessions. All panelists work at the Faculty of Technology Novi Sad, Serbia, and have wide expertise in the sensory evaluation of foods. Panelists were trained according to methods described in ISO 8586 [44], in a sensory laboratory equipped according to ISO 8589 [45]. Evaluation of sensory attributes (color, odor and flavor) was performed using the difference-from-control test [46]. Prior to analyses, sausages were equilibrated to room temperature for about 15 min. and marked with a three-digit sample number. The sausages were sliced into 2 mm thick pieces and placed on a white porcelain plate. Consumers were firstly questioned to evaluate the control sample (without JEO and with the corresponding contents of fat and nitrite) and afterward to determine how different the coded samples were from the control one. The difference was rated on a scale from 0 to 6, where 0 = no difference; 1 = very slight difference; 2 = slight/moderate difference; 3 = moderate difference; 4 = moderate/large difference; 5 = large difference; and 6 = very large difference.

2.7. Statistical Analysis

The statistical program STATISTICA 13.0 (TIBCO Software Inc., Palo Alto, CA, USA) was used for data analyses. The main effects (fat content, nitrite content, JEO content and storage day) were compared. All data were expressed as mean value with their standard deviation (Stdev). The two-way, three-way and four-way interactions between these effects were also tested. Differences among treatment means were compared according to t-test and Duncan's multiple range test ($p < 0.05$).

3. Results and Discussion

3.1. Chemical Profile of JEO

Chemical profile of JEO was determined by GC-MS and results are presented in Table 1.

It can be observed that β-myrcene (14.12%) was the predominant compound in JEO, obtained using the conventional technique of hydro distillation. Other compounds detected in JEO with relative percentage higher than 1% were: (1) monoterpene hydrocarbons: sabinene (9.51%), β-pinene (5.39%), α-terpinene (1.95%), *p*-cymene (3.92%), d,l-limonene (8.36%), γ-terpinene (3.38%) and α-terpinolene (2.80%); (2) oxygenated monoterpenes: 4-terpineol (6.88%); (3) sesquiterpene hydrocarbons: α-cubebene (1.22%), α-copaene (1.39%), β-elemene (3.38%), caryophyllene (3.94%), α-humulene (3.26%), germacrene D (3.81%), ledene (1.40%), α-muurolene (1.30%), α-amorphene (5.43%) and germacrene B (3.74%), while all other compounds were present in content less than 1%. Results suggested a majority of terpenoids with hydrocarbons, while a lower amount could be accounted for by the oxygenated monoterpenes and sesquiterpenes. Present results were in accordance with similar studies since it was reported that monoterpene hydrocarbons [α-pinene (31.1%), β-myrcene (16.3%), sabinene (7.5%), limonene (6.2%) and β-pinene (3.7%)] were the major compounds identified in commercial JEO [47]. Similar results were also reported by Radoukova et al. [48] and Zheljazkov et al. [49]. It should be highlighted that variations in chemical profile of JEO could be related to genetic properties, geographical origin and climate conditions. Besides conventional technique of hydro distillation, Orav et al. [50] and Marković et al. [51] investigated the possibility of using a novel technique of extraction for JEO recovery. Orav et al. [50] reported that the JEO obtained using supercritical fluid extraction (SFE) with carbon-dioxide had a lower content of monoterpenes and a higher content of sesquiterpenes compared to JEO obtained using conventional hydro distillation. On the contrary, Marković et al. [51] determined a similar chemical profile for JEOs obtained using conventional hydro distillation and novel microwave-assisted hydro distillation. Therefore, it could be assumed that supercritical fluid extraction will cause co-extraction of other lipophilic compounds which could further alter the bioactivity of these extracts. Besides that, juniper variety, geographical origin, climate and post-harvest processing could significantly affect JEO yield and chemical profile of terpenoids. Furthermore, the possibility of utilization of other juniper materials, such as the needles (leaves) during hydro distillation cannot be excluded [47].

Table 1. Chemical profile of JEO determined by GC-MS.

	Retention Time (min)	Relative Percentage (%)
Sabinene	4.37	9.51
β-Pinene	4.45	5.39
β-Myrcene	4.71	14.12
Phellandrene	5.02	0.46
Δ-3-Carene	5.14	0.22
α-Terpinene	5.29	1.95
p-Cymene	5.49	3.92
d,l-Limonene	5.58	8.36
γ-Terpinene	6.33	3.38
n.i. [1]	6.68	0.28
α-Terpinolene	7.12	2.80
Linalool	7.47	0.29
n.i.	7.66	0.12
n.i.	8.09	0.08
n.i.	8.20	0.58
trans-Pinocarvenol	8.60	0.47
n.i.	8.77	0.23
n.i.	8.89	0.25
Borneol	9.47	0.36
4-Terpineol	9.80	6.88
p-Cymen-8-ol	10.07	0.35
n.i.	10.20	1.46
Benihinal	10.31	0.24
Verbenone	10.72	0.39
n.i.	12.34	0.10
n.i.	12.71	0.27
Bornyl acetate	13.10	0.72
n.i.	13.32	0.09
n.i.	13.38	0.12
n.i.	14.33	0.22
n.i.	14.72	0.14
α-Cubebene	15.09	1.22
Ylangene	15.76	0.13
α-Copaene	15.91	1.39
n.i.	16.21	0.20
β-Elemene	16.45	3.38
Isoledene	16.72	0.35
Caryophyllene	17.26	3.94
Aromadendrene	17.54	0.29
α-Humulene	18.30	3.26
trans-β-Farnesene	18.44	0.86
Germacrene D	19.15	3.81
β-Selinene	19.29	0.17

Table 1. *Cont.*

	Retention Time (min)	Relative Percentage (%)
Ledene	19.54	1.40
α-Muurolene	19.72	1.30
α-Amorphene	20.44	5.43
γ-Selinene	20.72	0.55
Aristolene	20.81	0.48
Germacrene B	21.36	3.74
n.i.	21.63	0.19
Spathulenol	21.99	0.62
Caryophyllene oxide	22.10	0.51
Humulene oxide	22.86	0.31
n.i.	23.01	0.38
n.i.	23.40	0.24
tau-Muurolol	23.82	0.85
α-Cadinol	24.18	0.99
n.i.	25.58	0.16
n.i.	25.89	0.17
Total		100

[1] Not identified.

3.2. pH and Instrumental Parameters of Color of Dry Fermented Sausages

The pH values of dry fermented sausages are presented in Table 2.

Table 2. pH, instrumental parameters of color and 2-Thiobarbituric acid reactive substances (TBARS) values of dry fermented sausages.

	pH	L^*	a^*	b^*	TBARS (mg MDA/kg)
			FC (%)		
15	5.46 ± 0.10 [a]	47.8 ± 3.0 [b]	14.0 ± 1.7 [a]	8.01 ± 1.27 [a]	0.17 ± 0.12 [a]
25	5.33 ± 0.10 [b]	52.4 ± 3.4 [a]	12.8 ± 1.7 [b]	7.66 ± 1.13 [b]	0.15 ± 0.10 [a]
p	<0.001	<0.001	<0.001	<0.001	0.258
			NC (mg/kg)		
0	5.38 ± 0.12 [a]	50.7 ± 4.1 [a]	13.5 ± 1.9 [a]	7.76 ± 1.16 [a]	0.20 ± 0.12 [a]
75	5.41 ± 0.12 [a]	50.1 ± 3.7 [a,b]	13.4 ± 1.8 [a]	7.83 ± 1.22 [a]	0.14 ± 0.10 [b]
150	5.39 ± 0.12 [a]	49.5 ± 3.9 [b]	13.3 ± 1.7 [a]	7.91 ± 1.27 [a]	0.15 ± 0.10 [b]
p	0.362	0.019	0.680	0.484	<0.001
			JC (µL/g)		
0	5.37 ± 0.10 [a]	51.2 ± 4.2 [a]	13.3 ± 1.9 [a]	7.80 ± 1.13 [a]	0.20 ± 0.11 [a]
0.01	5.40 ± 0.12 [a]	49.9 ± 3.8 [b]	13.3 ± 1.9 [a]	7.83 ± 1.36 [a]	0.16 ± 0.11 [a,b]
0.05	5.40 ± 0.14 [a]	49.8 ± 3.7 [b]	13.4 ± 1.8 [a]	7.82 ± 1.22 [a]	0.14 ± 0.11 [b]
0.10	5.41 ± 0.11 [a]	49.4 ± 3.9 [b]	13.5 ± 1.7 [a]	7.89 ± 1.15 [a]	0.14 ± 0.10 [b]
p	0.316	0.001	0.759	0.924	0.003
			SD		
0	5.26 ± 0.08 [d]	50.8 ± 3.6 [a]	13.0 ± 1.8 [b]	7.59 ± 1.21 [b]	0.04 ± 0.03 [d]
75	5.47 ± 0.08 [a]	50.3 ± 4.1 [a,b]	13.2 ± 1.6 [b]	7.51 ± 1.11 [b]	0.12 ± 0.06 [c]
150	5.40 ± 0.10 [c]	49.7 ± 3.7 [b]	13.2 ± 1.8 [b]	7.72 ± 1.06 [b]	0.20 ± 0.05 [b]
225	5.44 ± 0.10 [b]	49.5 ± 4.3 [b]	14.2 ± 1.8 [a]	8.52 ± 1.22 [a]	0.28 ± 0.09 [a]
p	<0.001	0.020	<0.001	<0.001	<0.001

FC—fat content; NC—nitrite content; JC—*Juniperus communis* L. essential oil (JEO) content; SD—storage day; Means ± Stdev with different letters [a–d] in the same column are significantly different ($p < 0.05$).

The fat content and storage time had a significant ($p < 0.05$) effect on the pH values. The samples produced with 15% of fat had a higher pH value. Regarding storage time, it can be observed that pH values inconsistently increased throughout storage, probably as the result of formation of amino-compounds during the proteolysis in fermented sausages [52,53]. The two-way (SD × JC), three-way (FC × NC × SD, FC × SD × JC) and four-way (FC × NC × SD × JC) interactions had a significant ($p < 0.05$–0.001) effect on the pH values (Table S2—Supplementary material). Values of pH ranged from 5.11 (FC = 25%, NC = 0 mg/kg, SD = 0, JC = 0.05 µL/g) to 5.63 (FC = 15%, NC = 150 mg/kg, SD = 75, JC = 0.10 µL/g). Similar results were observed by Kurćubić et al. [52] and Ozaki et al. [54] in fermented meat products.

Color is one of the key quality parameters for meat and meat products [16]. The instrumental parameters of color (L^*, a^* and b^*) are displayed in Table 2. The contents of fat, nitrite and JEO, as well as storage time, had a significant ($p < 0.05$) effect on L^* values. As expected, the samples produced with 15% of fat had lower L^* values. Moreover, storage time had the effect of decreasing L^* values, according with the findings of Pateiro et al. [55]. Finally, the addition of JEO decreased the L^* value, probably as the result of interactions among bioactive compounds of JEO (phenolics, terpenes) and myoglobin [30]. The two-way (FC × SD) and four-way (FC × NC × SD × JC) interactions were also significant ($p < 0.05$) for L^* values (Table S2). The L^* values ranged across a wide interval from 43.92 (FC = 15%, NC = 150 mg/kg, SD = 225, JC = 0.10 µL/g) to 56.64 (FC = 25%, NC = 0 mg/kg, SD = 0, JC = 0.00 µL/g).

Fat content and storage time had a significant ($p < 0.05$) effect on the a^* values. As expected, the samples produced with 15% of fat had higher a^* values. Concerning storage time, the increasing of a^* values after the 150th day of storage can be noticed. This is in accordance with the findings of Pateiro et al. [55]. The increase of a^* values could be related to the growth of the *Staphylococcus* species [56]. Faustman and Cassens [56] reported that enzymes (NADH-cytochrome b5 reductase systems, metmyoglobin reductase and nitrate reductase) of *S. carnosus* or *S. xylosus* can alter metmyoglobin to form red myoglobin derivatives and enhance the color of meat products. Two-way (FC × NC) and three-way interactions (FC × NC × SD and FC × SD × JC) suggested a significant ($p < 0.05$–0.01) effect of using both sodium nitrite (150 mg/kg) and JEO (0.10 µL/g) for enhancing the redness of low-fat (15%) dry fermented sausages (Table S2). The lowest (9.38) and the highest (15.95) a^* values were determined in the samples: FC = 25%, NC = 0 mg/kg, JC = 0.00 µL/g, SD = 0; FC= 15%, NC = 150 mg/kg, JC = 0.10 µL/g SD = 225. No significant ($p > 0.05$) four factor interaction was detected for the a^* value.

Fat content and storage time had a significant ($p < 0.05$) effect on the b^* values. Surprisingly, the samples produced with 15 fat had higher b^* values. After the 150th day of storage, the increase in b^* value can also be noticed. Similar findings were observed by Rubio et al. [57] for comparable meat products. Two-way (FC × SD, FC × JC, NC × JC, SD × JC), three-way (FC × NC × SD, NC × SD × JC) and four-way (FC × NC × SD × JC) interactions were significant ($p < 0.05$–0.001) for b^* values (Table S2). The lowest (5.88) and the highest (10.17) b^* value was detected in the samples: FC = 25%, NC = 0 mg/kg, SD = 75, JC = 0.01 µL/g; FC = 15%, NC = 150 mg/kg, SD = 225, JC = 0.01 µL/g.

3.3. TBARS Values of Dry Fermented Sausages

Lipid oxidation is one the most important parameters of quality for meat and meat products [30]. TBARS values of dry fermented sausages are presented in Table 2. The contents of nitrite and JEO and storage time had a significant ($p < 0.05$) effect on TBARS values. The inclusion of sodium nitrite (75 and 150 mg/kg) decreased TBARS values. This was probably the result of the antioxidant activity of sodium nitrite [11]. Furthermore, Honikel [11] reported that antioxidant activity of nitrites is associated with the ability of NO to fix and stabilize heme iron (Fe) of meat myoglobin, making it unavailable to catalyze reactions of oxidation. Also, Karwowska et al. [58] reported that the reduction of nitrites, from 150 to 50 mg/kg, increased TBARS values in cooked meat products. Moreover, samples produced with the addition of JEO (0.05 and 0.10 µL/g) had lower TBARS values compared to samples produced without JEO. This is the consequence of the strong antioxidant potential of JEO. Höferl et al. [25] reported that juniper berry oil significantly prevented the formation of lipid peroxidation by-products caused by TBA. Certain compounds, such as α-terpinene, γ-terpinene and α-terpinolene exhibit strong antioxidant activity in prevention of lipid oxidation which could be compared with α-tocopherol [59]. On the other hand, certain compounds from JEO (pinene, sabinene and limonene) have a rather weak effect. Similar findings of the antioxidant effects of JEO in meat products were observed in our previous study [30]. As expected, storage time had a significant ($p < 0.05$) effect on increasing TBARS values, as the result of lipid oxidation [2]. The two-way interactions (FC × SD and NC × SD) were significant ($p < 0.05$) for TBARS values. Moreover, three-way (FC × NC × SD and FC × SD × JC) and four-way interactions had a significant ($p < 0.05$–0.001) effect on TBARS values (Table S2). The highest TBARS value (0.398 mg MDA/kg) was observed in the sample: FC = 15%, NC = 0 mg/kg, SD = 225, JC = 0.05 µL/g. At the same time, TBARS values in the samples were: FC = 25%, NC = 75 mg/kg, SD = 225, JC = 0.01 µL/g and FC = 25%, NC = 75 mg/kg, SD = 225, JC = 0.05 µL/g amounted 0.117 and 0.110 mg MDA/kg, respectively. According to Melton [60], the TBARS value of 0.3 mg MDA/kg is marked as the threshold for rancidity of meat products. The obtained results suggested that interaction between sodium nitrite (75 mg/kg) and JEO (0.01 and 0.05 µL/g) efficiently reduced the lipid oxidation in high-fat (25%) dry fermented sausages. Regarding the strong lipo-solubility of terpenoid compounds (e.g., β-myrcene, sabinene, β-pinene, limonene) JEO possessed a higher antioxidant potential in dry fermented sausages produced with higher fat content (25%).

3.4. Microbiological Analysis of Dry Fermented Sausages

Total plate count (TPC) and lactic acid bacteria (LAB) of dry fermented sausages are presented in Table 3.

Table 3. Microbiological quality of dry fermented sausages.

	TPC (log CFU/g)	LAB (log CFU/g)
	FC (%)	
15	5.55 ± 0.73 [a]	5.71 ± 0.87 [a]
25	5.43 ± 0.84 [a]	5.53 ± 0.87 [a]
p	0.459	0.336
	NC (mg/kg)	
0	5.51 ± 0.77 [a]	5.64 ± 0.75 [a]
75	5.57 ± 0.77 [a]	5.67 ± 0.80 [a]
150	5.40 ± 0.81 [a]	5.55 ± 1.04 [a]
p	0.705	0.851
	JC (µL/g)	
0	5.51 ± 0.96 [a]	5.62 ± 0.70 [a]
0.01	5.65 ± 0.66 [a]	5.68 ± 0.93 [a]
0.05	5.44 ± 0.68 [a]	5.55 ± 0.93 [a]
0.10	5.37 ± 0.68 [a]	5.63 ± 0.99 [a]
p	0.642	0.968
	SD	
0	5.10 ± 0.38 [c]	6.47 ± 0.51 [a]
75	4.74 ± 0.53 [d]	5.81 ± 0.67 [b]
150	6.23 ± 0.47 [a]	5.52 ± 0.44 [b]
225	5.91 ± 0.63 [b]	4.68 ± 0.71 [c]
p	<0.001	<0.001

TPC—total plate count; LAB—lactic acid bacteria; FC—fat content; NC—nitrite content; JC—JEO content; SD—storage day; Means ± Stdev with different letters [a–d] in the same column are significantly different ($p < 0.05$).

The contents of fat and nitrite did not exhibit a significant ($p > 0.05$) effect on TPC and LAB. In the case of JEO, the addition of this essential oil (≥ 0.05 μL/g) had a tendency to reduce the TPC (for 0.14 log cfu/g), but differences among the samples were not significant ($p > 0.05$). Hence, further optimization with a higher concentration is necessary. Moreover, in our previous study [30] we found that JEO addition (≥ 0.10 μL/g) efficiently reduced TPC in cooked pork sausages. On the contrary, Selim et al. [29] showed that JEO had no effect on the reduction of microbial growth in fresh beef meat. The antimicrobial potential of essential oil depends of its chemical shape. Generally, monoterpenes from the JEO (α-pinene, β-pinene, sabinene, γ-terpinene, β-myrcene, and limonene) are not efficient antimicrobials when applied singly [61]. However, a mixture of these compounds with the presence of other JEO constituents present as a minor content could result in additive or synergistic antimicrobial effects [62]. As mentioned, the utilization of a novel extraction technique (e.g., SFE) could be a good solution in order to improve the chemical profile of JEO [50]. Orav et al. [50] found that JEO obtained using SFE contained less monoterpenes (5.1%) and more sesquiterpenes and oxygenated sesquiterpenes (69.8%) with a higher antimicrobial potential. As expected, storage time significantly ($p < 0.05$) affected TPC and LAB. TPC decreased during the first 75 days of storage, then increased until the 150th day of storage and again decreased until the end of storage. This trend could be related to the reduction of LAB during storage, especially after the 150th day of storage. As mentioned, the population of LAB decreased throughout storage, probably as the consequences of low storage temperature (15 °C) and the exhaustion of sugar [63]. No significant ($p > 0.05$) two, three or four factor interactions were detected for both TPC and LAB (Table S2). It can also be noticed that foodborne pathogens (*Escherichia coli*, *Listeria monocytogenes*, *Salmonella* spp. and sulfite-reducing clostridia) were not detected in any sample throughout the storage. The obtained results suggested that all treatments provided a satisfactory microbiological quality according to EU regulation [64].

3.5. Texture Analysis of Dry Fermented Sausages

Results of instrumental determination of texture characteristics are shown in Table 4.

As expected, fat content significantly changed ($p < 0.05$) the texture parameters (hardness, springiness, cohesiveness and chewiness). Samples with lower fat content showed a higher value of hardness and chewiness. An increase in hardness and chewiness as fat content decreases in dry fermented sausages was also reported by other authors [65,66], probably as a result of a more pronounced moisture loss in sausages with higher proportions of lean meat [66]. During the storage, hardness and chewiness values showed significant ($p < 0.05$) increase until the 150th day of storage, followed by decrease till day 225. Springiness value showed constant increase, while cohesiveness showed constant decrease until the 150th day of storage. Rubio et al. [67] reported increase of hardness, springiness, cohesiveness and chewiness of dry fermented sausage over the whole storage period (till 210 days), while Severini et al. [68] reported decrease in firmness, due to proteolysis. In the case of nitrites, it can be observed that the addition of sodium nitrite significantly ($p < 0.05$) affected the cohesiveness of dry fermented sausages. A similar finding was observed by Dong et al. [69] in cooked pork sausages. Moreover, Villaverdre et al. [70] found that the sodium nitrite addition at the levels of 75 and 150 mg/kg increased the hardness of fermented sausages. This could be related to the ability of sodium nitrite to promote protein oxidation and Schiff base formation [70]. Regarding JEO, it has been noticed that JEO addition had no impact on the texture parameters (hardness, cohesiveness and chewiness) of dry fermented sausages. Similarly, Viuda-Martos et al. [71] reported that rosemary essential oil has no effect on texture parameters of cooked sausages. Two-way (FC × SD) and four-way interactions were also significant ($p < 0.05$–0.001) for all texture parameters (Table S2). The main texture parameters (hardness and chewiness) ranged in interval from 3539 (FC = 25%, NC = 75 mg/kg, SD = 0, JC = 0.00 μL/g) to 10,990 g (FC = 15%, NC = 0 mg/kg, SD = 150, JC = 0.05 μL/g) and from 926 (FC = 25%, NC = 0 mg/kg, SD = 225, JC = 0.05 μL/g) to 2759 g (FC = 15%, NC = 150 mg/kg, SD = 225, JC = 0.05 μL/g), respectively. Similar results for hardness of different dry fermented sausages were observed by Triki et al. [65] and Rubio et al. [67].

Table 4. Texture parameters of dry fermented sausages.

	Hardness (g)	Springiness	Cohesiveness	Chewiness (g)
	FC (%)			
15	7579 ± 1611 [a]	0.488 ± 0.05 [a]	0.511 ± 0.03 [b]	1902 ± 489 [a]
25	5282 ± 1020 [b]	0.505 ± 0.05 [b]	0.525 ± 0.04 [a]	1407 ± 321 [b]
p	<0.001	<0.001	<0.001	<0.001
	NC (mg/kg)			
0	6271 ± 1707 [a]	0.489 ± 0.05 [a]	0.504 ± 0.04 [c]	1552 ± 438 [b]
75	6542 ± 1874 [a]	0.500 ± 0.05 [a]	0.519 ± 0.04 [b]	1694 ± 508 [a]
150	6462 ± 1712 [a]	0.500 ± 0.05 [a]	0.531 ± 0.03 [a]	1713 ± 480 [a]
p	0.377	0.079	<0.001	0.005
	JC (µL/g)			
0	6269 ± 1752 [a]	0.508 ± 0.05 [a]	0.523 ± 0.04 [a]	1654 ± 449 [a]
0.01	6502 ± 1570 [a]	0.492 ± 0.05 [b]	0.516 ± 0.03 [a]	1671 ± 467 [a]
0.05	6672 ± 211 [a]	0.492 ± 0.05 [b]	0.512 ± 0.04 [a]	1675 ± 562 [a]
0.10	6264 ± 1556 [a]	0.494 ± 0.04 [b]	0.521 ± 0.04 [a]	1612 ± 437 [a]
p	0.219	0.022	0.093	0.739
	SD			
0	4730 ± 915 [c]	0.444 ± 0.04 [c]	0.553 ± 0.04 [a]	1153 ± 202 [c]
75	6649 ± 1341 [b]	0.505 ± 0.03 [b]	0.521 ± 0.03 [b]	1741 ± 323 [b]
150	7518 ± 1641 [a]	0.516 ± 0.03 [a]	0.503 ± 0.03 [c]	1940 ± 416 [a]
225	6801 ± 1718 [b]	0.521 ± 0.03 [a]	0.496 ± 0.03 [c]	1777 ± 506 [b]
p	<0.001	<0.001	<0.001	<0.001

FC—fat content; NC—nitrite content; JC—JEO content; SD—storage day; Means ± Stdev with different letters [a-c] in the same column are significantly different ($p < 0.05$).

3.6. Sensory Analysis of Dry Fermented Sausages

Results of sensory analysis are shown in Table 5.

The fat content, JEO content and storage time had a significant ($p < 0.05$) effect on the sensory attribute of color. Two-way (FC × NC, FC × SD, NC × SD, FC × JC, SD × JC), three-way (FC × NC × SD, FC × NC × JC, FC × SD × JC, NC × SD × JC) and four-way interactions were also significant ($p < 0.05$–0.001) for this sensory attribute (Table S2). Moreover, nitrite content, JEO content and storage had a significant ($p < 0.05$) effect on sensory attribute of odor. Furthermore, two-way (FC × NC, NC × JC, SD × JC) and three-way (FC × NC × JC, NC × SD × JC) interactions had a significant ($p < 0.05$–40.001) effect on odor (Table S2). It should also be noticed that the numerical data for the sensory attributes of color and odor did not exceed the values of 2.0 (slight differences, less than 1.67 (color) and 1.78 (odor)), in any samples. Hence, the obtained results suggested that the fluctuations of fat, sodium nitrite and JEO, as well as storage time had no negative impact on these sensory attributes. Nitrite content, JEO addition and storage time had a significant ($p < 0.05$) effect on sensory attribute of flavor. Two-way (FC × JC, NC × JC, SD × JC), three-way (FC × NC × JC, FC × SD × JC, NC × SD × JC) and four-way interactions had also a significant ($p < 0.05$–0.001) effect on the flavor (Table S2). The highest differences (>3, higher than moderate) of typical flavor were observed in the samples: FC = 25%, NC = 150 mg/kg, SD = 225, JC = 0.10 µL/g; FC = 15%, NC = 150 mg/kg, SD = 225, JC = 0.10 µL/g. Regarding JEO content of 0.05 µL/g, the highest difference (1.33) was observed in samples: FC = 15%, NC = 150 mg/kg, SD = 225, JC = 0.05 µL/g; FC = 15%, NC = 75 mg/kg, SD = 150, JC = 0.05 µL/g). This difference could be the result of interaction among the sodium nitrite and terpenoid-compounds of JEO. In our previous study we also determined that a high percentage of JEO had a significant effect on the strong aroma of cooked pork sausages [30]. Using novel extraction techniques (e.g., supercritical fluid extraction) at optimum conditions results in extracts which possess a strong antioxidant and antimicrobial potential, as well as mild flavor, which enables their application at lower concentration in meat processing [17,23].

Table 5. Sensory parameters of dry fermented sausages.

	Color	Odor	Flavor
	FC (%)		
15	0.32 ± 0.64 [b]	0.34 ± 0.67 [a]	0.89 ± 1.18 [a]
25	0.77 ± 0.77 [a]	0.29 ± 0.68 [a]	0.81 ± 1.12 [a]
p	<0.001	0.164	0.137
	NC (mg/kg)		
0	0.58 ± 0.78 [a]	0.18 ± 0.46 [b]	0.74 ± 1.06 [b]
75	0.57 ± 0.77 [a]	0.36 ± 0.73 [a]	0.85 ± 1.17 [a,b]
150	0.49 ± 0.67 [a]	0.41 ± 0.77 [a]	0.98 ± 1.20 [a]
p	0.063	<0.001	0.002
	JC (μL/g)		
0	0.00 ± 0.00 [c]	0.00 ± 0.00 [d]	0.00 ± 0.00 [d]
0.01	0.67 ± 0.72 [b]	0.10 ± 0.36 [c]	0.17 ± 0.49 [c]
0.05	0.72 ± 0.81 [b]	0.22 ± 0.50 [b]	0.80 ± 0.77 [b]
0.10	0.80 ± 0.79 [a]	0.94 ± 0.94 [a]	2.44 ± 0.86 [a]
p	<0.001	<0.001	<0.001
	SD		
0	0.34 ± 0.65 [c]	0.27 ± 0.68 [b]	0.70 ± 1.07 [b]
75	0.48 ± 0.67 [b]	0.15 ± 0.41 [c]	0.76 ± 1.09 [b]
150	0.91 ± 0.82 [a]	0.46 ± 0.77 [a]	0.94 ± 1.10 [a]
225	0.45 ± 0.69 [b]	0.38 ± 0.73 [a]	1.01 ± 1.29 [a]
p	<0.001	<0.001	<0.001

FC—at content; NC—nitrite content; JC—JEO content; SD—storage day; Means ± Stdev with different letters [a–d] in the same column are significantly different ($p < 0.05$).

4. Conclusions

Monoterpene hydrocarbon β-myrcene (14.12%) was the most abundant compound identified in JEO. The sausages produced with a lower fat content were significantly darker and redder ($p < 0.05$). Moreover, the values of hardness and chewiness were significantly ($p < 0.05$) higher in the samples produced with a lower fat content. The variations in the contents of nitrite and JEO had no negative impact on the color and texture parameters of dry fermented sausages. No foodborne pathogens were detected in any samples. The highest concentration of JEO (0.10 μL/g) had negative impact on flavor. The addition of JEO (0.01 and 0.05 μL/g) combined with reduced concentration of sodium nitrite (75 mg/kg) efficiently retarded the lipid oxidation of high-fat (25%) dry fermented sausages during 225 days of storage. Hence, JEO with evident antioxidant potential could be used as a partial replacement for sodium nitrite in fermented sausages processing. In order to enhance the antimicrobial potential of JEO, the usage of novel extraction technique (e.g., SFE) could be an effective solution. Further investigations are needed to analyze the synergistic effects of different natural extracts, isolated from various plant sources, on improving the quality and shelf-life of meat products.

Supplementary Materials: The following are available online at http://www.mdpi.com/2304-8158/9/8/1066/s1. Table S1. Proximate chemical composition of dry ferment sausages; Table S2. The effect of two-way, three-way and four-way interactions among processing parameters on the quality of dry fermented sausages expressed as *p*-value.

Author Contributions: Conceptualization, V.T.; methodology, V.T. and B.Š.; software, V.T.; formal analysis, B.Š., J.S., S.K.-T., B.P. and M.J.; investigation, V.T. and B.Š.; resources, J.S., N.P., D.V.; writing—original draft preparation, B.Š. and B.P.; writing—review and editing, V.T., V.Đ., N.P. and A.M.; supervision, D.V.; project administration, V.T. and A.M.; All authors have read and agreed to the published version of the manuscript.

Funding: This research was funded by the Ministry of Education, Science and Technological Development, Republic of Serbia, under Grant 451-03-68/2020-14/200134. Also, this research has been done in liaison with the activities defined by the grant for the establishment and implementation of the research-innovation-scientific program "Centre of Excellence (CoE) for digitalization of microbial food safety risk assessment and quality parameters for accurate food authenticity certification (FoodHub)", financed by the Ministry of Science of Montenegro, under the Grant No. 01-3660/2.

Conflicts of Interest: The authors declare no conflict of interest.

References

1. Chen, X.; Mi, R.; Qi, B.; Xiong, S.; Li, J.; Qu, C.; Qiao, X.; Chen, W.; Wang, S. Effect of proteolytic starter culture isolated from Chinese Dong fermented pork (Nan x Wudl) on microbiological, biochemical and organoleptic attributes in dry fermented sausages. *Food Sci. Hum. Wellness* **2020**, *3*. in press.

2. Šojić, B.V.; Petrović, L.S.; Mandić, A.I.; Sedej, I.J.; Džinić, N.R.; Tomović, V.M.; Jokanović, M.; Tasić, T.; Škaljac, S.; Ikonić, P.M. Lipid oxidative changes in traditional dry fermented sausage *Petrovská klobása* during storage. *Hem. Ind.* **2014**, *68*, 27–34. [CrossRef]

3. Van Ba, H.; Seo, H.W.; Cho, S.H.; Kim, Y.S.; Kim, J.H.; Ham, J.S.; Park, P.Y.; Pil-Nam, S. Effects of extraction methods of shiitake by-products on their antioxidant and antimicrobial activities in fermented sausages during storage. *Food Control* **2017**, *79*, 109–118. [CrossRef]

4. Oswell, N.J.; Harshavardhan, T.; Ronald, B.P. Practical use of natural antioxidants in meat products in the US: A review. *Meat Sci.* **2018**, *145*, 469–479. [CrossRef] [PubMed]

5. Tomović, V.; Jokanović, M.; Šojić, B.; Škaljac, S.; Ivić, M. Plants as natural antioxidants for meat products. In Proceedings of the 59th International Meat Industry Conference MEATCON2017, IOP Conference Series: Earth and Environmental Science, Zlatibor, Serbia, 1–4 October 2017.

6. Jayasena, D.D.; Cheorun, J. Essential oils as potential antimicrobial agents in meat and meat products: A review. *Trends Food Sci. Technol.* **2013**, *34*, 96–108. [CrossRef]

7. Gómez, B.; Barba, F.J.; Domínguez, R.; Putnik, P.; Bursać Kovačević, D.; Pateiro, M.; Toldrá, F.; Lorenzo, J.M. Microencapsulation of antioxidant compounds through innovative technologies and its specific application in meat processing. *Trends Food Sci. Technol.* **2018**, *82*, 135–147. [CrossRef]

8. Abd Hamid, N.F.H.; Khan, M.M.; Hoon, L.L. Assessment of nitrate, nitrite and chloride in selected cured meat products and their exposure to school children in Brunei Darussalam. *J. Food Compos. Anal.* **2020**, *91*, 103520. [CrossRef]

9. European Commission (EC). Directive 2006/52/EC of the European Parliament and of the Council of 5 July 2006 amending Directive 95/2/EC on food additives other than colours and sweeteners and Directive 94/35/EC on sweeteners for use in foodstuffs. *Off. J. Eur. Union* **2006**, *204*, 10–22.

10. Choi, J.H.; Song, D.H.; Hong, J.S.; Ham, Y.K.; Ha, J.H.; Choi, Y.S.; Kim, H.W. Nitrite scavenging impact of fermented soy sauce in vitro and in a pork sausage model. *Meat Sci.* **2019**, *151*, 36–42. [CrossRef]

11. Honikel, K.O. The use and control of nitrate and nitrite for the processing of meat products. *Meat Sci.* **2008**, *78*, 68–76. [CrossRef]

12. Jin, S.K.; Choi, J.S.; Yang, H.S.; Park, T.S.; Yim, D.G. Natural curing agents as nitrite alternatives and their effects on the physicochemical, microbiological properties and sensory evaluation of sausages during storage. *Meat Sci.* **2018**, *146*, 34–40. [CrossRef] [PubMed]

13. Herrmann, S.S.; Granby, K.; Duedahl-Olesen, L. Formation and mitigation of N-nitrosamines in nitrite preserved cooked sausages. *Food Chem.* **2015**, *174*, 516–526. [CrossRef] [PubMed]

14. Slima, S.B.; Ktari, N.; Trabelsi, I.; Triki, M.; Feki-Tounsi, M.; Moussa, H.; Salah, R.B. Effect of partial replacement of nitrite with a novel probiotic *Lactobacillus plantarum* TN8 on color, physico-chemical, texture and microbiological properties of beef sausages. *LWT Food Sci. Technol.* **2017**, *86*, 219–226. [CrossRef]

15. Posthuma, J.A.; Rasmussen, F.D.; Sullivan, G.A. Effects of nitrite source, reducing compounds, and holding time on cured color development in a cured meat model system. *LWT Food Sci. Technol.* **2018**, *95*, 47–50. [CrossRef]

16. Šojić, B.; Pavlić, B.; Ikonić, P.; Tomović, V.; Ikonić, B.; Zeković, Z.; Kocić-Tanackov, S.; Jokanović, M.; Škaljac, S.; Ivić, M. Coriander essential oil as natural food additive improves quality and safety of cooked pork sausages with different nitrite levels. *Meat Sci.* **2019**, *157*, 107879. [CrossRef]

17. Šojić, B.; Pavlić, B.; Tomović, V.; Ikonić, P.; Zeković, Z.; Kocić-Tanackov, S.; Đurović, S.; Škaljac, S.; Jokanović, M.; Ivić, M. Essential oil versus supercritical fluid extracts of winter savory (*Satureja montana* L.)—Assessment of the oxidative, microbiological and sensory quality of fresh pork sausages. *Food Chem.* **2019**, *287*, 280–286. [CrossRef]

18. Ribeiro-Santos, R.; Andrade, M.; de Melo, N.R.; Sanches-Silva, A. Use of essential oils in active food packaging: Recent advances and future trends. *Trends Food Sci. Technol.* **2017**, *61*, 132–140. [CrossRef]

19. Burt, S. Essential oils: Their antibacterial properties and potential applications in foods—A review. *Int. J. Food Microbiol.* **2004**, *94*, 223–253. [CrossRef]

20. Meira, N.V.; Holley, R.A.; Bordin, K.; de Macedo, R.E.; Luciano, F.B. Combination of essential oil compounds and phenolic acids against *Escherichia coli* O157: H7 in vitro and in dry-fermented sausage production. *Int. J. Food Microbiol.* **2017**, *260*, 59–64. [CrossRef]

21. Kocić Tanackov, S.; Dimić, G.; Đerić, N.; Mojović, L.; Tomović, V.; Šojić, B.; Đukić-Vuković, A.; Pejin, J. Growth control of molds isolated from smoked fermented sausages using basil and caraway essential oils, in vitro and in vivo. *LWT Food Sci. Technol.* **2020**, *123*, 109095. [CrossRef]

22. Soncu, E.D.; Özdemir, N.; Arslan, B.; Küçükkaya, S.; Soyer, A. Contribution of surface application of chitosan–thyme and chitosan–rosemary essential oils to the volatile composition, microbial profile, and physicochemical and sensory quality of dry-fermented sausages during storage. *Meat Sci.* **2020**, *166*, 108127. [CrossRef] [PubMed]

23. Šojić, B.; Pavlić, B.; Tomović, V.; Kocić-Tanackov, S.; Đurović, S.; Zeković, Z.; Belović, M.; Torbica, A.; Jokanović, M.; Uromović, N.; et al. Tomato pomace extract and organic peppermint essential oil as effective sodium nitrite replacement in cooked pork sausages. *Food Chem.* **2020**, *330*, 127202. [CrossRef] [PubMed]

24. Huang, L.; Zeng, X.; Sun, Z.; Wu, A.; He, J.; Dang, Y.; Pan, D. Production of a safe cured meat with low residual nitrite using nitrite substitutes. *Meat Sci.* **2020**, *162*, 108027. [CrossRef] [PubMed]

25. Höferl, M.; Stoilova, I.; Schmidt, E.; Wanner, J.; Jirovetz, L.; Trifonova, D.; Krastev, L.; Krastanov, A. Chemical composition and antioxidant properties of Juniper berry (*Juniperus communis* L.) essential oil. Action of the essential oil on the antioxidant protection of Saccharomyces cerevisiae model organism. *Antioxidants* **2014**, *3*, 81–98. [CrossRef] [PubMed]

26. Pandey, S.; Tiwari, S.; Kumar, A.; Niranjan, A.; Chand, J.; Lehri, A.; Chauhan, P.S. Antioxidant and anti-aging potential of Juniper berry (*Juniperus communis* L.) essential oil in Caenorhabditis elegans model system. *Ind. Crop. Prod.* **2018**, *120*, 113–122. [CrossRef]

27. Vasilijević, B.; Mitić-Ćulafić, D.; Djekic, I.; Marković, T.; Knežević-Vukčević, J.; Tomasevic, I.; Nikolić, B. Antibacterial effect of *Juniperus communis* and *Satureja montana* essential oils against *Listeria monocytogenes* in vitro and in wine marinated beef. *Food Control* **2019**, *100*, 247–256. [CrossRef]

28. Schelz, Z.; Molnar, J.; Hohmann, J. Antimicrobial and antiplasmid activities of essential oils. *Fitoterapia* **2006**, *77*, 279–285. [CrossRef]

29. Selim, S. Antimicrobial activity of essential oils against Vancomycin-Resistant enterococci (VRE) and *Escherichia coli* O157: H7 in feta soft cheese and minced beef meat. *Braz. J. Microbiol.* **2011**, *42*, 187–196. [CrossRef]

30. Šojić, B.; Tomović, V.; Jokanović, M.; Ikonić, P.; Džinić, N.; Kocić-Tanackov, S.; Popović, L.; Tasić, T.; Đuranović, J.; Živković Šojić, N. Antioxidant activity of *Juniperus communis* L. essential oil in cooked pork sausages. *Czech J. Food Sci.* **2017**, *35*, 189–193.

31. Pavlić, B.; Bera, O.; Teslić, N.; Vidović, S.; Parpinello, G.; Zeković, Z. Chemical profile and antioxidant activity of sage herbal dust extracts obtained by supercritical fluid extraction. *Ind. Crop. Prod.* **2018**, *120*, 305–312. [CrossRef]

32. ISO. *Meat and Meat Products—Determination of Moisture Content (Reference Method)*; ISO 1442:1997; International Organization for Standardization: Geneva, Switzerland, 1997.

33. ISO. *Meat and Meat Products—Determination of Nitrogen Content*; ISO 937:1978; International Organization for Standardization: Geneva, Switzerland, 1978.

34. ISO. *Meat and Meat Products—Determination of Total Fat Content*; ISO 1443:1973; International Organization for Standardization: Geneva, Switzerland, 1973.

35. ISO. *Meat and Meat Products—Determination of Total Ash*; ISO 936:1998; International Organization for Standardization: Geneva, Switzerland, 1998.

36. Ikonić, P.; Jokanović, M.; Petrović, L.; Tasić, T.; Škaljac, S.; Šojić, B.; Džinić, N.; Tomović, V.; Tomić, J.; Danilović, B.; et al. Effect of starter culture addition and processing method on proteolysis and texture profile of traditional dry-fermented sausage *Petrovská klobása*. *Int. J. Food Proper.* **2016**, *19*, 1924–1937. [CrossRef]

37. Botsoglou, N.A.; Fletouris, D.J.; Papageorgiou, G.E.; Vassilopoulos, V.N.; Mantis, A.J.; Trakatellis, A.G. Rapid, sensitive and specific thiobarbituric acid method for measurement of lipid peroxidation in animal tissue, food and feedstuff samples. *J. Agric. Food Chem.* **1994**, *42*, 1931–1937. [CrossRef]

38. ISO. *Microbiology of the Food Chain—Horizontal Method for the Enumeration of Microorganisms—Part 1: Colony Count at 30 °C by the Pour Plate Technique*; ISO 4833-1:2013; International Organization for Standardization: Geneva, Switzerland, 2013.

39. ISO. *Microbiology of Food and Animal Feeding Stuffs—Horizontal Method for the Enumeration of Mesophilic Lactic Acid Bacteria—Colony-Count Technique at 30 °C*; ISO 15214:1998; International Organization for Standardization: Geneva, Switzerland, 1998.

40. ISO. *Microbiology of Food and Animal Feeding Stuffs—Horizontal Method for the Enumeration of beta-glucuronidase-positive Escherichia coli—Part 2: Colony-Count Technique at 44 Degrees C Using 5-bromo-4-chloro-3-indolyl beta-D-glucuronide*; ISO 16649-2:2001; International Organization for Standardization: Geneva, Switzerland, 2001.

41. ISO. *Microbiology of the Food Chain—Horizontal Method for the Detection and Enumeration of Listeria Monocytogenes and of Listeria spp.—Part 2: Enumeration Method*; ISO 11290-2:2017; International Organization for Standardization: Geneva, Switzerland, 2017.

42. ISO. *Microbiology of the food chain—Horizontal Method for the Detection, Enumeration and Serotyping of Salmonella—Part 1: Detection of Salmonella spp*; ISO 6579-1:2017; International Organization for Standardization: Geneva, Switzerland, 2017.

43. ISO. *Microbiology of Food and Animal Feeding Stuffs—Horizontal Method for the Enumeration of Sulfite-Reducing Bacteria Growing Under Anaerobic Conditions*; ISO 15213:2003; International Organization for Standardization: Geneva, Switzerland, 2003.

44. ISO. *Sensory Analysis—General Guidelines for The Selection, Training And Monitoring of Selected Assessors and Expert Sensory Assessors*; ISO 8586:2012; International Organization for Standardization: Geneva, Switzerland, 2012.

45. ISO. *Sensory Analysis—General Guidance for the Design of Test Rooms*; ISO 8589: 2007; International Organization for Standardization: Geneva, Switzerland, 2007.

46. Meilgaard, M.; Civille, G.V.; Carr, B.T. Selection and training of panel members. In *Sensory Evaluation Techniques*, 3rd ed.; Taylor & Francis Group: Boca Raton, FL, USA, 1999; pp. 174–176.

47. Falcão, S.; Bacém, I.; Igrejas, G.; Rodrigues, P.J.; Vilas-Boas, M.; Amaral, J.S. Chemical composition and antimicrobial activity of hydrodistilled oil from juniper berries. *Ind. Crop. Prod.* **2018**, *124*, 878–884. [CrossRef]

48. Radoukova, T.; Zheljazkov, V.D.; Semerdjieva, I.; Dincheva, I.; Stoyanova, A.; Kačániová, M.; Salamon, I. Differences in essential oil yield, composition, and bioactivity of three juniper species from Eastern Europe. *Ind. Crop. Prod.* **2018**, *124*, 643–652. [CrossRef]

49. Zheljazkov, V.D.; Astatkie, T.; Jeliazkova, E.A.; Heidel, B.; Ciampa, L. Essential oil content, composition and bioactivity of Juniper species in Wyoming, United States. *Nat. Prod. Commun.* **2017**, *12*, 1934578X1701200215. [CrossRef]

50. Orav, A.; Koel, M.; Kailas, T.; Müürisepp, M. Comparative analysis of the composition of essential oils and supercritical carbon dioxide extracts from the berries and needles of Estonian juniper (*Juniperus communis* L.). *Proc. Chem.* **2010**, *2*, 161–167. [CrossRef]

51. Marković, M.S.; Radosavljević, D.B.; Pavićević, V.P.; Ristić, M.S.; Milojević, S.Ž.; Bošković-Vragolović, N.M.; Veljković, V.B. Influence of common juniper berries pretreatment on the essential oil yield, chemical composition and extraction kinetics of classical and microwave-assisted hydrodistillation. *Ind. Crop. Prod.* **2018**, *122*, 402–413. [CrossRef]

52. Kurćubić, V.S.; Mašković, P.Z.; Vujić, J.M.; Vranić, D.V.; Vesković-Moračanin, S.M.; Okanović, Đ.G.; Lilić, S.V. Antioxidant and antimicrobial activity of *Kitaibelia vitifolia* extract as alternative to the added nitrite in fermented dry sausage. *Meat Sci.* **2014**, *97*, 459–467. [CrossRef]

53. Lashgari, S.S.; Noorolahi, Z.; Sahari, M.A.; Ahmadi Gavlighi, H. Improvement of oxidative stability and textural properties of fermented sausage via addition of pistachio hull extract. *Food Sci. Nutr.* **2020**, in press. [CrossRef]

54. Ozaki, M.M.; Munekata, P.E.; de Souza Lopes, A.; do Nascimento, M.D.S.; Pateiro, M.; Lorenzo, J.M.; Pollonio, M.A.R. Using chitosan and radish powder to improve stability of fermented cooked sausages. *Meat Sci.* **2020**, *167*, 108165. [CrossRef]

55. Pateiro, M.; Bermúdez, R.; Lorenzo, J.M.; Franco, D. Effect of addition of natural antioxidants on the shelf-life of *"Chorizo"*, a Spanish dry-cured sausage. *Antioxidants* **2015**, *4*, 42–67. [CrossRef] [PubMed]

56. Faustman, C.; Cassens, R.G. The biochemical basis for discoloration in fresh meat: A review. *J. Musc. Foods* **1990**, *1*, 217–243. [CrossRef]

57. Rubio, B.; Martínez, B.; García-Cachán, M.D.; Rovira, J.; Jaime, I. Effect of the packaging method and the storage time on lipid oxidation and colour stability on dry fermented sausage *salchichón* manufactured with raw material with a high level of mono and polyunsaturated fatty acids. *Meat Sci.* **2008**, *80*, 1182–1187. [CrossRef]

58. Karwowska, M.; Kononiuk, A.; Wójciak, K.M. Impact of Sodium Nitrite Reduction on Lipid Oxidation and Antioxidant Properties of Cooked Meat Products. *Antioxidants* **2020**, *9*, 9. [CrossRef] [PubMed]

59. Ruberto, G.; Baratta, M.T. Antioxidant activity of selected essential oil components in two lipid model systems. *Food Chem.* **2000**, *69*, 167–174. [CrossRef]

60. Melton, S.L. Methodology for following lipid oxidation in muscle foods. *Food Technol.* **1983**, *38*, 105–111.

61. Dorman, H.J.D.; Deans, S.G. Antimicrobial agents from plants: Antibacterial activity of plant volatile oils. *J. Appl. Microbiol.* **2000**, *88*, 308–316. [CrossRef]

62. Ložienė, K.; Venskutonis, P.R. Juniper (*Juniperus communis* L.) oils. In *Essential Oils in Food Preservation, Flavor and Safety*; Academic Press: Cambridge, MA, USA, 2016; pp. 495–500.

63. Hussein, F.H.; Razavi, S.H. Physicochemical properties and sensory evaluation of reduced fat fermented functional beef sausage. *Appl. Food Biotechnol.* **2017**, *4*, 93–102.

64. European Community (EC). Commission regulation (EC) No. 2073/2005. Microbiological criteria for foodstuffs. *Off. J. Eur. Union* **2005**, *338*, 1–26.

65. Triki, M.; Herrero, A.M.; Rodríguez-Salas, L.; Jiménez-Colmenero, F.; Ruiz-Capillas, C. Chilled storage characteristics of low-fat, n-3 PUFA-enriched dry fermented sausage reformulated with a healthy oil combination stabilized in a konjac matrix. *Food Control* **2016**, *31*, 158–165. [CrossRef]

66. Fonseca, S.; Gómez, M.; Domínguez, R.; Lorenzo, J.M. Physicochemical and sensory properties of Celta dry-ripened *"salchichón"* as affected by fat content. *Grasas y Aceites* **2015**, *66*, e059.

67. Rubio, B.; Martínez, B.; Sánchez, M.J.; García-Cachán, M.D.; Rovira, J.; Jaime, I. Study of the shelf life of a dry fermented sausage *"salchichon"* made from raw material enriched in monounsaturated and polyunsaturated fatty acids and stored under modified atmospheres. *Meat Sci.* **2007**, *76*, 128–137. [CrossRef] [PubMed]

68. Severini, C.; De Pilli, T.; Baiano, A. Partial substitution of pork backfat with extra-virgin olive oil in 'salami' products: Effects on chemical, physical and sensorial quality. *Meat Sci.* **2003**, *64*, 323–331. [CrossRef]

69. Dong, Q.L.; Tu, K.; Guo, L.Y.; Yang, J.L.; Wang, H.; Chen, Y.Y. The effect of sodium nitrite on the textural properties of cooked sausage during cold storage. *J. Texture Stud.* **2007**, *38*, 537–554. [CrossRef]

70. Villaverde, A.; Ventanas, J.; Estévez, M. Nitrite promotes protein carbonylation and Strecker aldehyde formation in experimental fermented sausages: Are both events connected? *Meat Sci.* **2014**, *98*, 665–672. [CrossRef] [PubMed]

71. Viuda-Martos, M.; Ruiz-Navajas, Y.; Fernández-López, J.; Pérez-Álvarez, J.A. Effect of added citrus fibre and spice essential oils on quality characteristics and shelf-life of *mortadella*. *Meat Sci.* **2010**, *85*, 568–576. [CrossRef]

 *foods*

Article

Phosphate Elimination in Emulsified Meat Products: Impact of Protein-Based Ingredients on Quality Characteristics

Olivier Goemaere, Seline Glorieux, Marlies Govaert, Liselot Steen and Ilse Fraeye *

Research Group for Technology and Quality of Animal Products, Department of Microbial and Molecular Systems, Leuven Food Science and Nutrition Research Centre (LFoRCe), KU Leuven Ghent Technology Campus, Gebroeders De Smetstraat 1, B-9000 Gent, Belgium; olivier.goemaere@kuleuven.be (O.G.); seline.glorieux@tradelio.eu (S.G.); marliesgovaert@hotmail.com (M.G.); Liselot.Steen@solina-group.eu (L.S.)
* Correspondence: ilse.fraeye@kuleuven.be; Tel.: +32-9-331-66-17

Abstract: The addition of phosphates to meat products improves the emulsifying and gelling properties of meat proteins, in turn enhancing overall product quality. The current market trend towards additive-free products and the health issues related to phosphate challenge the industry to develop phosphate-free meat products. The aim of this study was to evaluate the potential of seven protein-based ingredients (pea, blood plasma, gelatin, soy, whey, egg, and potato) to remediate quality losses of emulsified meat products (cooked sausages) upon phosphate elimination. First, the intrinsic gelling and emulsifying characteristics of the proteins were assessed. Next, the proteins were added to phosphate-free sausages, of which quality characteristics during production (viscoelastic behavior and emulsion stability) and of the final products (texture, cooking loss, and pH) were screened. Blood plasma and soy were superior in phosphate-free cooked sausages, as no significant differences in hardness, cooking yield, or stability were found compared to phosphate-containing sausages. Egg and pea also improved the previously mentioned quality characteristics of phosphate-free sausages, although to a lesser extent. These insights could not entirely be explained based on the intrinsic gelling and emulsifying capacity of the respective proteins. This indicated the importance of a well-defined standardized meat matrix to determine the potential of alternative proteins in meat products.

Keywords: phosphate elimination; emulsified meat products; proteins; standardized meat matrix

Citation: Goemaere, O.; Glorieux, S.; Govaert, M.; Steen, L.; Fraeye, I. Phosphate Elimination in Emulsified Meat Products: Impact of Protein-Based Ingredients on Quality Characteristics. *Foods* **2021**, *10*, 882. https://doi.org/10.3390/foods10040882

Academic Editor: Gonzalo Delgado-Pando

Received: 18 March 2021
Accepted: 12 April 2021
Published: 17 April 2021

Publisher's Note: MDPI stays neutral with regard to jurisdictional claims in published maps and institutional affiliations.

1. Introduction

Food phosphates exist in different types (mono-, di-, tri-, and polyphosphates) and are often used in the meat industry due to their impact on pH, chelation, ionic strength, and antibacterial activity. They fulfill several functional properties in meat products such as a good buffering capacity (monophosphates) and the ability to dissociate the actomyosin complex of meat (diphosphates) and activate the meat proteins by chelating the protein-bound Mg^{2+} and Ca^{2+}, leading to increased solubilization of the meat proteins and depolymerization of the thick and thin filaments (tri- and polyphosphates) [1,2]. Due to these effects, meat proteins can maximally exert their emulsifying and gelling properties, which are very important with regard to water holding capacity (WHC) and fat emulsification. In addition, most di- or polyphosphates contribute to an increase in pH or ionic strength, respectively. Both effects result in increased electrostatic repulsion and consequently more space to bind water and fat between the meat proteins, which again contributes to increased water and fat stabilization. The different phosphate types (or blends) in meat products can be added to a maximum amount of 0.5% (expressed as P_2O_5) according to European legislation [3]. However, in a former study of Ritz, et al. [4], an association was found between a high intake of phosphate additives and cardiovascular morbidity and mortality. This health issue was already recognized for chronic kidney disease patients, but questions arose with regard to the general population. The EFSA Panel on Food Additives and Flavorings further investigated the matter and provided a scientific opinion re-evaluating the safety of phosphates as food additives in 2019. They considered phosphates to be of low acute oral toxicity, and there was no concern with respect to genotoxicity and carcinogenicity. Furthermore, the Panel considered an acceptable daily intake (ADI) of 40 mg/kg body weight per day. However, this ADI does not apply to humans with a reduction in renal function. Ten percent of the general population might have chronic kidney disease with reduced renal function and they may not tolerate the proposed ADI [5].

In a recent study, it was shown that the current amount of P_2O_5 added to emulsified meat products (cooked sausages) can be strongly reduced with minimal loss in product quality [6]. Nevertheless, the market trend towards additive-free products [2,7] and the negative effect of phosphates on human health for certain population groups justify attempts to develop phosphate-free emulsified meat products. Unfortunately, phosphate elimination results in decreased meat protein functionality, which causes quality defects such as compromised water and fat stabilization [6]. Therefore, alternative ingredients or even innovative technologies are needed to compensate for this functionality loss. These include pH improving ingredients, starches, hydrocolloids, or the use of high-pressure technology [2,8–11]. Additionally, proteins, from both animal and vegetable sources, can act as enhancers to compensate for the loss of functionalized meat proteins due to phosphate elimination in meat products. This is mainly related to their gelling and emulsification properties. They have already been proven useful to boost the quality characteristics of meat products related to water and fat binding properties, gel network formation, texture, and/or sensorial properties. In this respect, they have been successfully deployed as fat-replacers, processing aids of low-cost meat products, and substitutes for meat proteins [12–20].

However, only a limited number of studies aimed to investigate the opportunities of reduction/replacement of phosphate in meat products [11], especially with regard to the use of alternative proteins. Hurtado, et al. [21] concluded that porcine blood plasma was a useful functional ingredient to replace phosphate and caseinate in frankfurters. Pereira, et al. [22] stated that the addition of collagen fibers improved cooking yield and hardness in phosphate-free sausages. Enhancement of water holding capacity, sensorial attributes, color, and microbial stability could also be achieved by replacing phosphates with a purified beef collagen powder in injected beef strip loins [23]. Furthermore, Muguruma, et al. [24] stated that the addition of biopolymers containing soybean and milk proteins may permit a reduction in phosphate content without a loss of the texture of chicken sausages.

In summary, alternative proteins have been proven to function as functional ingredients in different meat systems on account of their gelling and emulsifying capacities. In contrast, the more specific ability to act as a phosphate replacer, in order to manage the loss of functionalized meat proteins, has only been studied for a limited number of proteins. Furthermore, standardized comparison between proteins remains difficult, because these surveys were conducted on different meat matrices (difference in meat product class, composition of recipes, processing conditions) and different analyzing techniques were applied, making it impossible to identify the most promising protein. Xiong [25] stated that a valid comparison between proteins is only possible if the screening is made under identical processing and storage conditions. Therefore, the aim of this study was to evaluate the potential of seven protein-based ingredients, from both animal and vegetable sources, to counter the loss of quality due to phosphate elimination in emulsified meat products (cooked sausages). In the first stage, the intrinsic protein characteristics (gelation and emulsification potential), related to improving the quality of meat products, are studied. In the second stage, the ability of the proteins to enhance the properties of phosphate-free cooked sausages (viscoelastic and textural properties, emulsion stability, cooking loss, and pH) is evaluated. This research is of important industrial relevance, since phosphate elimination in emulsified meat products will decrease potential health concerns and is a promising step towards clean-label products.

2. Materials and Methods

2.1. Determination of the Intrinsic Characteristics of Selected Proteins

The proteins discussed in this study are egg white (Pulviver), pea (Nutralys), potato (KMC), soy concentrate (Pulviver), blood plasma (Veos), gelatin (Rousselot), and whey (Caldic) protein. Proteins were selected based on their industrial relevance. In addition, a balanced distribution between animal and vegetable proteins was envisioned. In order to learn more about their intrinsic properties related to enhancing meat product quality, their gelation and emulsification potential were studied in a watery environment. It is well described that the salt level and acidity of a medium have an important impact on protein characteristics [12,26,27]. In order to create an aqueous medium that reflects the composition of emulsified meat products, proteins were suspended in a 0.05 M Na phosphate buffer (pH = 6) containing 3.5% NaCl and stirred (800 rpm) for 105 min at room temperature before analysis. The applied protein concentration was dependent on the screening technique used, as described below. The concentration of salt corresponds to its quantity in the cooked sausage model (see Section 2.2) expressed in the water phase, and the pH value is in the range of common meat products.

2.1.1. Gelation Potential

The gelation properties of the proteins were determined through rheological measurements using an AR2000ex stress-controlled rheometer (TA instruments, New Castle, DE, USA) equipped with a 40-mm parallel plate system. A crosshatched upper plate and a lower plate were used to prevent slippage of the sample. The gap was set at 500 μm. The AR2000ex was supplemented with an efficient Peltier temperature control system and an upper heated plate (TA Instruments) to control the sample temperatures accurately. Temperature sweeps were conducted to investigate structural changes of the protein suspensions (4.5% protein) during a heating and cooling process, representative of the manufacturing process of emulsified meat products. The following profile was applied: (1) a heating step from 20 to 76 °C at a constant heating rate of 2 °C/min; (2) an isothermal heating step at 76 °C for 3 min; (3) a cooling step from 76 to 20 °C at a constant cooling rate of 2 °C/min. Oscillation measurements during the entire process were performed at a fixed frequency of 1 Hz and a strain of 0.025, a value found to be within the linear viscoelastic region based on preliminary experiments. The storage modulus (G') and phase angle (δ, with δ of 90° representing a fully viscous material and δ of 0° representing a fully elastic material) at the end of the isothermal heating step and cooling step (G'$_{76\,°C,suspension}$, $\delta_{76\,°C,suspension}$ and G'$_{end,suspension}$, $\delta_{end,suspension}$, respectively) were derived from the temperature sweep profiles using the software (Rheology Advantage Data Analysis, v. 5.7.0, TA Instruments). All G'-values are expressed logarithmically. Protein suspensions were made in duplicate for each protein, and determination of the gelling potential was performed in duplicate per suspension.

2.1.2. Emulsification

The emulsifying properties of the proteins were screened according to the procedure described by Steen, et al. [28], which was based upon the turbidimetric method of Pearce and Kinsella [29]. Emulsions were prepared by mixing 2.0 mL sunflower oil and 8.0 mL protein suspensions (0.15% protein) for 1 min at a speed of 12,000 rpm and room temperature using an Ultra-Turrax homogenizer (model T25, IKA-Werke GmbH, Staufen, Germany). Immediately and 10 min after emulsion formation, 50 μL of the emulsion was taken and diluted with 5 mL of 0.1% sodium dodecyl sulfate (SDS) solution. Absorbance values were measured at 500 nm (A_{500}) and used to calculate the emulsifying activity index (EAI, m²/g) according to the equations below.

$$\mathrm{EAI}\left(\mathrm{m^2/g}\right) = \frac{2xTxF}{\varphi\,C} \tag{1}$$

$$T = \frac{2.303 x A500}{L} \tag{2}$$

where A_{500} represents the absorbance at 500 nm, L the light path length ($L = 0.01$ m), φ the volume fraction (v/v) of the dispersed phase ($\varphi = 0.20$), C the protein concentration ($C = 1500$ g/m^3) before emulsification, T the turbidity, and F the dilution factor ($F = 100$). The emulsion activity index immediately after emulsion formation is represented by the abbreviation EAI0. Emulsion stability (ES) was the percentage of emulsion turbidity remaining after 10 min. Emulsions were made in duplicate for each protein, and determination of the emulsifying properties was performed in duplicate for each emulsion.

2.2. Manufacturing of Cooked Sausage

Cooked sausages were prepared in the pilot plant of the research group "Technology and Quality of Animal Products" (KU Leuven Technology Campus Gent, Belgium). Raw materials (pork shoulder and pork backfat) were obtained from a local industrial meat supplier (De Lausnay Rene bvba, Destelbergen, Belgium), chopped, homogenized to generate one batch, vacuum-packed, and stored at −18 °C until preparation. Cooked sausages contained pork shoulder (35/100 g), pork backfat (35/100 g), and ice (30/100 g), together with nitrite curing salt (1.5/100 g), sugar (0.5/100 g), white pepper (0.2/100 g), foil (0.05/100 g), ascorbic acid (0.05/100 g), glutamate (0.05/100 g), coriander (0.025/100 g), and cardamom (0.025 g/100 g). All non-meat ingredients were purchased from Solina Group (Eke-Nazareth, Belgium). Ingredients were calculated relative to the total mass of meat raw materials (pork shoulder and pork backfat) and ice. First, a standardized reference treatment was prepared, containing 0.32/100 g tetrasodium pyrophosphate (TSPP) (Solina Group, Eke-Nazareth, Belgium), which is equal to 0.171% P$_2$O$_5$, a standard amount used in the meat industry for emulsified meat products. The reference containing phosphate will be referred to as M+TSPP. Secondly, TSPP was eliminated and standardized phosphate-free cooked sausages were prepared. These cooked sausages will be referred to as M-TSPP. Finally, the seven above-described protein-based ingredients were added to the phosphate-free treatment. All proteins were added in a mass fraction of 2/100 g, a commonly used dosage [17]. Proteins were calculated relative to the total mass of meat raw materials (pork shoulder and pork backfat) and ice. These phosphate-free treatments containing protein-based ingredients will be referred to as M-TSPP+"corresponding protein source". During manufacturing of the M+TSPP preparations, the raw lean meat was first pre-chopped together with ice, salt, and TSPP in a bowl cutter for 7 min and 30 s (Stephan cutter UM12, Hameln, Germany), corresponding to a final temperature of 5 °C. Next, the pork backfat was added to the meat batter together with the remaining food ingredients. The total mass was ground under vacuum for 4 min and 30 s to obtain a homogenous batter. The temperature did not exceed 14 °C during processing to avoid protein denaturation and fat coalescence. Phosphate-free sausages were prepared the same way, but without the addition of TSPP. When phosphate-free sausages containing protein-based ingredients were prepared, these proteins were added during the first grinding step of the manufacturing process instead of TSPP. Part of the batter, prepared in duplicate per treatment, was immediately analyzed for dynamic viscoelastic properties (Section 2.3) and emulsion stability (Section 2.4). In order to standardize the cooking process, the remainder of the batter was filled into cans of standardized dimensions (diameter 7 cm, height 5 cm, mass ± 250 g, Crown Verpakking België NV, Hoboken, Belgium), cooked at 76 °C (core temperature 72 °C) for 90 min in a cooking chamber (Rational Climaplus Combi CPC 61, Claes Machines, Paal, Belgium) and finally cooled to 4 °C. The resulting meat products, cooked in cans, served as a model product for cooked sausage and will be referred to as "cooked sausages". Each treatment, including the reference products with or without TSPP, was manufactured in duplicate. One week after the manufacturing process, three sausages per replicate were analyzed for cooking loss (Section 2.5), pH (Section 2.6), and textural properties (Section 2.7). The number of measuring points is described in the respective analyses below.

2.3. Dynamic Viscoelastic Properties

The dynamic viscoelastic properties of the batters were analyzed using the same equipment as described in Section 2.1.1. The gap was set at 1000 μm for both rheological procedures described below (stress sweep and temperature sweeps).

Stress sweeps were conducted at a temperature of 13 °C, between 0.1 and 1000 Pa, and at a fixed frequency of 1 Hz to determine the linear viscoelastic region (LVR). Hereby, parameters G', G'' (storage and loss modulus respectively), and δ were directly obtained from the software. The complex modulus (G^*), representing the materials' overall rigidity or resistance to deformation, was calculated by the following formula,

$$G* = \sqrt{G'^2 + G''^2} \tag{3}$$

The LVR represents the stress range within which G^* (and thus G',G'') is independent of the imposed stress amplitude and is determined according to Glorieux, Goemaere, Steen and Fraeye [6]. LVR is determined in duplo per replicate of each treatment and referred to as LVR_{batter}. Furthermore, the corresponding G^*_{batter}, expressed logarithmically, and δ_{batter} within the LVR are reported.

Temperature sweeps were conducted to investigate the impact of phosphate elimination and use of alternative proteins on the structure formation of meat batters during a heating and cooling procedure, representative of the manufacturing process of cooked sausages. Similar profiles and conditions were applied as described in Section 2.1.1., except for the initial (before heating) and final (end of cooling) temperatures, which were both set at 13 °C in accordance with the final temperature of the raw batter at the end of the cutter process. The parameters G' and δ at the end of the isothermal heating step and cooling step ($G'_{76°C,batter}$, $\delta_{76°C, batter}$ and $G'_{end,batter}$ $\delta_{end,batter}$, respectively) were derived from the temperature sweep profiles using the software. G'-values are expressed logarithmically. All rheological parameters ($G'_{76°C,batter}$, $\delta_{76°C,batter}$, $G'_{end,batter}$, $\delta_{end,batter}$) were determined in duplo per replicate of each treatment.

2.4. Emulsion Stability

Emulsion stability of the meat batter was determined immediately after the grinding process, according to Glorieux, Goemaere, Steen and Fraeye [6] with slight modifications. Summarized, emulsion stability is expressed as drip loss upon heating (30 min, 70 °C) and centrifugation at $4230\times g$ (6000 rpm in a rotor Cat. No. 1620 A, Hettich, Germany) at 25 °C for 3 min, of a pre-weighed amount of raw batter. The percentage of total expressible fluid (TEF) was expressed as follows:

$$\text{TEF (\%)} = \frac{drip\ loss\ meat\ batter}{initial\ weight\ meat\ batter} \times 100 \tag{4}$$

Furthermore, the relative amount of water, next to the fat in the drip, was determined. Therefore drip loss after centrifugation was weighed before and after drying in an oven (Typ U 40, Memmert, Germany) for 24 h. The relative amount of water in the drip loss was expressed as follows:

$$\text{Relative amount of } H_2O \text{ in drip, (\%)} = \frac{drip\ before\ drying - drip\ after\ drying}{drip\ before\ drying} \times 100 \tag{5}$$

TEF and Relative amount of H_2O in drip were determined six times per replicate of each treatment.

2.5. Cooking Loss

Cooking loss (CL) of the cooked sausages of each treatment was measured according to Glorieux, Goemaere, Steen and Fraeye [6]. CL was calculated as follows:

$$CL\ (\%) = \frac{drip\ loss\ sausage}{initial\ weight\ sausage} \times 100 \tag{6}$$

Measurements were determined in triplicate per replicate of each treatment.

2.6. pH Measurement

The pH of the cooked sausages was measured three times on three different sausages (nine measurements) per replicate of each treatment, according to the methods described in Glorieux, Goemaere, Steen and Fraeye [6].

2.7. Texture

The hardness of the cooked sausages was analyzed using a Lloyd Texture Analyzer (Model LF plus, Lloyd Instruments, Bognor Regis, UK) and expressed as the maximum force (N) to penetrate the sample, as described in Glorieux, Goemaere, Steen and Fraeye [6]. Per replicate of each treatment, hardness was measured three times on three different sausages (nine measurements).

2.8. Statistical Analysis

Results are expressed as mean $\pm$ standard deviation. All results were evaluated by one-way ANOVA. A Tukey's post hoc test was performed with a significance level of $p < 0.05$ to identify significant differences. Statistical analysis was performed using the software IBM SPSS Statistics 25 (IBM, Armonk, NY, USA).

3. Results and Discussion

3.1. Intrinsic Characteristics of Selected Proteins

Screening of the intrinsic characteristics of functional ingredients is often executed in watery media. It is a rather quick and easy method to evaluate ingredient functionality that requests no specific and often expensive process equipment to imitate industrial food products. Moreover, it can provide a broad view of the application potential of the ingredients in several food products. Food proteins are mainly applied in meat products in relation to their gelling and emulsifying properties, enabling them to improve overall meat product quality. Results regarding these intrinsic characteristics are described below in Sections 3.1.1 and 3.1.2.

3.1.1. Gelation Potential

The gelling characteristics of proteins are one of the key reasons they are applied for meat product improvement. Figure 1 shows the gelling properties of the proteins upon heating, with the exception of gelatin. Gelatin is a cold-gelling protein that solubilizes during heating [30] and can therefore only participate in gel network formation at sufficiently low temperatures. The critical temperature below which gelling can occur is dependent on gelatin concentration, cooling rate, and maturing temperature [31]. The applied thermal processing and used gelatin (concentration, source) did not allow the expression of the cold gelling character of gelatin. For all other protein suspensions, the heating and subsequent cooling process caused in general an overall increase in G' and a decrease of δ. This suggests the formation of a gel-like structure and increased elastic behavior. $G'_{76°C,suspension}$ and $G'_{end,suspension}$ are highest for potato and egg white protein, indicating the strongest gelling potential of all screened proteins. The irreversible heat coagulation of egg white proteins involves the formation of spherical aggregates via hydrophobic interactions, which are further stiffened through sulfhydryl–disulfide reactions to finally give rise to a gel, which explains the rather high values of $G'_{76°C,suspension}$. Furthermore, the G'-values of egg white protein still increase ($G'_{76°C,suspension}$ vs. $G'_{end,suspension}$) during cooling, which can be attributed to the numerous hydrogen bonds that are formed at lower temperatures [32]. The suspension of potato protein also exerted very good gelling properties upon heating. The low denaturation temperature of patatin, one of the main potato protein fractions, may be partially responsible for this. The denaturation temperature is roughly 20 °C lower compared to common food proteins as ovalbumin (egg) or soy glycinin [33,34]. Figure 1 reveals no significant difference between $G'_{76°C,suspension}$ of the egg white protein and $G'_{76°C,suspension}$ of soy concentrate, indicating good gelation characteristics of the latter. Pea proteins are mainly composed of globulins. Pea globulins are recognized for their lower gelling ability compared to their soy counterparts. This can also be observed in Figure 1, where $G'_{76°C,suspension}$ and $G'_{end,suspension}$ of pea proteins are significantly lower than the values of soy concentrate. The gelation of pea proteins appeared to be governed mainly by nonspecific interactions, whereas the involvement of disulfide bonds was reported for soy proteins [35]. Furthermore, high temperatures are required to induce the gelation of the pea proteins because of their high denaturation temperature (>85 °C) [36]. The applied thermal processing in this research was therefore not sufficient to obtain proper gelling of pea proteins. Whey protein suspensions start to form gels at concentrations higher than 80 mg whey protein/g H_2O when heated above 75 °C [37]. The rather short heating time above 75 °C and applied concentration could therefore explain the somewhat low values of $G'_{76°C,suspension}$, and $G'_{end,suspension}$ of whey protein. Blood plasma also showed relatively low values of both $G'_{76°C,suspension}$, and $G'_{end,suspension}$ and is probably attributed to the same reasons as described for whey protein. Research stated that heating to 75 °C was a necessity to create strong gels from 4% w/v plasma protein solutions [38]. Other sources claimed that suspensions containing 4–5% blood plasma already tend to form firm and irreversible gels when temperatures over 70 °C are applied [39].

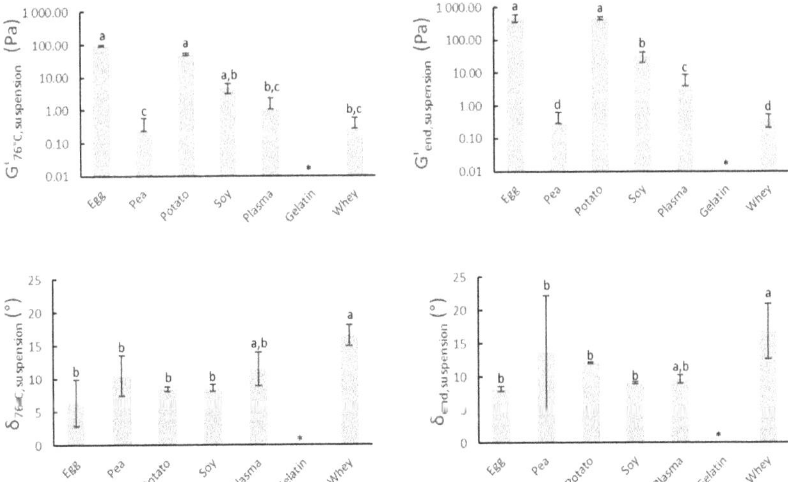

Figure 1. $G'_{76°C,suspension}$ and $\delta_{76°C,suspension}$ represent the elasticity modulus and phase angle of protein suspensions at the end of the isothermal heating at 76 °C for 3 min. $G'_{end,suspension}$ and $\delta_{end,suspension}$ represent the elasticity modulus and phase angle of protein suspensions after further cooling from 76 to 20 °C. (*) = no value of gelatin could be obtained. Mean values and standard deviations are presented (n = 4). Letters a–d: different letters indicate significant differences ($p < 0.05$) between different proteins.

Based upon the gathered data and literature study, potato and egg white protein show the most potential for use in meat products. Their gelling properties may lead to a better structure formation of the meat gel and as a consequence to improved water binding or texture of the sausages.

3.1.2. Emulsification

In addition to their gelation potential, proteins are also of interest to the meat industry because of their ability to stabilize emulsions. The intrinsic emulsifying and emulsion-stabilizing properties of the proteins can be derived from Figure 2. The EAI0 indicates the area of interface stabilized per unit weight of protein (m^2/g) and is associated with the ability of the protein to coat the water–oil interface immediately after emulsion formation. ES represents the percentage of emulsion turbidity remaining after 10 min and therefore refers to the ability of an emulsion to resist changes in its properties over time, e.g., droplet coalescence, creaming, and/or flocculation [28].

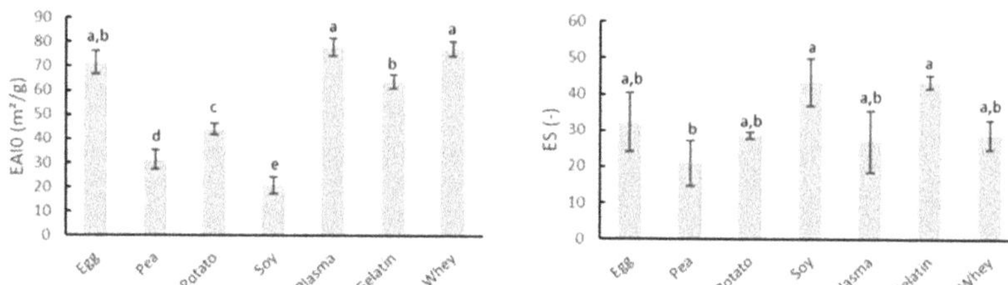

Figure 2. Comparison of emulsifying activity index immediately after emulsion formation (EAI0, m^2/g) and emulsion stability (ES) of several protein sources. Mean values and standard deviations are presented (n = 4). Letters a–e: different letters indicate significant differences ($p < 0.05$) between different proteins.

Gelatin possesses good emulsifying properties, as can be noticed by the high values of both EAI0 and ES in Figure 2. Gelatin is capable of reducing the surface tension of aqueous environments and forming the necessary identically charged film around the fat droplets of the dispersed phase. Therefore, the isoelectric point (IEP) is of great importance in the surface activity effects of the used gelatin [40]. The protein carried a net negative charge under the conditions in which this analysis was performed. Whey proteins are well-known for their ability to stabilize interfaces, explaining their great emulsifying properties, as seen in the present research [41]. Figure 2 also shows that blood plasma and egg white protein exerted excellent emulsifying properties. Research by Rodriguez Furlán, et al. [42] confirmed the good emulsifying properties of blood plasma. Yet, literature stated that ovalbumin, the major protein in egg white, may perform good emulsifying ability and stability under extreme acidic conditions, which is in contrast to the watery suspensions applied in this research, while under neutral and alkaline pH the stability of egg white emulsions was limited [43]. The emulsifying capacity of soy concentrate was rather limited, as indicated by the low value of EAI0 in Figure 2. The study of Amine, et al. [44] also indicated soy protein as a poor emulsifier for oil in water emulsions, based upon the measurement of oil droplet particle sizes. The same research presented potato protein as the better emulsifier compared to soy and pea proteins, as was the case in this study. Pea protein also exhibited poor emulsifying properties, as seen in Figure 2. Several studies concluded that pea proteins are usually inferior to traditional emulsifiers such as milk and egg proteins [45].

Results indicated the use of egg white protein, blood plasma, gelatin, or whey protein may be more beneficial in stabilizing meat emulsion regarding water and fat binding compared to the other screened proteins because of their high initial emulsion activity in combination with their good emulsion stability.

3.2. Impact of Seven Different Protein-Based Ingredients on the Quality Characteristics of Cooked Sausage

The results presented in the following sections deal with the impact of the selected proteins on several quality characteristics of phosphate-free sausage.

3.2.1. Dynamic Viscoelastic Properties of Meat Batters Influenced by Protein Source

Stress sweeps were performed to study the structure of the raw meat batter immediately after the grinding process, prior to thermal processing. Data (Table 1, Stress sweeps) indicated that the LVR, the stress range in which the structure of the sample remains intact, significantly ($p < 0.05$) increased when TSPP was eliminated (M-TSPP) compared to the model preparation containing phosphate (M+TSPP). Since TSPP has the ability to dissociate the actomyosin complex [1], the M+TSPP batter was presumably more sensitive to external deformation. This is reflected in a significantly ($p < 0.05$) lower δ_{batter} value compared to the M-TSPP sample, the latter having more "solid-like" behavior. In parallel, the G^*_{batter} of M-TSPP was significantly higher compared to M+TSPP, indicating that M-TSPP showed high resistance to deformation. A higher LVR_{batter}, lower δ_{batter}, and higher G^*_{batter} as a result of phosphate elimination were also seen in our previous study [6].

The addition of protein-based ingredients to phosphate-free raw sausage batters did not affect the LVR_{batter} or G^*_{batter} compared to the M-TSPP preparations, with the exception of the preparation with gelatin (M-TSPP+gelatin) and egg white protein (M-TSPP+egg). The addition of gelatin to phosphate-free raw sausage batter (M-TSPP+gelatin) significantly increased the G^*_{batter}, which can possibly be attributed to the cold gelling capacity of the protein [46]. Raw phosphate-free sausage batter containing 2% egg white protein (M-TSPP+egg) gave rise to a significantly lower LVR_{batter}, and, at the same time, a remarkably high G^*_{batter} compared to M-TSPP. An explanation of this striking observation is given in Appendix A.

To study the rheological properties of the sausage batters during thermal processing, all samples were subjected to a temperature sweep as described in Section 2.3. The heat causes the myofibrillar proteins to unfold and/or dissociate, followed by association and aggregation, resulting in a gelled system in which water and fat are entrapped [47,48]. The high G'_{batter}-values in Figure 3 confirmed the formation of gel structures. $\delta_{76°C,batter}$ is lower than 10° for all batters, indicating a strong elastic behavior of the formed network. Significant differences in $\delta_{76°C,batter}$ between batters have little relevance.

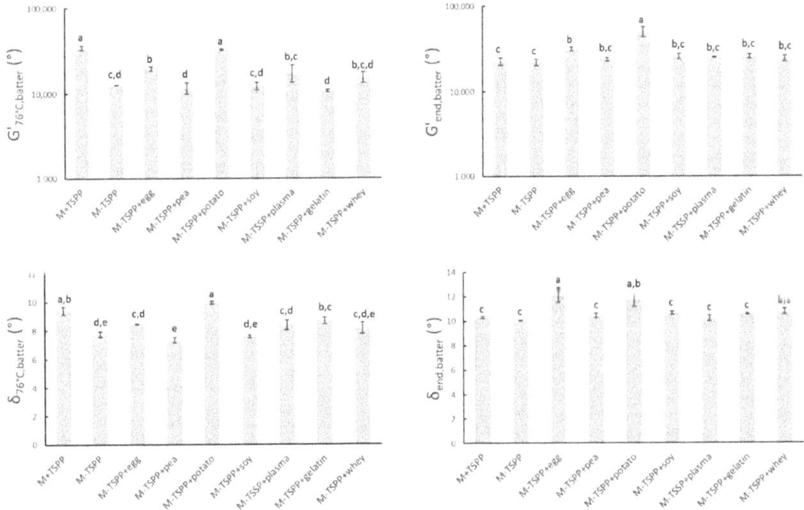

Figure 3. $G'_{76°C,batter}$ and $\delta_{76°C,batter}$ represent the elasticity modulus and phase angle, respectively, of sausage batters during rheological measurements at the end of isothermal heating at 76 °C. $G'_{end,batter}$ and $\delta_{end,batter}$ represent the elasticity modulus and phase angle of sausage batters after further cooling from 76 to 20 °C. Mean values and standard deviations are presented (n = 2). Abbreviations used: M (model) and TSPP (tetrasodium pyrophosphate). Letters a–e: different letters indicate significant differences ($p < 0.05$) between treatments.

Table 1. Structural parameters of sausage batters (stress sweeps) and hardness, pH, water, and fat binding characteristics (CL, TEF, and relative amount of H_2O) of cooked sausages with phosphate (M+TSPP), without phosphate (M-TSPP), and several protein-based ingredients instead of phosphate (M-TSPP+protein). Mean values and standard deviations are presented (n = 2). Different letters indicate significant differences at $p < 0.05$. Abbreviations used: M (Model), TSPP (tetrasodium pyrophosphate), LVR_{batter} (linear viscoelastic region of the sausage batter), G^*_{batter} (complex modulus of the sausage batter), δ_{batter} (phase angle of the sausage batter), TEF (total expressible fluid), and CL (cooking loss). Superscripts a–e: different letters indicate significant differences ($p < 0.05$) between different treatments.

	M+TSPP	M-TSPP	M-TSPP +egg	M-TSPP +pea	M-TSPP +potato	M-TSPP +soy	M-TSPP +plasma	M-TSPP +gelatin	M-TSPP +whey
Stress sweeps									
LVR_{batter}/Pa	25 ± 4 [bc]	56 ± 9 [a]	7 ± 7 [c]	53 ± 8 [a]	45 ± 17 [a,b]	54 ± 5 [a]	48 ± 4 [a]	56 ± 6 [a]	42 ± 0 [a,b]
G^*_{batter}/10^2Pa	12 ± 1 [d]	36 ± 0.1 [c]	227 ± 67 [a]	46 ± 2 [c]	38 ± 2 [c]	52 ± 0.6 [b,c]	39 ± 4 [c]	77 ± 5 [b]	45 ± 1 [c]
δ_{batter}/°	37.2 ± 0.1 [a]	11.3 ± 0.2 [c,d]	11.2 ± 1.3 [c,d]	11.1 ± 0.0 [c,d]	12.0 ± 0.0 [b,c]	10.8 ± 0.2 [d]	10.7 ± 0.3 [d,e]	9.8 ± 0.2 [e]	12.8 ± 0.5 [b]
Stability									
TEF/%	1.9 ± 0.4 [c]	5.2 ± 0.6 [a]	3.2 ± 0.6 [b,c]	2.9 ± 0.5 [c]	4.8 ± 0.7 [a,b]	1.9 ± 0.6 [c]	2.5 ± 0.6 [c]	6.5 ± 0.7 [a]	5.4 ± 0.3 [a]
Relative amount of H_2O/%	89.1 ± 1.2 [a,b,c]	90.8 ± 0.4 [a,b]	89.2 ± 1.0 [a,b,c]	88.8 ± 0.3 [a,b,c]	91.8 ± 0.3 [a]	85.7 ± 1.1 [c,d]	86.1 ± 1.5 [c,d]	87.5 ± 0.1 [b,c,d]	84.1 ± 1.3 [d]
CL/%	0.6 ± 0.2 [d]	5.7 ± 0.6 [a]	2.1 ± 0.2 [c,d]	3.2 ± 0.1 [b,c]	4.2 ± 0.1 [a,b]	1.9 ± 0.4 [c,d]	0.9 ± 0.1 [d]	5.4 ± 1.0 [a]	1.2 ± 0.1 [d]
pH/(-)	6.98 ± 0.09 [a]	6.62 ± 0.00 [b,c]	6.59 ± 0.01 [b,c]	6.52 ± 0.07 [b,c]	6.46 ± 0.07 [c]	6.44 ± 0.04 [c]	6.69 ± 0.08 [b]	6.41 ± 0.04 [c]	6.43 ± 0.01 [c]
Hardness/N	4.3 ± 0.0 [c]	4.8 ± 0.1 [b,c]	5.6 ± 0.4 [a,b]	5.3 ± 0.2 [a,b]	5.2 ± 0.1 [a,b]	5.0 ± 0.4 [a,b,c]	5.0 ± 0.2 [a,b,c]	5.8 ± 0.2 [a]	5.1 ± 0.2 [a,b,c]

Comparison between M+TSPP and M-TSPP indicated that the elimination of phosphate had an effect on the viscoelastic behavior of the meat batter during heating ($G'_{76°C,batter}$ $\delta_{76°C,batter}$). Values of $G'_{76,batters}$ revealed that phosphate elimination significantly reduced gel strength at the end of heating prior to cooling. On the other hand, upon subsequent cooling, no significant differences could be observed anymore between M+TSPP and M-TSPP ($G'_{end,batter}$ and $\delta_{end,batter}$). The stronger increase of G' during heating upon the addition of phosphates was possibly caused by conformation transitions, exposure of hydrophobic groups, and the formation of more disulfide bonds of the meat proteins [49]. In other words, TSPP promoted gelation, as it aids in the extraction of myofibrillar proteins that will subsequently aggregate and gel upon thermal processing [48,50–52]. However, Sun and Holley [48] also reported that it was possible that polyphosphates do not influence myofibrillar gel strength, as this is dependent on the applied protein source and preparation and gelation conditions that are used.

The strong gelling properties of potato protein nullified the drop in $G'_{76°C,batter}$ due to phosphate elimination (M-TSPP+potato). A similar observation can be made when egg white protein (M-TSPP+egg) is used, although the total impact of phosphate elimination on $G'_{76,batters}$ could not be compensated, since a significant difference in $G'_{76°C,batters}$ between M+TSPP and M-TSPP+egg remained. Furthermore, $G'_{end,batter}$ and $\delta_{end,batter}$ significantly increased ($p < 0.05$) with the addition of egg white (M-TSPP+egg) and potato proteins (M-TSPP+potato) compared to M-TSPP. Potato and egg white proteins probably formed additional protein networks or improved interactions for gel formation compared to the other proteins, leading to increased structure formation. Studies on the impact of egg albumin on the thermal gelation of myofibrillar proteins are contradictory. Some authors concluded egg proteins caused disruption of the meat gel by interfering with the gelling process of the myofibrillar proteins or by the formation of mixed egg–myofibrillar protein gels, while others reported egg proteins participated in meat gel network formation [53]. Hunt, et al. [54] also observed a positive effect on gelation characteristics of Alaska pollock fish protein upon the addition of dried egg white protein. No significant difference in $G'_{76°C,batters}$, $G'_{end,batter}$, or $\delta_{end,batter}$ between M-TSPP and preparations with pea, gelatin, whey, blood plasma, or soy concentrate (M-TSPP+pea, M-TSPP+gelatin, M-TSPP+whey, M-TSPP+plasma, and M-TSPP+soy, respectively) could be observed. On the other hand, the studies of Wang, et al. [55] and Li, et al. [56] claimed an improvement of the gelling characteristics and structural strength of myofibrillar protein gels upon the addition of soy protein. Additionally, the addition of blood plasma has been shown to affect the thermal gelation of myofibrils and therefore influence the final gel strength [57,58]. Sun and Holley [48] stated that due to a lack of interaction between nonmeat and muscle proteins, it is possible that texture is negatively affected by interference with the gelation of the myofibrillar proteins. This could not be deducted from Figure 3, as the final gel strength ($G'_{end,batter}$) of all phosphate-free batches with different proteins is similar or higher on average compared to M-TSPP.

The rather strong gelation potential of egg white and potato protein in the meat matrix during heating was also seen in the watery medium (Figure 1), as described in Section 3.1.1. Despite this similarity, batter parameters $G'_{76°C,batter}$ and $G'_{end,batter}$ were significantly higher for potato protein compared to egg white protein, which was not the case for $G'_{76°C, suspension}$ and $G'_{end,suspension}$ of the same proteins. Furthermore, blood plasma and whey protein resulted in similar values of $G'_{76°C,batter}$ and $G'_{end,batter}$ compared to egg white protein, which was not observed in $G'_{76°C, suspension}$ and $G'_{end,suspension}$ of the same proteins. In contrast, soy concentrate resulted in a significantly lower value of $G'_{76°C,batter}$ compared to egg white protein, while this was not the case for $G'_{76°C,suspension}$. On the other hand, $G'_{76°C,batter}$ was lowest for preparations with pea protein and gelatin, which was identically reflected in $G'_{76°C,suspension}$.

Evaluating these insights, it seems that the gelation potential of the different proteins, as determined in an aqueous medium during thermal processing (Section 3.1.1), was not always clearly noticeable in a meat system. This suggests the importance of a well-defined meat matrix, imitating industrial meat products, to determine and understand the impact of ingredient functionality. A food environment is a more complex system, where ingredients and other components (i.e., salts, lipids, and proteins) may interact, thus modifying the added value to the product quality of the functional ingredient.

3.2.2. Emulsion Stability of Meat Batters and Cooking Loss of Cooked Sausages Influenced by Protein Source

Significant ($p < 0.05$) differences in emulsion stability and cooking loss (CL) were found between the different preparations (Table 1). Elimination of TSPP (M-TSPP) resulted in a significant increase of total expressible fluid (TEF) and thus lower emulsion stability, and increased CL compared to M+TSPP. These findings are in line with our former study [6]. It is known from the literature that TSPP is able to dissociate the actomyosin complex, releasing myosin, which can act as a natural emulsifier. Additionally, more myofibrillar proteins are extracted by TSPP, helping to stabilize the protein matrix in which water and fat are entrapped [1].

Preparations containing egg white protein (M-TSPP+egg), pea (M-TSPP+pea), soy concentrate (M-TSPP+soy), and blood plasma proteins (M-TSPP+plasma) significantly ($p < 0.05$) reduced TEF compared to M-TSPP and even resulted in similar percentages of TEF as the preparation containing phosphate (M+TSPP), indicating an equal stabilization of moisture and fat in the meat matrix. These proteins were thus able to compensate for the decreased emulsion stability due to phosphate elimination. On the other hand, there was no significant difference in TEF between M-TSPP and preparations with the addition of potato (M-TSPP+potato), whey proteins (M-TSPP+whey), and gelatin (M-TSPP+gelatin). Furthermore, the use of some proteins also caused a shift in composition (water vs. fat) of the drip loss. The relative amount of fat in the drip loss was significantly higher when adding blood plasma (M-TSPP+plasma), gelatin (M-TSPP+gelatin), soy concentrate (M-TSPP+soy), or whey protein (M-TSPP+whey) to phosphate-free sausages (M-TSPP). This could mean that fat stabilization in the meat matrix could be altered by using additional proteins, which could affect the final product characteristics such as texture or mouthfeel [59].

In almost all cases, CL significantly ($p < 0.05$) decreased with the addition of protein-based ingredients compared to M-TSPP. The addition of blood plasma (M-TSPP+plasma), whey proteins (M-TSPP+whey), egg white proteins (M-TSPP+egg), and soy concentrate (M-TSPP+soy) even resulted in similar CL as the cooked sausages containing phosphate (M+TSPP). Blood plasma proteins are good emulsifiers [9] and were found to be a useful substitute for polyphosphate in frankfurters, as they did not affect the water holding capacity and cooking losses compared to frankfurters containing 0.5% sodium tripolyphosphate [21]. Research by Prabhu [60] also indicated blood plasma was suitable to improve the emulsion stability, texture, flavor, and juiciness of comminuted meat products. Additionally, the use of pea protein (M-TSPP+pea) could significantly decrease CL compared to M-TSPP, although to a lesser extent than the previously mentioned proteins. On the other hand, the addition of gelatin (M-TSPP+gelatin) or potato protein (M-TSPP+potato) did not change CL compared to M-TSPP. This was in contrast with the study by Nieto, Castillo, Xiong, Álvarez, Payne and Garrido [20] in which cooking losses were reduced when 2.5% hydrolyzed potato proteins were added to phosphate-free meat emulsions.

Comparison between the intrinsic properties of the proteins discussed in Section 3.1 and their impact on fat and water binding characteristics of cooked sausages indicated limited analogy. Stronger gelation potential of the protein samples as measured in the watery medium would suggest better water and especially fat binding in meat products. Furthermore, proteins with good emulsifying capacities are expected to be able to stabilize emulsified meat products to a greater extent and contribute to reducing cooking loss (especially fat release). While potato and egg white protein both showed very good gelation properties, only the latter could positively improve the cooking yield. In contrast, pea proteins showed both low emulsifying capacity and gelling behavior in the watery medium, while in the cooked sausages, they could reduce cooking loss and TEF. Results even surpassed those of potato protein. Blood plasma proteins, showing an average gelation potential and good attribution to emulsion stability in the watery environment, outperformed the other proteins, with the exception of soy concentrate, regarding water and fat binding in cooked sausages. Again, these results underline the importance of a well-defined meat matrix, close to industrial meat products, to determine and fully understand the impact of ingredient functionality.

3.2.3. pH of Cooked Sausages Influenced by Alternative Protein Source

Data (Table 1) showed that the elimination of TSPP (M-TSPP) resulted in significantly lower pH values compared to the model system containing TSPP (M+TSPP), which was in line with our former study [6]. The pH of TSPP (1% solution) is equal to 10.2 [1], which explains the pH difference between preparations M-TSPP and M+TSPP. Due to phosphate elimination, the pH of the meat product was decreased and was closer to the iso-electric pH of the myofibrillar proteins. This led to a reduction in their net charge and repulsion between proteins, causing a negative impact on water and fat binding, as seen in Section 3.2.2 [61]. The decrease in pH by phosphate elimination could not be compensated by the addition of protein-based ingredients, as seen in Table 1. Velemir, et al. [62] determined no significant difference in pH upon the addition of 1.5% whey or soy protein to sausages. Blood plasma, despite its higher pH, could also not remediate the lower pH of phosphate-free sausages, which was also seen in the research of Hurtado, Saguer, Toldrà, Parés and Carretero [21]. The proteins could therefore not contribute to water binding by generating a higher concentration of negative meat protein charges.

3.2.4. Textural Properties of Cooked Sausages Influenced by Protein Type

The differences in hardness of the different preparations are limited (Table 1). Phosphate elimination (M-TSPP) did not significantly affect the hardness of the cooked sausages, which was in line with our former study [6]. A lower hardness might be expected when TSPP is eliminated, since TSPP dissociates the actomyosin complex, resulting in more proteins being available for emulsification and the formation of a more stable gel matrix during heating. Yet, gel strength at the end of thermal processing (see Section 3.2.1) also revealed no difference in $G'_{end,batter}$ between M+TSPP and M-TSPP. On the other hand, the increase in CL when phosphate is eliminated could lead to a firmer meat product. The addition of gelatin (M-TSPP+gelatin) increased hardness compared to M-TSPP, despite no significant difference in CL being measured. Therefore, it could be concluded that gelatin itself had an impact on the final hardness of the phosphate-free cooked sausage, which could probably be attributed to its cold gelling properties [46]. The addition of the other protein-based ingredients did not significantly affect hardness compared to M-TSPP, but a significant increase in hardness compared to the reference sausage containing phosphate (M+TSPP) was determined upon the addition of pea, potato, egg white protein, and again gelatin. Nieto, Castillo, Xiong, Álvarez, Payne and Garrido [20] found that the addition of 2.5% hydrolyzed potato proteins had no effect on the hardness of phosphate-free frankfurters, which was also in line with our results. Furthermore, Youssef and Barbut [14] concluded that soy protein could increase or decrease the product texture depending on the type of soy used. The impact of whey proteins on hardness is linked to their degree of denaturation, which is dependent on their production process. In general, undenatured whey protein preparations deteriorate textural properties, while partially denatured whey concentrates enhance the binding and texture of sausages and other comminuted meat products [25]. This could possibly explain the mild impact on the observed hardness of whey proteins. Fernandez, et al. [63] also found no difference in hardness when 2% of dried egg white was added to chicken meat batters. Cofrades, Guerra, Carballo, Fernández-Martín and Colmenero [19] noted an increase in the product hardness of Bologna sausages when blood plasma was applied. This observation could not be established in this research.

4. Conclusions

The elimination of phosphate had a negative impact on several quality characteristics of cooked sausages. Next to an increase in cooking loss and reduced emulsion stability, a change in gel network formation during thermal processing could be observed, although the final gel strength was not influenced. The cause of these quality losses is mainly related to the reduced functionality of the myofibrillar proteins due to phosphate elimination. This research indicated that the addition of specific proteins could remediate the negative impact of phosphate elimination. However, it is important to keep in mind that different protein sources exhibit varying potential in this respect. Hereby, it is crucial to evaluate the potential of the proteins in a well-defined standardized meat matrix. The intrinsic protein properties, gelation and emulsification, related to improving meat quality are often evaluated in aqueous media. This study showed that protein characteristics determined in this manner did not entirely reflect their capacity to enhance the characteristics of phosphate-free emulsified meat products.

In phosphate-free cooked sausages, blood plasma and soy protein overall showed the most promising results, as no significant differences in terms of product hardness, cooking yield, or emulsion stability could be found compared to standard phosphate-containing sausages. These proteins may therefore provide an added value for the meat industry to further reduce E-numbers and contribute to the healthy image of meat products. Other screened proteins, such as egg white, pea, and whey protein, also proved to be beneficial, yet the quality level of the phosphate-containing sausages could not be equaled. Potato protein and gelatin showed the least improvement to the phosphate-free cooked sausages.

Future research can be conducted on the use of combinations of different protein sources or mixtures of proteins with certain hydrocolloids to further remediate the loss of quality due to phosphate elimination in emulsified meat products.

Author Contributions: Conceptualization: O.G. and I.F.; formal analysis: S.G.; funding acquisition: I.F.; investigation: O.G., M.G., and L.S.; methodology: O.G. and I.F.; project administration. I.F.; resources: I.F.; supervision: I.F.; validation: M.G. and L.S.; visualization: O.G.; writing—original draft: O.G. and S.G.; writing—review and editing: M.G., L.S., and I.F. All authors have read and agreed to the published version of the manuscript.

Funding: This research was 80% funded by Flanders Food and the Agency for Innovation by Science and Technology (IWT), grant number IWT 130500, and 20% co-funded by the Flemish meat and ingredient producing industry.

Institutional Review Board Statement: Not applicable.

Informed Consent Statement: Not applicable.

Data Availability Statement: Not applicable.

Conflicts of Interest: Participating companies in the research (co-funding partners) had the opportunity to propose ingredients to be included in the study. However, the final decision of ingredient choice was solely and independently made by the researchers. Protein-based ingredients were chosen to obtain a good mix of different animal and vegetable sources and with relevance to the broad industry. Furthermore, they had no role in the design of the study; in the collection, analyses, or interpretation of data; in the writing of the manuscript; or in the decision to publish the results.

Appendix A

Raw phosphate-free sausage batter containing 2% egg white protein (M-TSPP+egg) gave rise to a significantly lower LVR_{batter}, and, at the same time, a remarkably high G^*_{batter} compared to M-TSPP, as seen in Table 1 (main text). In order to gain insight into this striking observation, the stress sweeps as seen in Figure A1 were evaluated.

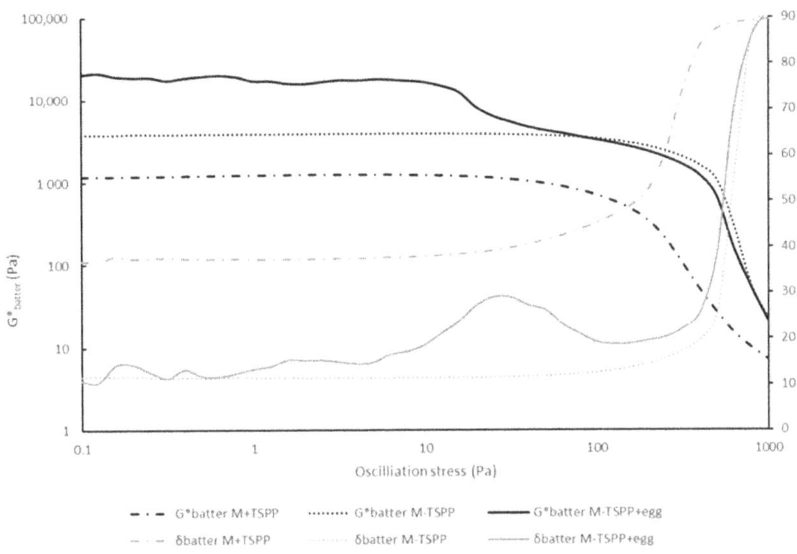

Figure A1. Stress sweep curves of meat batters containing phosphate (M+TSPP, dash-dot line), without phosphate (M-TSPP, dotted line), and without phosphate-containing egg white proteins (M-TSPP+egg, full line). The complex modulus (G*, black color) and phase angle (δ, grey color) are displayed on the primary and secondary y-axis, respectively. Confidence intervals are not included to enhance the readability of the graph (n = 2).

According to Glorieux, Goemaere, Steen and Fraeye [6], the LVR is calculated as the stress level at which G* deviates more than 5% from a constant G* (plateau) value and indicates irreversible structure breakdown. However, the raw sausage batter containing egg white proteins (M-TSPP+egg white) contained two plateau regions in which G′ and G″ (and thus G*) were independent of the applied stress amplitude. The first plateau was characterized by a high G_{batter}^* value and ranged up to ±10 Pa, the stress value at which structure breakdown occurred. However, from around a stress value of 30 Pa, the structure stabilized again, resulting in another plateau that reached stress values of ±200 Pa until irreversible structure breakdown occurred (Figure A1). The two LVR regions could possibly be explained by the presence of two distinct protein structures. The first plateau is characterized by G_{batter}^* higher than G_{batter}^* of M-TSPP, which may be attributed to the presence of the egg white proteins. The second plateau is characterized by G_{batter}^* values in the same order of magnitude as M-TSPP, and the LVR ends at a comparable stress value, presumably indicating that this part of the LVR was stabilized independently of the added protein based ingredient.

References

1. Long, N.H.B.S.; Gál, R.; Buňka, F. Use of phosphates in meat products. *Afr. J. Biotechnol* **2011**, *10*, 19874–19882. [CrossRef]
2. Balestra, F.; Petracci, M. Technofunctional Ingredients for Meat Products: Current Challenges. In *Sustainable Meat Production and Processing*; Galanakis, C.M., Ed.; Academic Press: Cambridge, MA, USA, 2019; pp. 45–68. [CrossRef]
3. European Food Safety Authority. Assessment of one published review on health risks associated with phosphate additives in food. *EFSA J.* **2013**, *11*, n. [CrossRef]
4. Ritz, E.; Hahn, K.; Ketteler, M.; Kuhlmann, M.K.; Mann, J. Phosphate Additives in Food-a Health Risk. *Dtsch. Arztebl. Int.* **2012**, *109*, 49–55. [CrossRef]
5. Younes, M.; Aquilina, G.; Castle, L.; Engel, K.-H.; Fowler, P.; Frutos Fernandez, M.J.; Fürst, P.; Gürtler, R.; Husøy, T.; Mennes, W.; et al. Re-evaluation of phosphoric acid–phosphates – di-, tri- and polyphosphates (E 338–341, E 343, E 450–452) as food additives and the safety of proposed extension of use. *EFSA J.* **2019**, *17*. [CrossRef]
6. Glorieux, S.; Goemaere, O.; Steen, L.; Fraeye, I. Phosphate Reduction in Emulsified Meat Products: Impact of Phosphate Type and Dosage on Quality Characteristics. *Food Technol. Biotechnol.* **2017**, *55*, 390–397. [CrossRef] [PubMed]

7. van Gunst, A.; Roodenburg, A.J.C. Consumer Distrust about E-numbers: A Qualitative Study among Food Experts. *Foods* **2019**, *8*, 178. [CrossRef]
8. Sen, A.R.; Naveena, B.M.; Muthukumar, M.; Babji, Y.; Murthy, T.R.K. Effect of chilling, polyphosphate and bicarbonate on quality characteristics of broiler breast meat. *Br. Poult. Sci.* **2005**, *46*, 451–456. [CrossRef] [PubMed]
9. Petracci, M.; Bianchi, M.; Mudalal, S.; Cavani, C. Functional ingredients for poultry meat products. *Trends Food Sci. Technol.* **2013**, *33*, 27–39. [CrossRef]
10. Solomon, M.B.; Liu, M.N.; Patel, J.; Paroczay, E.; Eastridge, J.; Coleman, S.W. Tenderness improvement in fresh or frozen/thawed beef steaks treated with hydrodynamic pressure processing. *J. Muscle Foods* **2008**, *19*, 98–109. [CrossRef]
11. Thangavelu, K.P.; Kerry, J.P.; Tiwari, B.K.; McDonnell, C.K. Novel processing technologies and ingredient strategies for the reduction of phosphate additives in processed meat. *Trends Food Sci. Technol.* **2019**, *94*, 43–53. [CrossRef]
12. Phillips, G.O.; Williams, P.A. *Handbook of Food Proteins*; Elsevier Science & Technology: Cambridge, UK, 2011. [CrossRef]
13. Youssef, M.K.; Barbut, S. Effects of caseinate, whey and milk proteins on emulsified beef meat batters prepared with different protein levels. *J. Muscle Foods* **2010**, *21*, 785–800. [CrossRef]
14. Youssef, M.K.; Barbut, S. Effects of two types of soy protein isolates, native and preheated whey protein isolates on emulsified meat batters prepared at different protein levels. *Meat Sci.* **2011**, *87*, 54–60. [CrossRef]
15. Egbert, W.R.; Payne, C.T. Plant Proteins. In *Ingredients in Meat Products: Properties, Functionality and Applications*; Tarté, R., Ed.; Springer: New York, NY, USA, 2009; pp. 111–129. [CrossRef]
16. Tarté, R. Meat-Derived Protein Ingredients. In *Ingredients in Meat Products: Properties, Functionality and Applications*; Tarté, R., Ed.; Springer: New York, NY, USA, 2009; pp. 145–171. [CrossRef]
17. Feiner, G. Additives: Proteins, carbohydrates, fillers and other additives. In *Meat Products Handbook*; Feiner, G., Ed.; Woodhead Publishing: Cambridge, UK, 2006; pp. 89–141. [CrossRef]
18. Choe, J.-H.; Kim, H.-Y.; Lee, J.-M.; Kim, Y.-J.; Kim, C.-J. Quality of frankfurter-type sausages with added pig skin and wheat fiber mixture as fat replacers. *Meat Sci.* **2013**, *93*, 849–854. [CrossRef]
19. Cofrades, S.; Guerra, M.A.; Carballo, J.; Fernández-Martín, F.; Colmenero, F.J. Plasma Protein and Soy Fiber Content Effect on Bologna Sausage Properties as Influenced by Fat Level. *J. Food Sci.* **2000**, *65*, 281–287. [CrossRef]
20. Nieto, G.; Castillo, M.; Xiong, Y.L.; Álvarez, D.; Payne, F.A.; Garrido, M.D. Antioxidant and emulsifying properties of alcalase-hydrolyzed potato proteins in meat emulsions with different fat concentrations. *Meat Sci.* **2009**, *83*, 24–30. [CrossRef] [PubMed]
21. Hurtado, S.; Saguer, E.; Toldrà, M.; Parés, D.; Carretero, C. Porcine plasma as polyphosphate and caseinate replacer in frankfurters. *Meat Sci.* **2012**, *90*, 624–628. [CrossRef]
22. Pereira, A.G.T.; Ramos, E.M.; Teixeira, J.T.; Cardoso, G.P.; Ramos, A.d.L.S.; Fontes, P.R. Effects of the addition of mechanically deboned poultry meat and collagen fibers on quality characteristics of frankfurter-type sausages. *Meat Sci.* **2011**, *89*, 519–525. [CrossRef]
23. Lowder, A.C.; Goad, C.L.; Lou, X.; Morgan, J.B.; Mireles DeWitt, C.A. Evaluation of a dehydrated beef protein to replace sodium-based phosphates in injected beef strip loins. *Meat Sci.* **2011**, *89*, 491–499. [CrossRef]
24. Muguruma, M.; Tsuruoka, K.; Katayama, K.; Erwanto, Y.; Kawahara, S.; Yamauchi, K.; Sathe, S.K.; Soeda, T. Soybean and milk proteins modified by transglutaminase improves chicken sausage texture even at reduced levels of phosphate. *Meat Sci.* **2003**, *63*, 191–197. [CrossRef]
25. Xiong, Y.L. Dairy Proteins. In *Ingredients in Meat Products: Properties, Functionality and Applications*; Tarté, R., Ed.; Springer: New York, NY, USA, 2009; pp. 131–144. [CrossRef]
26. Dàvila, E.; Parés, D.; Cuvelier, G.; Relkin, P. Heat-induced gelation of porcine blood plasma proteins as affected by pH. *Meat Sci.* **2007**, *76*, 216–225. [CrossRef]
27. Sun, X.D.; Arntfield, S.D. Dynamic oscillatory rheological measurement and thermal properties of pea protein extracted by salt method: Effect of pH and NaCl. *J. Food Eng.* **2011**, *105*, 577–582. [CrossRef]
28. Steen, L.; Glorieux, S.; Goemaere, O.; Brijs, K.; Paelinck, H.; Foubert, I.; Fraeye, I. Functional Properties of Pork Liver Protein Fractions. *Food Bioprocess. Tech.* **2016**, *9*, 970–980. [CrossRef]
29. Pearce, K.N.; Kinsella, J.E. Emulsifying properties of proteins: Evaluation of a turbidimetric technique. *J. Agric. Food Chem.* **1978**, *26*, 716–723. [CrossRef]
30. Sultana, S.; Ali, M.E.; Ahamad, M.N.U. Gelatine, collagen, and single cell proteins as a natural and newly emerging food ingredients. In *Preparation and Processing of Religious and Cultural Foods*; Ali, M.E., Nizar, N.N.A., Eds.; Woodhead Publishing: Cambridge, UK, 2018; pp. 215–239. [CrossRef]
31. Haug, I.J.; Draget, K.I. 5-Gelatin. In *Handbook of Food Proteins*; Phillips, G.O., Williams, P.A., Eds.; Woodhead Publishing: Cambridge, UK, 2011; pp. 92–115. [CrossRef]
32. Strixner, T.; Kulozik, U. 7-Egg proteins. In *Handbook of Food Proteins*; Phillips, G.O., Williams, P.A., Eds.; Woodhead Publishing: Cambridge, UK, 2011; pp. 150–209. [CrossRef]
33. Alting, A.C.; Pouvreau, L.; Giuseppin, M.L.F.; van Nieuwenhuijzen, N.H. 12-Potato proteins. In *Handbook of Food Proteins*; Phillips, G.O., Williams, P.A., Eds.; Woodhead Publishing: Cambridge, UK, 2011; pp. 316–334. [CrossRef]
34. Katzav, H.; Chirug, L.; Okun, Z.; Davidovich-Pinhas, M.; Shpigelman, A. Comparison of Thermal and High-Pressure Gelation of Potato Protein Isolates. *Foods* **2020**, *9*, 1041. [CrossRef]

35. Mession, J.-L.; Sok, N.; Assifaoui, A.; Saurel, R. Thermal denaturation of pea globulins (Pisum sativum L.)-molecular interactions leading to heat-induced protein aggregation. *J. Agric. Food Chem.* **2013**, *61*, 1196–1204. [CrossRef] [PubMed]
36. Zhang, S.; Huang, W.; Feizollahi, E.; Roopesh, M.S.; Chen, L. Improvement of pea protein gelation at reduced temperature by atmospheric cold plasma and the gelling mechanism study. *Innovative Food Sci. Emerg. Technol.* **2021**, *67*. [CrossRef]
37. Boye, J.I.; Alli, I.; Ismail, A.A.; Gibbs, B.F.; Konishi, Y. Factors affecting molecular characteristics of whey protein gelation. *Int. Dairy J.* **1995**, *5*, 337–353. [CrossRef]
38. O'Riordan, D.; Kinsella, J.E.; Mulvihill, D.M.; Morrissey, P.A. Gelation of plasma proteins. *Food Chem.* **1989**, *33*, 203–214. [CrossRef]
39. Tarté, R. Meat protein ingredients. In *Handbook of Food Proteins*; Phillips, G.O., Williams, P.A., Eds.; Woodhead Publishing: Cambridge, UK, 2011; pp. 56–91. [CrossRef]
40. Schrieber, R.; Gareis, H. From Collagen to Gelatine. In *Gelatine Handbook*; Wiley-VCH Verlag GmbH: Weinheim, Germany, 2007; pp. 45–117. [CrossRef]
41. Boland, M. Whey proteins. In *Handbook of Food Proteins*; Phillips, G.O., Williams, P.A., Eds.; Woodhead Publishing: Cambridge, UK, 2011; pp. 30–55. [CrossRef]
42. Rodriguez Furlán, L.; Rinaldoni, A.N.; Padilla, A.; Campderros, M. Assessment of Functional Properties of Bovine Plasma Proteins Compared with Other Protein Concentrates, Application in a Hamburger Formulation. *Am. J. Food Technol.* **2011**, *6*, 717–729. [CrossRef]
43. Chang, C.; Li, X.; Li, J.; Niu, F.; Zhang, M.; Zhou, B.; Su, Y.; Yang, Y. Effect of enzymatic hydrolysis on characteristics and synergistic efficiency of pectin on emulsifying properties of egg white protein. *Food Hydrocoll.* **2017**, *65*, 87–95. [CrossRef]
44. Amine, C.; Dreher, J.; Helgason, T.; Tadros, T. Investigation of emulsifying properties and emulsion stability of plant and milk proteins using interfacial tension and interfacial elasticity. *Food Hydrocoll.* **2014**, *39*, 180–186. [CrossRef]
45. Burger, T.G.; Zhang, Y. Recent progress in the utilization of pea protein as an emulsifier for food applications. *Trends Food Sci. Technol.* **2019**, *86*, 25–33. [CrossRef]
46. Gómez-Guillén, M.C.; Giménez, B.; López-Caballero, M.E.; Montero, M.P. Functional and bioactive properties of collagen and gelatin from alternative sources: A review. *Food Hydrocoll.* **2011**, *25*, 1813–1827. [CrossRef]
47. Totosaus, A.; Montejano, J.G.; Salazar, J.A.; Guerrero, I. A review of physical and chemical protein-gel induction. *Int. J. Food Sci. Tech.* **2002**, *37*, 589–601. [CrossRef]
48. Sun, X.D.; Holley, R.A. Factors Influencing Gel Formation by Myofibrillar Proteins in Muscle Foods. *Compr. Rev. Food Sci. Food Saf.* **2011**, *10*, 33–51. [CrossRef]
49. Shan, L.; Li, Y.; Wang, Q.; Wang, B.; Guo, L.; Sun, J.; Xiao, J.; Zhu, Y.; Zhang, X.; Huang, M.; et al. Profiles of gelling characteristics of myofibrillar proteins extracted from chicken breast: Effects of temperatures and phosphates. *LWT* **2020**, *129*, 109525. [CrossRef]
50. Cao, Y.; Ma, W.; Huang, J.; Xiong, Y.L. Effects of sodium pyrophosphate coupled with catechin on the oxidative stability and gelling properties of myofibrillar protein. *Food Hydrocoll.* **2020**, *104*, 105722. [CrossRef]
51. Liu, G.; Xiong, Y.L. Gelation of Chicken Muscle Myofibrillar Proteins Treated with Protease Inhibitors and Phosphates. *J. Agric. Food Chem.* **1997**, *45*, 3437–3442. [CrossRef]
52. Çarkcioğlu, E.; Rosenthal, A.J.; Candoğan, K. Rheological and Textural Properties of Sodium Reduced Salt Soluble Myofibrillar Protein Gels Containing Sodium Tri-Polyphosphate. *J. Texture Stud.* **2016**, *47*, 181–187. [CrossRef]
53. Pietrasik, Z. Binding and textural properties of beef gels processed with κ-carrageenan, egg albumin and microbial transglutaminase. *Meat Sci.* **2003**, *63*, 317–324. [CrossRef]
54. Hunt, A.; Park, J.W.; Handa, A. Effect of Various Types of Egg White on Characteristics and Gelation of Fish Myofibrillar Proteins. *J. Food Sci.* **2009**, *74*, C683–C692. [CrossRef] [PubMed]
55. Wang, Z.; Liang, J.; Jiang, L.; Li, Y.; Wang, J.; Zhang, H.; Li, D.; Han, F.; Li, Q.; Wang, R.; et al. Effect of the interaction between myofibrillar protein and heat-induced soy protein isolates on gel properties. *CYTA J. Food* **2015**, *13*, 527–534. [CrossRef]
56. Li, J.; Chen, Y.; Dong, X.; Li, K.; Wang, Y.; Wang, Y.; Du, M.; Zhang, J.; Bai, Y. Effect of chickpea (Cicer arietinumL.) protein isolate on the heat-induced gelation properties of pork myofibrillar protein. *J. Sci. Food Agric.* **2020**. [CrossRef]
57. AUTIO, K.; MIETSCH, F. Heat-induced Gelation of Myofibrillar Proteins and Sausages: Effect of Blood Plasma and Globin. *J. Food Sci.* **1990**, *55*, 1494–1496. [CrossRef]
58. Sun, X.D.; Arntfield, S.D. Molecular forces involved in heat-induced pea protein gelation: Effects of various reagents on the rheological properties of salt-extracted pea protein gels. *Food Hydrocoll.* **2012**, *28*, 325–332. [CrossRef]
59. Gordon, A.; Barbut, S.; Schmidt, G. Mechanisms of meat batter stabilization: A review. *Crit. Rev. Food Sci. Nutr.* **1992**, *32*, 299–332. [CrossRef] [PubMed]
60. Prabhu, G.A. Utilizing functional meat-based proteins in processed meat applications. In Proceedings of the 55th Annual Reciprocal Meat Conference, American Meat Science Association, Savoy, IL, USA, 29 July 2002; pp. 29–34.
61. Offer, G.; Trinick, J. On the mechanism of water holding in meat: The swelling and shrinking of myofibrils. *Meat Sci.* **1983**, *8*, 245–281. [CrossRef]
62. Velemir, A.; Mandić, S.; Vučić, G.; Savanović, D. Effects of non-meat proteins on the quality of fermented sausages. *Foods Raw Mater.* **2020**, *8*, 259–267. [CrossRef]
63. Fernandez, P.; Cofrades, S.; Solas, M.T.; Carballo, J.; Colmenero, F.J. High Pressure-Cooking of Chicken Meat Batters with Starch, Egg White, and Iota Carrageenan. *J. Food Sci.* **1998**, *63*, 267–271. [CrossRef]

 *foods*

Article

Replacement of Sodium Tripolyphosphate with Iota Carrageenan in the Formulation of Restructured Ostrich Ham [†]

Sumari Schutte [1,2], Jeannine Marais [2], Magdalena Muller [2] and Louwrens C. Hoffman [1,3,*]

[1] Department of Animal Sciences, University of Stellenbosch, Private Bag X1, Matieland, Stellenbosch 7602, South Africa; sumari.milner@gmail.com

[2] Department of Food Science, University of Stellenbosch, Private Bag X1, Matieland, Stellenbosch 7602, South Africa; jeanninemarais@sun.ac.za (J.M.); mm7@sun.ac.za (M.M.)

[3] Centre for Nutrition and Food Sciences, Queensland Alliance for Agriculture and Food Innovation (QAAFI), The University of Queensland, Health and Food Sciences Precinct, 39 Kessels Rd., Coopers Plains 4108, Australia

*   Correspondence: Louwrens.hoffman@uq.edu.au

[†] The article is derived from the MSc thesis of the first author "Development of value added ostrich (Struthio camelus) meat products".

Citation: Schutte, S.; Marais, J.; Muller, M.; Hoffman, L.C. Replacement of Sodium Tripolyphosphate with Iota Carrageenan in the Formulation of Restructured Ostrich Ham. *Foods* **2021**, *10*, 535. https://doi.org/10.3390/foods10030535

Academic Editors: Gonzalo Delgado-Pando, Tatiana Pintado and Andrea Garmyn

Received: 28 December 2020
Accepted: 2 March 2021
Published: 5 March 2021

Publisher's Note: MDPI stays neutral with regard to jurisdictional claims in published maps and institutional affiliations.

Abstract: The influence of iota carrageenan (iota-CGN) as a partial replacement of sodium tripolyphosphate (STPP) was investigated on the physical (pH, yield, instrumental color, texture profile analysis), chemical (moisture, protein, total fat, ash, phosphate) and sensory (descriptive analysis, acceptance testing) quality of restructured ostrich ham (95% lean meat plus fat). Treatments consisted of five decreasing levels of STPP (0.70%, 0.53%, 0.35%, 0.18% and 0%) that were simultaneously substituted with five increasing levels of iota-CGN (0%, 0.1%, 0.2%, 0.3% and 0.4%). Cooked yield, hardness, cohesiveness, and gumminess of restructured ostrich ham increased ($p \leq 0.05$) with decreasing levels of STPP (and increased levels of iota-CGN). No significant trend in instrumental color measurements or springiness were observed between treatments. Ostrich ham with 0.35% STPP and lower had increased ostrich meat aroma and flavor, while spicy aroma and flavor, mealiness and consumer acceptance decreased. Iota carrageenan can be substituted for STPP (up to 0.35% STPP and 0.2% iota-CGN) to produce reduced STPP ham.

Keywords: iota carrageenan; chemical composition; consumer acceptance; descriptive analysis; ham; ostrich; phosphate; polysaccharide; processed; restructured meat; sensory profile

1. Introduction

Restructured ham is usually prepared from large pieces of meat that are molded together to resemble a whole muscle meat product after cooking. The actual binding of adjacent meat pieces relies on extraction of myofibrillar proteins by salt (NaCl), phosphate and mechanical action (massaging or tumbling). During subsequent heating, the latter proteins, of which myosin is the major protein, coagulate and act as a bonding agent holding the meat pieces together [1–5]. The binding properties of restructured ham are essential to produce a uniformly attractive product with desirable slicing characteristics. The most desirable properties of high-quality cooked ham are cohesiveness, textural firmness, and juiciness.

Polyphosphates are used extensively in restructured meat products due to their functional properties of increasing binding strength, water holding capacity and yield [4,6–10]. Polyphosphate action is ascribed to the increase in the pH and ionic strength in meat products [11,12]. Tripolyphosphates (TPP) are the most widely used of all the phosphates utilized in meat processing and are typically permitted up to 3.5% of final product weight. However, there is an increase in the demand for meat products with reduced phosphate levels [13].

The presence of excessive amounts of phosphates in the diet may influence the calcium, iron, and magnesium balance in the human body, and can increase the risk of bone diseases [14–16]. Furthermore, consumers and retailers generally associate polyphosphates with cost reduction and lower quality products. Consumers also seem to associate the term 'polyphosphates' with non-food applications, viewing them as 'chemical products'. The former indicates an opportunity for the use of alternatives to phosphates in restructured cooked meat products [3,5,13,17]. Numerous non-meat functional ingredients, mainly proteins and polysaccharides, have been applied as binders, fillers, and extenders to improve the quality of restructured meat products [4,5,18,19]. These ingredients are primarily used for their water binding ability and texture modification functionality [20].

Hydrocolloids with their unique characteristics in building texture, stability and emulsification are of great interest in the low-fat processed meat area due to their ability to bind water and form gels [21]. Carrageenan (CGN), a sulphated polysaccharide extracted from seaweed, is a hydrocolloid used extensively in the food industry in a broad range of applications because of its water binding, thickening and gelling properties [22,23]. There are three major types: kappa (κ, gelling); iota (ι, gelling); and lambda-CGN (λ, non-gelling). These differ in degree and manner of sulphation, the position of the 3–6 anhydrogalactose residues, their pyranose ring conformations, and the cations associated with the sulphate groups [23]. Carrageenans, alone or combined with other ingredients, have been used extensively in restructured meat products [24–29] for their ability to form gels, retain water and to provide a desirable texture [30,31]. Various levels of ingredients in combination with CGN have been studied; for example, the use of 1.5% salt with iota-CGN improved the cooking yield, juiciness, and tenderness of restructured pork nuggets [25]. Kappa-CGN favorably affected hydration properties and thermal stability, yielding lower cooking loss, purge, and expressible moisture of beef gels [27], whilst kappa-CGN increased the sliceability and rigidity in roasted turkey breasts [24], and improved the adhesion in pork hams [26].

Ostrich meat is frequently marketed as a healthy alternative to other red meats as it has a favorable fatty acid profile and a low intramuscular fat content [32–34]. Ostrich meat has a high ultimate pH of ca. 6.0 [35] and should by implication have a high-water binding capacity and thus be able to retain high levels of moisture. Therefore, moisture-retaining agents, such as phosphates, in restructured ostrich meat products could be reduced.

To maintain the health characteristics of ostrich meat, it is suggested that an alternative ingredient, that mimics the textural, functional and flavor characteristics of phosphate, be introduced in the formulation of restructured meat products. Therefore, the aim of this study was to investigate the effect of replacement of sodium tripolyphosphate (STPP) with iota carrageenan (iota-CGN) on the physical, chemical, sensory characteristics, and consumer acceptability of restructured cooked ostrich ham.

2. Materials and Methods

2.1. Ham Manufacture

Five different ham formulations with decreased levels of STPP and increased levels of iota-CGN were produced (Table 1). Each treatment was formulated to contain a 95% Total Meat Equivalent (TME) on chemical analysis (lean meat and fat). Brine ingredients, expressed as percentage in the brine, consisted of 9% NaCl, 0.25% sodium erythorbate, 1% curing salt (NaCl + 0.6% nitrite), 20% starch (corn flour), 1% ground garlic, 1% ground ginger, STPP (3.5%, 2.63%, 1.75%, 0.88% and 0%, respectively), iota-CGN (0%, 0.5%, 1.0%, 1.5% and 2.0%, respectively), and water (64.25%, 64.62%, 65.00%, 65.37% and 65.75%, respectively). The corn flour was added to the brine and the meat after the first tumble cycle.

Table 1. Formulation of five ostrich ham treatments.

	Sodium Tripolyphosphate/Iota Carrageenan Levels				
Ingredients (%)	**0.70%/0.0%**	**0.53%/0.1%**	**0.35%/0.2%**	**0.18%/0.3%**	**0.00%/0.4%**
Sodium tripolyphosphate	0.70	0.53	0.35	0.18	0.00
Iota carrageenan	0.00	0.10	0.20	0.30	0.40
Additives *	6.45	6.45	6.45	6.45	6.45
Water	12.85	12.92	13.00	13.07	13.15
Brine	20.00	20.00	20.00	20.00	20.00
Meat	80.00	80.00	80.00	80.00	80.00
Total	100.00	100.00	100.00	100.00	100.00

* Salt (1.8%), curing salt (0.2%), sodium erythorbate (0.05%), ginger (0.2%), garlic (0.2%), starch (4%).

Ostrich (*Struthio camelus* var. *domesticus*) fan fillets (n = 40 different birds; 1–1.5 kg weight per fan fillet) were obtained from a local European Union approved abattoir, Mosstrich (Mossdustria, Mossel Bay, South Africa), with all the muscles being randomly sampled from one day's kill. The meat was vacuum packed and frozen before being transported to Stellenbosch; where it was stored at −20 °C until used. Iota-CGN (GENU® texturizer type MB-150F) from Tranarc (Tranarc Holdings Pty Ltd., Benmore, South Africa) was used. All the remaining ingredients were provided by a single provider, Deli Spices (Epping, Cape Town, South Africa).

The thawed (24 h at 4 °C) ostrich fan fillets (n = 10 fillets per batch) were cut into fist sized pieces (±100 g per piece) and mixed in a container. The meat structure was subsequently further disrupted by the mild shearing action of passing through a meat mincing machine without any cutting blades or plates. The latter opened the meat structure to facilitate brine penetration and protein extraction, without reducing particle size. The meat from each batch was then divided into five smaller batches—one batch per treatment. The brine mixture for each treatment was then added to the meat and the latter mixture was tumbled (Biro VTS-41) under vacuum (25 kPa) for 6 h (4 °C) with a cycle of 20 min tumble and 10 min rest. After tumbling, the ham mixtures were vacuum stuffed (Talsa Model T0101, Germany) into impermeable plastic casings. The above-mentioned procedures were followed four times to produce four replications per treatment. Each replicate sample weighed approx. 1.5 kg and was 30 cm in length and 12 cm in diameter. Each stuffed casing within each treatment was weighed and cooked in a water bath until a core temperature of 72 °C was reached (approximately 1 h). The internal temperature of the ham was measured using a thermocouple probe inserted into the center of the product. After cooking, the hams were immediately immersed in cold water containing ice for 15 min before refrigeration at 4 °C prior to subsequent analyses.

2.2. Chemical Analyses

Homogenized samples of the five ham treatments (of a randomly selected ham within each treatment) were analyzed in duplicate for total percentages of moisture, ash, and phosphorus (according to AOAC Official Methods 934.01, 942.05, and 960.03, respectively) [36]. The total crude protein content was determined on dried (60 °C for 24 h), defatted and ground (with a pestle and mortar to a fine powder) samples (0.1 mg) encapsulated in LecoTM foil sheets and analyzed using a Leco Protein Analyzer (FP-528, Leco Corporation). An EDTA calibration sample (Leco Corporation, St. Joseph, HI, USA, Part number 502–092, lot number 1038) was analyzed before and after every 10 samples, with the intention of ensuring the accuracy and recovery rate of each sample. A Nitrogen conversion factor of 6.25 was used to determine the total protein content. The total fat content was determined by extracting the fat with a 2:1 mixture of chloroform:methanol [37]. The laboratory at the Department of Animal Sciences, Stellenbosch University, is accredited by the Agricultural Laboratory Association of South Africa (AgriLASA) to perform accurate and reliable proximate analyses. For validation of accuracy and repeatability, the laboratory partakes in the monthly National Inter-laboratory Scheme where blind tests are conducted. The lean meat equivalent (LME) was calculated using a conversion factor of 30 to convert protein to lean meat and the total meat equivalent (TME) was obtained through the summation of the LME and fat.

2.3. Physical Analyses

The pH of the refrigerated (4 °C) cooked hams was measured with the use of a calibrated (standard buffers pH 4.0 and 7.0) portable Testo 502 pH-meter. Cooked yield, color (CIE lightness L*, a* and b* color coordinates) and Texture Profile Analysis (TPA) measurements were recorded on each of the four ham replicates per treatment. Cooking yield was expressed as follows:

Cooked yield (%) = (W1 − W2) × 100 where W1 = ham weight after cooking and W2 = ham weight before cooking

The weight of the cooked product was recorded after 24 h chilling (4 °C), when the products were removed from the casings, touch dried with absorbent paper, and casing weight recorded, separate from product weight. Product weight losses occurred primarily during thermal processing; weight loss due to the exudate remaining in the tumbler was small (about 1%) as the tumbler surfaces had been scraped with a spatula to reclaim as much exudate as possible.

Instrumental color measurements of cooked ham were recorded on three slices obtained from each of the four ham replicates per treatment [38]. A color-guide 45°/O° colorimeter (Cat no: 6805; BYK-Gardner, BYK-Instruments, Orlando, FL, USA) was used; the colorimeter was calibrated using the supplied calibration white tile according to the supplier's instruction before and between every 10 samples. Three ham slices (1.5 to 2.0 cm thick) of each treatment were allowed to "bloom" for 30 min at ambient temperature (*ca.* 20 °C) prior to color measurements. Four color measurements were recorded for each slice at randomly selected positions and expressed by the coordinated L*, a* and b* of the CIELab colorimetric space. In the color space L* indicates lightness and a* and b* are the chromaticity coordinates, where a* is the red-green range, and b* the yellow-blue range of the color spectrum.

Instrumental textural properties were analyzed using the Instron Universal Testing Machine (UTM, model 3344, 825 University Ave, Norwood, MA, 02062–2643, USA). Texture Profile Analysis (TPA) was performed on five cores (2.5 cm height and 2 cm diameter) per slice (two slices of each of the four replicates within the five treatments = 40 measurements per treatment). The cores were placed on the platform of the UTM. A circular plate of 2.5 cm diameter was attached to a 50 N load cell and the sample was compressed to 50% of its original height at a cross head speed of 200 mm/min twice in two cycles [39]. Hardness (N), springiness (mm), cohesiveness (ratio) and gumminess (N) were calculated for each sample [39].

2.4. Sensory Evaluation

2.4.1. Descriptive Sensory Analysis

Descriptive sensory analysis (DSA) was conducted to determine the effect of STPP reduction on the sensory quality characteristics of all five treatments of ostrich ham [40]. For each treatment four replicate encased hams were produced. The encased hams (stored at 4 °C) were opened 2 h prior to sensory analysis, sliced into 3.5 mm thick slices and vacuum packed (Multivac C200, Bahnhofstraße 4, D-87787 Wolfertschwenden, Germany). Four slices were placed next to each other and the slices did not overlap when vacuum packed.

A panel of assessors (n = 8), with extensive experience in DSA of meat, was trained in two interactive sessions to familiarize them with the treatments and to identify the aroma, flavor and mouthfeel characteristics associated with the respective treatments. Reference standards were also used to enable the assessors to calibrate their sensory perception during training, thereby allowing them to recognize and score all the characteristics tested in the respective treatments. The reference standards included commercial beef fillet, ostrich fan fillet and pork ham, resembling the meaty, ostrich meat and spicy aroma and flavor notes, respectively. Beef liver was used to illustrate a mealy meat texture. The questionnaire was compiled during the first training session and refined and tested during the second training session. Unstructured 100-point line scales were used to analyze the sensory characteristics. Table 2 depicts the sensory characteristics and definitions used.

Table 2. Definitions of sensory characteristics for descriptive sensory analysis of five ostrich ham treatments.

Characteristics	Definition	Scale
Mealy aroma	The intensity of an overall meaty aroma, perceived by sniffing	0 = None; 100 = Strong
Ostrich meat aroma	The intensity of an ostrich meat aroma, perceived by sniffing	0 = None; 100 = Strong
Spicy aroma	The intensity of a spicy aroma, derived from ginger and garlic content, perceived by sniffing	0 = None; 100 = Strong
Meaty flavor	The intensity of an overall meaty flavor, perceived by tasting	0 = None; 100 = Strong
Ostrich meat flavor	The intensity of an ostrich meat flavor, perceived by tasting	0 = None; 100 = Strong
Spicy flavor	The intensity of a spicy flavor, derived from ginger and garlic content, perceived by tasting	0 = None; 100 = Strong
Mealiness	The degree of mealiness in the mouth, indicative of cohesiveness of sample, perceived by tasting	0 = None; 100 = Prominent

Sensory testing was performed in individual booths fitted with Compusense® software (Compusense, Guelph, ON, Canada) in a temperature—(20 °C) and light-controlled (equivalent to daylight) sensory evaluation area. A sample of each of the five treatments was served to the assessors in a randomized order in four replicate test sessions (two sessions per day). The sample size per treatment per test session was one slice, with each assessor receiving an eighth ($^1/_8$) of a slice. Each sample was coded with a three-digit blinding code and served at a refrigeration temperature of ca. 6–10 °C. Assessors were provided with distilled water, dried apple pieces and water biscuits as palate cleansers.

2.4.2. Acceptance Testing

Sensory acceptance testing was conducted with a hundred target consumers (79 females, 21 males) recruited among staff and students at Stellenbosch University, Stellenbosch, South Africa. The consumers tested three of the treatments (STPP levels 0.70%, 0.35% and 0.00%), without any knowledge of the formulation of the products. The sample size per treatment per consumer was an eighth ($^1/_8$) of a slice. Samples were coded with three-digit blinding codes and served in a random order to each consumer at a refrigeration temperature of ca. 6–10 °C. Testing was done in a temperature- (20 °C) and light-controlled (equivalent to daylight) sensory evaluation area. Consumer acceptance testing was tested using the traditional nine-point hedonic scale ranging from 1 (dislike extremely) to 9 (like extremely) [40].

2.5. Statistical Analysis

The experimental design consisted of five treatments and four replicates per treatment. One-way analysis of variance (ANOVA) was performed to compare treatment means in terms of chemical, physical and sensory data, using SAS version 9.1 statistical software [41]. The Shapiro–Wilk test was performed to test for non-normality [42]. In some cases, deviations from normality were the cause of one or two outliners, which were removed before the final analysis [43]. Student's t-Least Significant Difference (LSD) was calculated at a 5% significant level to compare treatment means. Pearson correlation coefficients were also calculated to measure the strength and direction of the linear relationship between selected variables.

For the consumer data, hedonic score values of three of the treatments were subjected to one-way ANOVA. Student's t-Least Significant Difference (LSD) was calculated at a 5% significant level to compare treatment means.

3. Results and Discussion

3.1. Chemical and Physical Characteristics

The chemical composition, total meat equivalent (TME), product pH, cooking yield, textural properties, and instrumental color of the five ham treatments with decreasing levels of STPP are presented in Table 3.

Table 3. Means (±SD) of the chemical and physical characteristics of five ostrich hams manufactured with decreasing Sodium tripolyphosphate levels (n = 4 per treatment) *.

	Sodium Tripolyphosphate/Iota Carrageenan Levels					LSD
	0.70%/0.0%	0.53%/0.1%	0.35%/0.2%	0.18%/0.3%	0.00%/0.4%	
Chemical composition						
Moisture (%)	73.2 [b] ± 0.0	73.4 [b] ± 0.1	73.8 [ab] ± 0.1	74.3 [a] ± 0.6	73.4 [b] ± 0.0	0.78
Fat (%)	2.9 [a] ± 0.1	2.8 [a] ± 0.3	2.5 [a] ± 0.2	2.8 [a] ± 0.3	2.7 [a] ± 0.2	0.61
Protein (%)	19.4 [a] ± 0.3	19.6 [a] ± 0.4	19.4 [a] ± 0.0	18.9 [a] ± 0.8	19.6 [a] ± 0.1	1.07
Ash (%)	4.0 [a] ± 0.0	3.7 [ab] ± 0.0	3.4 [bc] ± 0.3	3.3 [bc] ± 0.1	3.2 [c] ± 0.1	0.42
Phosphorus (%)	1.42	1.03	0.78	0.76	0.51	n/a
TME (calculated)[¤]	97.00	96.79	95.87	93.28	96.78	n/a
Product pH	6.24	6.23	6.26	6.21	6.20	n/a
Cooked yield (%)	86.0 [d] ± 0.9	88.1 [c] ± 0.2	91.9 [b] ± 2.4	94.1 [a] ± 1.5	92.5 [ab] ± 1.2	2.0
Instrumental color						
Lightness (L*)	48.1 [c] ± 1.9	49.4 [bc] ± 2.3	51.7 [a] ± 1.2	48.6 [c] ± 1.5	50.8 [ab] ± 2.2	1.53
Redness (a*)	9.8 [a] ± 0.6	9.1 [b] ± 0.7	8.3 [c] ± 0.5	9.5 [ab] ± 0.8	9.5 [ab] ± 0.9	0.59
Yellowness (b*)	11.4 [b] ± 0.5	12.4 [a] ± 1.2	12.7 [a] ± 1.2	12.6 [a] ± 0.9	13.0 [a] ± 0.7	0.77
Instrumental textural properties						
Hardness (N)	18.9 [c] ± 4.2	21.2 [c] ± 2.3	29.5 [b] ± 5.1	30.8 [b] ± 4.2	35.1 [a] ± 3.3	3.55
Cohesiveness (ratio)	0.42 [c] ± 0.64	0.44 [bc] ± 0.05	0.46 [abc] ± 0.03	0.49 [ab] ± 0.07	0.49 [a] ± 0.07	0.05
Gumminess (N)	8.3 [c] ± 2.0	10.9 [bc] ± 2.5	11.6 [bc] ± 6.5	14.3 [ab] ± 4.1	15.5 [a] ± 3.6	3.64
Springiness (mm)	5.3 [c] ± 0.6	5.1 [c] ± 0.5	5.6 [bc] ± 0.5	6.5 [a] ± 0.6	5.9 [b] ± 0.6	0.52

* Statistical analyses were performed on all data except for phosphorus, TME (Total Meat Equivalent: % Lean Meat Equivalent + % Total Fat) and pH, as these were measured/calculated only once per treatment; SD, Standard Deviation; LSD, Least Significant Difference ($p = 0.05$); [a–d] Means within the same row with different superscripts differ significantly ($p \leq 0.05$), where L* represents white (100) to black (0), a* represents green ($-$ve values) to red (+ve values) and b* represents blue ($-$ve values) to yellow (+ve values).

3.1.1. Chemical Composition

The ham formulated with 0.18% STPP presented the highest moisture content of 74.3% that differed ($p \leq 0.05$) from the hams formulated with 0.70%, 0.53% and 0% STPP (Table 3). As expected, since no fat was added during the manufacturing process, there were no differences ($p > 0.05$) in the lipid and protein content between the five ham treatments. In an earlier study [44], the lipid content of restructured pork shoulder was found to be in a range of 23% to 25%. This is much higher than the lipid content (2.5% to 2.9%) in the present study, which could be attributed to the low intramuscular fat content of ostrich meat [32]. The ash content decreased ($p \leq 0.05$) with decreased levels of STPP; the ham formulated with 0.70% STPP had the highest ash content (4.0%) whilst the ham formulated with 0% STPP had the lowest (3.2%). As the spice content was kept constant, the decrease in ash content may be attributed to the decreasing STPP levels. As expected, the phosphorus content in the hams also decreased with decreasing levels of STPP. However, the phosphorus content measured in the end-product proved to be much higher than the expected calculated phosphate content. These elevated values could be due to the natural phosphorus content (0.51%) of the meat as reflected in the ham formulated with no STPP added to the brine. Since a constant amount of phosphate was incrementally decreased in the formulation, it could be assumed that the discrepancies in the elevated phosphorus values were due to either sampling error or increased phosphorus content of a specific batch. Decreasing levels of STPP were found to have no effect on the pH of the cooked product.

3.1.2. Total Meat Equivalent (TME)

In this study the TME values of the hams formulated with 0.70%, 0.53% and 0% STPP were higher than the targeted value of 95% and therefore exceeded legal requirements, whereas the TME value of the 0.18% STPP level ham was lower (93.28%) (Table 3). Once more, the reason for this variation is unknown but may be linked to the latter sample having a lower protein and higher ($p \leq 0.05$) moisture content thus resulting in the calculated difference.

3.1.3. Cooked Yield

The decrease in STPP levels with a concomitant increase in iota-CGN levels resulted in an increase ($p \leq 0.05$) in the cooked yield of the restructured ostrich ham (Table 3). The latter can be attributed to the gelling properties and increased water binding capacity of the increased iota-CGN content [4]. During cooking, water and water-soluble components are released from myofibrils caused by the heat denaturation of the muscle proteins. Carrageenan develops a gel layer on the surface of the ham, which has a sealing effect, thereby decreasing the loss of the internal components [4]. The cooked yield levels observed in this experiment (86.0% to 94.1%) are substantially lower than that reported by Fisher and co-workers [44], who found that an ostrich ham-like product formulated with 0.3% and 1.5% phosphate produced a cooking yield of 99.21% and 99.42%, respectively. This difference could be due to different processing techniques, i.e., Fisher and co-workers [44] tumbled the meat for 20 min, whereas in this study, the meat was tumbled for 6 h.

3.1.4. Instrumental Color

The lightness (L* value) of the samples ranged between 48.1 and 51.7, redness (a* value) between 8.3 and 9.8 and yellowness (b* values) between 11.4 and 13.0 units (Table 3). The ham formulated with 0.35% STPP, was found to be the lightest (51.7) and least red (8.3) in color. However, the instrumental color measurements of the different ostrich ham samples revealed no pattern with relation to the decrease in STPP levels. This result is supported by a visually observed variation in the composition of each of the sample slices. Ostrich meat is known to have a darker color than other red meat types [45]. This is also evident in this study where the range of a* values (redness) in ostrich ham (8.3 to 9.8) are much higher than that of, for example, restructured beef steaks (3.82 to 5.94) [46]. Though not measured, it was observed that storage of the chilled (<4 °C) ham under lighting conditions (exposure of ham to light) between manufacture and consumption (over a 2-week period) led to browning of the product (decrease in redness). Light has a pro-oxidant effect that provokes a decrease in a* values due to oxidation and degradation of the nitroso-pigment [47,48]. This rapid oxidation warrants further investigation as does the use of higher nitrite levels to minimize this phenomenon.

3.1.5. Instrumental Texture Properties

The effect of the variation of the composition within each sample slice was reflected in the results for instrumental texture as no significant pattern was observed with the incremental decrease in the STPP levels (Table 3). However, significant differences in hardness, cohesiveness and gumminess were only observed with relation to the extreme manipulation of STPP (0.70% and 0%) during this experiment. The 0.53%, 0.35% and 0.18% STPP levels did not have a significant effect on the mentioned characteristics. The observed increase in the measured textural properties may be the results of increased levels of iota-CGN that forms a firm cohesive gel structure during cooling. These findings agree with results by Ulu [49], who studied the effect of carrageenan on the cooking and textural properties of low-fat meatballs.

3.2. Sensory Characteristics and Consumer Acceptance

The effect of reduced STPP on the sensory profile of five ham treatments is shown in Table 4. A meaty aroma was found to be the highest in the ham formulated with 0.35% (30.9), followed by 0.18% and 0.70% (25.7 and 25.4, respectively) STPP. Additionally, the ham formulated with 0.35% STPP was found to have the strongest ($p \leq 0.05$) meaty flavor, compared to the other ham treatments. All ham treatments illustrated perceptible meaty aromas and flavors, irrespective of STPP level. Ostrich meat aroma and flavor for the ham formulated with 0.18% and 0% STPP was found to be much stronger ($p \leq 0.05$) than the other ham treatments. The assessors were not able to discriminate ($p > 0.05$) between the ham treatments formulated with 0.70%, 0.53% and 0.35% STPP in terms of ostrich meat aroma and flavor. Therefore, a STPP level in ostrich ham of 0.18% and lower, does not conceal the typical aroma and flavor of ostrich meat even though spices were included at a constant level in all five treatments. Ginger and garlic were included in the formulae to mask the typical ostrich meat aroma and flavor. The sensory assessors perceived a slight spicy aroma and flavor in all ham treatments, which was perceived at lower intensities in the ham treatments with lower STPP levels (0.18% and 0%). Mealiness was defined as the mouthfeel experienced when the meat pieces separate upon chewing. This perception is indicative of the degree of cohesion between the meat pieces of the restructured ham. It seemed that STPP levels of 0.35% and higher resulted in increased mealiness ($p \leq 0.05$), significantly more than STPP levels 0.18% and 0.00%. Mealiness also correlated negatively ($r > -0.9$; $p \leq 0.05$) with the instrumental textural properties, particularly with the instrumental variables, hardness, and cohesiveness [49,50]. This increased mealiness could also be attributed to the increased cooking loss (Table 3) experienced in the higher % STPP inclusion treatments. This indicates that decreasing levels of STPP (coupled with increasing levels of iota-CGN) has a negative impact on the textural quality of the product as perceived by a trained taste panel.

Table 4. Means (±SD) for the sensory characteristics and hedonic scores (±SE) of five ostrich hams manufactured with decreasing Sodium tripolyphosphate levels (n = 4 per treatment).

	Sodium Tripolyphosphate/Iota Carrageenan Levels					LSD
	0.70%/0.0%	0.53%/0.1%	0.35%/0.2%	0.18%/0.3%	0.00%/0.4%	
Sensory characteristics						
Meaty aroma	25.4 [ab] ± 12.5	23.0 [b] ± 10.3	30.9 [a] ± 15.8	25.7 [ab] ± 14.3	23.6 [b] ± 15.3	5.88
Ostrich meat aroma	2.9 [b] ± 6.6	4.2 [b] ± 7.8	4.6 [b] ± 7.4	14.5 [a] ± 13.0	16.0 [a] ± 14.4	4.41
Spicy aroma	18.1 [ab] ± 17.6	19.2 [a] ± 16.8	13.1 [b] ± 11.8	4.6 [c] ± 8.2	6.3 [c] ± 10.7	5.48
Meaty flavor	26.8 [b] ± 14.4	25.5 [b] ± 14.5	40.2 [a] ± 18.5	22.1 [b] ± 15.4	22.2 [b] ± 16.4	5.08
Ostrich meat flavor	3.4 [b] ± 8.7	4.8 [b] ± 7.9	2.3 [b] ± 5.0	14.0 [a] ± 14.9	16.2 [a] ± 18.1	4.29
Spicy flavor	18.2 [a] ± 14.8	19.1 [a] ± 13.2	10.8 [b] ± 9.9	3.8 [c] ± 7.5	6.6 [bc] ± 11.0	5.38
Mealiness	17.5 [a] ± 14.4	18.8 [a] ± 16.2	11.8 [b] ± 10.0	3.5 [c] ± 4.2	5.7 [c] ± 8.8	4.29
Consumer preference						
Degree of liking	6.5 [a] ± 1.4	NE	6.4 [a] ± 1.4	NE	5.4 [b] ± 1.4	0.40

[a–c] Means within the same row with different superscripts differ significantly ($p \leq 0.05$); SD, Standard Deviation; SE, Standard Error; LSD, Least Significant Difference ($p = 0.05$); NE, Not Evaluated. Sensory characteristics were scored on 100-point scales, whereas the 9-point hedonic scale was used to score consumer preference.

Table 4 illustrates the degree of liking, as perceived by a group of target consumers, for three of the ostrich ham treatments. This group of consumers equally liked the ham formulated with 0.70% and 0.35% STPP ($p > 0.05$). However, the ostrich ham prepared with 0% STPP was found to be significantly ($p \leq 0.05$) less liked (an average value of 5.4 translates to neither like nor dislike on the nine-point hedonic scale). Therefore, it can be concluded that the STPP level in ostrich ham can be successfully reduced to an acceptable level of

0.35%. These results serve as a further confirmation that further product development is necessary to produce a feasible phosphate-free ostrich ham to the consumer [51].

4. Conclusions

The results from this study indicate that the production of a reduced STPP ostrich ham is a viable option for the ostrich meat industry. Due to the variation in the composition within the replicate samples of each treatment, no significant tendency was found with decreasing levels of STPP with relation to the chemical composition and physical properties measured. However, decreasing levels of STPP showed significant increases in the cooked yield, which could be attributed to the water binding ability of the increased levels of iota-CGN. The low-fat content of ostrich ham makes it a healthy option for the consumer. Descriptive sensory analysis and consumer acceptance results revealed that the STPP level in ostrich ham could be reduced to an acceptable level of 0.35%. Further research should investigate the use of other alternatives to substitute phosphate compounds and focus on optimizing the processing technique (i.e., tumbling time) for optimum myofibrillar protein extraction to manufacture a product with optimum textural and sensory quality. Further research should also include the use of antioxidants to control color changes and shelf-life studies of the product.

Author Contributions: Conceptualization and investigation, L.C.H., M.M. and S.S.; supervision, L.C.H. and M.M.; writing—review and editing, L.C.H., M.M., S.S. and J.M.; funding acquisition, L.C.H. All authors have read and agreed to the published version of the manuscript.

Funding: This research is supported by the South African Research Chairs Initiative (SARChI) and partly funded by the South African Department of Science and Technology (UID number: 84633), as administered by the National Research Foundation (NRF) of South Africa and by Stellenbosch University. Any opinions, findings and conclusions or recommendations expressed in this material are that of the author(s) and the National Research Foundation does not accept any liability in this regard.

Institutional Review Board Statement: Ethical review and approval were waived for this study as the meat products were processed according to South African legislation in an approved food grade facility.

Informed Consent Statement: Informed consent was obtained from all subjects involved in sensory and acceptance tests.

Acknowledgments: The inputs of Francois Mellett are appreciated.

Conflicts of Interest: The authors wish to declare no conflict of interest. The funders had no role in the design of the study; collection, analysis, and interpretation of data; writing of the manuscript, or in the decision to publish the research.

References

1. Macfarlane, J.J.; Schmidt, G.R.; Turner, R.H. Binding of meat pieces: A comparison of myosin, actomyosin and sarcoplasmic proteins as binding agents. *J. Food Sci.* **1977**, *42*, 1603. [CrossRef]
2. Siegel, D.G.; Theno, D.M.; Schmidt, G.R.; Norton, H.W. Meat massaging: The effects of salt, phosphate and massaging on cooking loss, binding strength and exudate composition in sectioned and formed ham. *J. Food Sci.* **1978**, *43*, 331–333. [CrossRef]
3. Resconi, V.C.; Keenan, D.F.; Garćia, E.; Allen, P.; Kerry, J.P.; Hamill, R.M. The effects of potato and rice starch as substitutes for phosphate in and degree of comminution on the technological, instrumental and sensory characteristics of restructured ham. *Meat Sci.* **2016**, *121*, 127–134. [CrossRef]
4. Sun, X.D. Utilization of restructuring technology in the production of meat products: A review. *CYTA J. Food.* **2009**, *7*, 153–162. [CrossRef]
5. Resconi, V.C.; Keenan, D.F.; Barahona, M.; Guerrero, L.; Kerry, J.P.; Hamill, R.M. Rice starch and fructo-oligosaccharides as substitutes for phosphate and dextrose in whole muscle cooked hams: Sensory analysis and consumer preferences. *LWT Food Sci. Technol.* **2016**, *66*, 284–292. [CrossRef]
6. Lee, J.B.; Hendricks, D.G.; Cornforth, D.P. Effect of sodium phytate, sodium pyrophosphate and sodium tripolyphosphate on physico-chemical characteristics of restructured beef. *Meat Sci.* **1998**, *50*, 273–283. [CrossRef]
7. Moiseev, I.V.; Cornforth, D.P. Sodium hydroxide and sodium tripolyphosphate effects on bind strength and sensory characteristics of restructured beef rolls. *Meat Sci.* **1997**, *45*, 53–60. [CrossRef]

8. Nielsen, G.S.; Petersen, B.R.; Møller, A.J. Impact of salt, phosphate and temperature on the effect of a transglutaminase (F XIIIa) on the texture of restructured meat. *Meat Sci.* **1995**, *41*, 293–299. [CrossRef]

9. Pepper, F.H.; Schmidt, G.R. Effect of blending time, salt, phosphate and hot-boned beef on binding strength and cook yield of beef rolls. *J. Food Sci.* **2006**, *40*, 227–230. [CrossRef]

10. Shults, G.W.; Wierbicki, E. Effects of sodium chloride and condensed phosphates on the water-holding capacity, pH and swelling of chicken muscle. *J. Food Sci.* **1973**, *38*, 991–994. [CrossRef]

11. Dziezak, J.D. Phosphates improve many foods. *Food Technol.* **1990**, *44*, 80–92.

12. Young, O.A.; Zhang, S.X.; Farouk, M.M.; Podmore, C. Effects of pH adjustment with phosphates on attributes and functionalities of normal and high pH beef. *Meat Sci.* **2005**, *70*, 133–139. [CrossRef]

13. Ruusunen, M.; Vainionpää, J.; Puolanne, E.; Lyly, M.; Lähteenmäki, L.; Niemistö, M.; Ahvenainen, R. Physical and sensory properties of low-salt phosphate-free frankfurters composed with various ingredients. *Meat Sci.* **2003**, *63*, 9–16. [CrossRef]

14. Calvo, M.S.; Park, Y.K. Changing phosphorus content of the US diet: Potential for adverse effects on bone. *J. Nutr.* **1996**, *126*, 1168S–1180S. [CrossRef]

15. Sandberg, A.S.; Brune, M.; Carlsson, N.G.; Hallberg, L.; Skoglund, E.; Rossander-Hulthén, L. Inositol phosphates with different numbers of phosphate groups influence iron absorption in humans. *Am. J. Clin. Nutr.* **1999**, *70*, 240–246. [CrossRef]

16. Norma, M.S.; Steinhardt, B.; Soullier, B.A.; Zemel, M.B. Effect of level and form of phosphorus and level of calcium intake on zinc, iron and copper bioavailability in man. *Nutr. Res.* **1984**, *4*, 371–379.

17. Flores, N.C.; Boyle, E.A.E.; Kastner, C.L. Instrumental and consumer evaluation of pork restructured with activa™ or with fibrimex™ formulated with and without phosphate. *LWT Food Sci. Technol.* **2007**, *40*, 179–185. [CrossRef]

18. Mittal, G.S.; Usborne, W.R. Meat emulsion extenders. *Food Technol.* **1985**, *39*, 121–130.

19. Ramírez, J.; Uresti, R.; Téllez, S.; Vázquez, M. Using salt and microbial transglutaminase as binding agents in restructured fish products resembling hams. *J. Food Sci.* **2002**, *67*, 1778–1784. [CrossRef]

20. Comer, F.W.; Dempster, S. Functionality of fillers and meat ingredients in comminuted meat products. *Can. Inst. Food Technol. J.* **1981**, *14*, 295–303. [CrossRef]

21. Candogan, K.; Kolsarici, N. Storage stability of low-fat beef frankfurters formulated with carrageenan or carrageenan with pectin. *Meat Sci.* **2003**, *64*, 207–214. [CrossRef]

22. DeFreitas, Z.; Sebranek, J.G.; Olson, D.G.; Carr, J.M. Carrageenan effects on salt-soluble meat proteins in model systems. *J. Food Sci.* **1997**, *62*, 539–543. [CrossRef]

23. Necas, L.; Bartosikova, L. Carrageenan: A review. *Vet. Med.* **2013**, *58*, 187–205. [CrossRef]

24. Bater, B.; Descamps, O.; Maurer, A.J. Quality characteristics of cured turkey thigh meat with added hydrocolloids. *Poult. Sci.* **1993**, *72*, 349–354. [CrossRef]

25. Berry, B.W.; Bigner, M.E. Use of carrageenan and konjac flour gel in low-fat restructured pork nuggets. *Food Res. Int.* **1996**, *29*, 355–362. [CrossRef]

26. Motzer, E.A.; Carpenter, J.A.; Reynolds, A.E.; Lyon, C.E. Quality of restructured hams manufactured with PSE pork as affected by water binders. *J. Food Sci.* **1998**, *63*, 1007–1011. [CrossRef]

27. Pietrasik, Z. Binding and textural properties of beef gels processed with κ-carrageenan, egg albumin and microbial transglutaminase. *Meat Sci.* **2003**, *63*, 317–324. [CrossRef]

28. Shand, P.J.; Sofos, J.N.; Schmidt, G.R. Kappa-carrageenan, sodium chloride and temperature affect yield and texture of structured beef rolls. *J. Food Sci.* **1994**, *59*, 282–287. [CrossRef]

29. Tsai, S.-J.; Unklesbay, N.; Unklesbay, K.; Clarke, A. Water and absorptive properties of restructured beef products with five binders at four isothermal temperatures. *LWT Food Sci. Technol.* **1998**, *31*, 78–83. [CrossRef]

30. Trudso, J.E. Increasing yields with carrageenan. *Meat Process.* **1985**, *24*, 37–39.

31. Verbeken, D.; Neirinck, N.; Van Der Meeren, P.; Dewettinck, K. Influence of κ-carrageenan on the thermal gelation of salt-soluble meat proteins. *Meat Sci.* **2005**, *70*, 161–166. [CrossRef]

32. Sales, J.; Marais, D.; Kruger, M. Fat content, caloric value, cholesterol content, and fatty acid composition of raw and cooked ostrich meat. *J. Food Compost. Anal.* **1996**, *9*, 85–89. [CrossRef]

33. Sales, J. Fatty acid composition and cholesterol content of different ostrich muscles. *Meat Sci.* **1998**, *49*, 489–492. [CrossRef]

34. Hoffman, L.C.; Jones, M.; Muller, N.; Joubert, E.; Sadie, A. Lipid and protein stability and sensory evaluation of ostrich (*Struthio camelus*) droëwors with the addition of rooibos tea extract (*Aspalathus linearis*) as a natural antioxidant. *Meat Sci.* **2014**, *96*, 1289–1296. [CrossRef] [PubMed]

35. Hoffman, L.C.; Wolmarans, W.J.; Smith, C.; Brand, T.S. Effect of transportation on ostrich (*Struthio camelus*) weight loss and meat quality. *Anim. Prod. Sci.* **2012**, *52*, 1153–1162. [CrossRef]

36. AOAC. *Official Method of Analysis*, 17th ed.; Association of Official Analytical Chemists: Gaithersburg, MD, USA, 2002.

37. Lee, C.M.; Trevino, B.; Chaiyawat, M. A simple and rapid solvent extraction method for determining total lipids in fish tissue. *J. AOAC Int.* **1996**, *79*, 487–492. [CrossRef]

38. Honikel, K.O. Reference methods for the assessment of physical characteristics of meat. *Meat Sci.* **1998**, *49*, 447–457. [CrossRef]

39. Bourne, M.C. Texture profile analysis. *Food Technol.* **1978**, *33*, 62–66, 72.

40. Lawless, H.T.; Heymann, H. *Sensory Evaluation of Food: Principals and Practices*, 2nd ed.; Chapman and Hall: New York, NY, USA, 2010.

41. SAS Institute. *SAS/STAT User's Guide*, 1st ed.; Version 9; SAS Institute Inc.: Cary, NC, USA, 2004.
42. Shapiro, S.S.; Wilk, M.S. An analysis of variance test for normality (complete samples). *Biometrika* **1965**, *52*, 591–611. [CrossRef]
43. Glass, G.V.; Peckham, P.D.; Sanders, J.R. Consequence of failure to meet assumptions underlying the fixed effects analyses of variance and covariance. *Rev. Educ. Res.* **1972**, *43*, 237–288. [CrossRef]
44. Fisher, P.; Hoffman, L.C.; Mellett, F.D. Processing and nutritional characteristics of value added ostrich products. *Meat Sci.* **2000**, *55*, 251–254. [CrossRef]
45. Hoffman, L.C.; Fisher, P.P. Comparison of meat quality characteristics between young and old ostriches. *Meat Sci.* **2001**, *59*, 335–337. [CrossRef]
46. Colmenero, F.J.; Serrano, A.; Ayo, J.; Solas, M.T.; Cofrades, S.; Carballo, J. Physiochemical and sensory characteristics of restructured beef steak with added walnuts. *Meat Sci.* **2003**, *65*, 1391–1397. [CrossRef]
47. Fernández-Ginés, J.M.; Fernández-Lopez, J.; Sayas-Barbera, E.; Sendre, E.; Pérez-Alvarez, J.A. Effects of storage conditions on quality characteristics of bologna sausage made with citrus fiber. *J. Food Sci.* **2003**, *68*, 710–715. [CrossRef]
48. Neethling, N.E.; Sigge, G.O.; Hoffman, L.C.; Suman, S.P.; Hunt, M.C. Exogenous and endogenous factors influencing color of fresh meat from ungulates. *Meat Muscle Biol.* **2017**, *1*, 253–275. [CrossRef]
49. Ulu, H. Effects of carrageenan and guar gum on the cooking and textual properties of low fat meatballs. *Food Chem.* **2006**, *95*, 600–605. [CrossRef]
50. Barbieri, S.; Soglia, S.; Palagano, R.; Tesini, F.; Bendini, A.; Petracci, M.; Cavani, C.; Toschi, T.G. Sensory and rapid instrumental methods as a combined tool for quality control of cooked ham. *Heliyon* **2016**, *2*, e00202. [CrossRef] [PubMed]
51. Steen, L.; Neyrinck, E.; De Mey, E.; De Grande, A.; Telleir, D.; Raes, K.; Paelinck, H.; Fraey, I. Impact of raw ham quality and tumbling time on the technological properties of polyphosphate-free cooked ham. *Meat Sci.* **2020**, *164*, 108093. [CrossRef] [PubMed]

 *foods*

Article

Effects of Garlic Powder and Salt on Meat Quality and Microbial Loads of Rabbit Burgers

Simone Mancini [1,*], Simona Mattioli [2], Roberta Nuvoloni [1,3], Francesca Pedonese [1,3], Alessandro Dal Bosco [2] and Gisella Paci [1,3]

[1]   Department of Veterinary Sciences, University of Pisa, Viale delle Piagge 2, 56124 Pisa, Italy; roberta.nuvoloni@unipi.it (R.N.); francesca.pedonese@unipi.it (F.P.); gisella.paci@unipi.it (G.P.)

[2]   Department of Agricultural, Food and Environmental Science, University of Perugia, Borgo XX Giugno 74, 06100 Perugia, Italy; simona.mattioli@hotmail.it (S.M.); alessandro.dalbosco@unipg.it (A.D.B.)

[3]   Interdepartmental Research Center "Nutraceuticals and Food for Health", University of Pisa, Via del Borghetto 80, 56124 Pisa, Italy

*   Correspondence: simone.mancini@unipi.it

Received: 9 July 2020; Accepted: 28 July 2020; Published: 31 July 2020

Abstract: The aim of the research study was to evaluate the effects of a common culinary spice such as garlic powder and salt addition on the quality and microbial shelf life of rabbit meat burgers. Rabbit burgers were evaluated for pH, the colour parameters, the water holding capacity and microbial loads during storage time of seven days at 4 °C. Four different formulations of burgers ($n = 180$ in total) were tested as control samples (only meat, C), burgers with garlic powder (at 0.25%, G), burgers with salt (at 1.00%, S) and burgers with both garlic powder and salt (0.25% and 1.00%, respectively, GS). As results, it was highlighted that garlic powder and salt addition significant affected pH, water holding capacity and some colour parameters of burgers. In particular, salt affected the pH of the raw burgers, leading to lower values that partially influenced all the colour parameters with higher a* values of S burgers. The mix of garlic powder and salt (GS burgers) showed mixed effects even if more closed to the G burgers than S ones. Salt expressed its properties of binding water molecules reducing drip and cooking losses in S and GS burgers. No variations in microbial loads were highlighted in relation to the formulations. Storage time affected all the parameters, highlighting a deterioration of the burgers' quality and an increase of the microbial loads.

Keywords: spice; ingredient; colour; ready-to-cook; meat preparation

1. Introduction

Decreasing rabbit meat consumption is spreading around the world, and the Mediterranean basin in particular, where rabbit meat was historically consumed, displays a decreasing per-capita consumption [1,2]. Other kinds of meat and the low amount of time available for cooking are affecting rabbit meat consumption, mostly among young consumers looking for more approachable kinds of food [3]. Indeed, rabbits are normally sold as pre-packed whole carcasses or cut-up (such as hind legs and loin) [4,5]. Production of meat products with rabbit meat could be a response to increase consumers' willingness to purchase this product.

Burgers are extensively consumed as fast meals and are recognizable and well known worldwide. Burgers represent an appetizing and easy to cook protein food that does not require long cooking processes or culinary preparation. This popularity and high rate of consumption have driven the lifestyle changes that drive consumers to also prefer ready-to-cook products in meat sector.

Burgers, and in general meat products, could represent a way to reintroduce rabbit meat to daily consumption. Indeed, in the last years attention has been paid to several different types of meat used as a basis for protein burgers as this type of product could be a solution to meet the consumers' willingness to purchase. Moreover, these combinations raise rabbit meat to a right market level of importance also in relation to its nutritional value, especially for children and the elderly [2].

Normally, burgers are sold as a ready-to-cook product, and several different recipes are available on market shelfs to attract consumers' attention. Burgers are sold as meat mixed with spices and other ingredients among which salt is always present. Spices and herbs are used as flavours, colours and aroma enhancers and also as preservatives due to their phytochemicals [6]. Salt (NaCl, sodium chloride) due to its chemical properties, mostly the ability to reduce water activity, can also play a role as a bacteriostatic in meat products that contain a low level of salt, typically 1–3%, increasing the overall safety of the food product [7].

In this study we tested garlic (*Allium sativum* L.), a widely used spice in the Mediterranean basin, in rabbit recipes. Garlic, besides its importance as flavour carrier and sensory characteristics, can also play an import role as an antimicrobial [8,9]. In this work burgers added with salt alone and garlic powder and salt mixes were tested in order to identify more products that could reliably meet consumers' liking. Although rabbit burgers are already on the market in some European countries where cuniculture and rabbit meat consumption have a historical tradition, such as Italy and Spain, the products have a low market penetration.

Therefore, the aim of this study was to evaluate different effects of garlic powder and salt, alone and in combinations, on burgers quality and microbial loads during refrigerated storage in order to increase our knowledge about rabbit meat products.

2. Materials and Methods

2.1. Experimental Design and Burgers Manufacture

Frozen rabbit hind legs (−20 °C, 1 month of frozen storage), derived from hybrid rabbits reared under intensive conditions (ninety day-old, 2.7 ± 0.30 kg) and fed commercial pelleted feed, were thawed for 18 h at 4 °C and then deboned. Meat was finely ground with a DN30323 meat mincer (DiNa Professional, Catania, Italy) into twelve meat batches, as previously reported [10,11]. Four different formulations (three batches per formulation) were performed as: control (C, only meat); meat supplemented with 0.25% of garlic powder (G); meat supplemented with 1.00% of salt (S); meat supplemented with 0.25% of garlic powder and 1.00% of salt (GS). Garlic powder and salt were purchased as food ingredients (garlic powder produced by Drogheria e Alimentari S.p.A., Florence, Italy, batch number: L010545; salt was sea salt type, NaCl).

Each batch was hand mixed and fifteen burgers of 100 g were formatted with a DN8097 forming machine (DiNa Professional, Catania, Italy; diameter 100 mm), for a total of forty-five burgers per formulation and 180 burgers in total. Burgers were then packaged in single Styrofoam trays, overwrapped with polyethylene film and stored raw at 4 ± 0.5 °C. At the fixed storage times (day 0, 4 and 7 of storage; D0, D4 and D7) burgers from each batch were tested as raw samples and as cooked samples.

Raw and cooked burgers were analysed at D0, D4 and D7 for the determination of the pH, the colour parameters, the water holding capacity and microbial loads (performed only on raw samples).

2.2. pH

The pH was measured using a pH meter (Eutech pH2700 Meter, Eutech Instruments Pte Ltd., Singapore) equipped with a XS Sensor Standard S7 (XS Sensor, Modena, Italy) and an automatic temperature compensator. pH meter was calibrated before each session with buffer solutions at pH 4.01 and 7.01 (HI7004L and HI7007L Hanna Instruments, Padova, Italy).

2.3. Colour

Colour was expressed as L*(lightness), a* (redness), and b* (yellowness) according to the CIElab system [12]. Colour parameters were measured using a Minolta CR300 chroma meter (Minolta, Osaka, Japan) with an aperture size of 8 mm, illuminant D65 and incidence angle of 0°. Before each session, the colorimeter was calibrated with a white tile (L* = 98.14, a* = −0.23 and b* = 1.89). Each data point was the mean of three replications measured on the surface of the burgers at randomly selected locations. Chroma (C*) and Hue angle (h*) were calculated as function of a* and b* following the formulas:

$$C^* = \sqrt{\left(a^{*2} + b^{*2}\right)}$$

$$h^* = \tan^{-1}\left(\frac{b^*}{a^*}\right)$$

The total colour difference (ΔE) was calculated as proposed by Sharma and Bala [13] between two different formulations at the same storage time or between two different storage times for the same formulation following the formula:

$$\Delta E_{\alpha-\beta} = \sqrt{(L^*_\alpha - L^*_\beta)^2 + (a^*_\alpha - a^*_\beta)^2 + (b^*_\alpha - b^*_\beta)^2}$$

where α and β subscripts of L*, a* and b* referred to two different formulations at the same storage time or two different storage times for the same formulation. Cooking effect on colour were determined for each F at each storage time (ST). As proposed by Sharma and Bala [13], the threshold of a human noticeable difference was fixed at 2.3 points.

2.4. Water Holding Capacity

Drip loss was calculated as proposed by Lundström and Malmfors [14] within the F between D0-D4 and D0-D7. Cooking loss was calculated as percentage of the decrease of weight before and after cooking in a preheated oven at 163 °C to an internal temperature of 71 °C and were turned every 4 min to prevent excess surface crust formation [15].

2.5. Microbial Quantifications

Ten grams of sample were aseptically removed and homogenised in a Stomacher 400 Circulator Lab Blender (Seward, Worthing, UK) with 90 mL of 0.1% peptone salt solution. Further serial dilutions were made in the same diluent and used for standard plate enumerations.

Total aerobic mesophilic and psychrotrophic bacteria were determined on Plate Count Agar (pour plate method) with incubation at 30 °C for 72 h, and 7 °C for 10 days, respectively; Enterobacteriaceae on Violet Red Bile Glucose Agar at 37 °C for 24 h; *Escherichia coli* on Tryptone Bile X Glucuronide Medium (TBX) at 44 °C for 24 h, lactic acid bacteria on MRS Agar (pour plate method) in anaerobiosis (Anaerogen 2.5L) at 30 °C for 72 h; *Brochothrix thermosphacta* on Streptomycin Thallous Acetate (STA) agar with STA selective supplement at 25 °C for 48 h; *Pseudomonas* spp. on Pseudomonas Agar base with CFC supplement at 25 °C for 72 h; yeasts and moulds on Yeast Extract Glucose Chloramphenicol Agar (pour plates method) after incubation at 25 °C for 5 days. Where not specified, spread plate method was used. All cultural media and supplements were from Oxoid (Basingstoke, UK). The bacterial counts were expressed as log Colony-Forming Units (CFU) per gram of sample.

2.6. Statistical Analysis

The effects of the formulation (F), of the storage time (ST) and their interaction (F × ST) on the burger parameters were analysed through a two-way ANOVA using the R software Version 1.2.5019 (R Core Team, The R Foundation for Statistical Computing, Vienna, Austria) [16]. The significance level was set at 5% (statistically significant for $p < 0.05$), and if statistical significance was found, the differences were assessed using Tukey's test ($p < 0.05$). When the interaction F × ST was not significant the results are reported as the mean of the fixed effects F and ST; the variability is expressed as root mean square error (RMSE).

3. Results and Discussion

Results of the pH, the colour parameters and the water holding capacity of the raw burgers are reported in Table 1. The formulation (F) significatively effected all the tested parameters. Furthermore, also storage time influenced quite all the parameters, indeed, only L* and b* coordinates did not show significant differences for ST ($p = 0.099$ and $= 0.066$, respectively). The addition of salt lead to lower pH values in average, as evidenced by the pH of S and GS formulations in relation to C and G burgers ($p = 0.002$). These differences in pH values might played a role in the colour values; indeed L* of S and GS burgers were lower than C and G ones ($p < 0.001$), as pH and lightness are linked by a negative correlation. Additions of garlic powder modified the redness (a*) and yellowness (b*) coordinates ($p < 0.001$ and $p = 0.013$, respectively). Natural pigments in garlic powder lead to a pale-yellow raw product than when processed turn to a strong yellow colour turning more vivid. Due to degradation processes the garlic pigments could produce a green-yellow tone [17–19]. Garlic natural pigments showed their effect in G and GS burgers with the reduction of redness value, as well as an increase of yellowness value. Additions of ingredients, both salt and garlic powder and its mix, decreased the chroma index in relation to the C burgers, beside S and GS burgers showed a greater decrement in chroma as salt addition affected negatively both a* and b* coordinates ($p < 0.001$). On the other hand, garlic induced an increase in h* leading to light-yellow burgers in colour due to the increase of b* value at the expense of a* value ($p < 0.001$).

Table 1. pH, colour parameters and water holding capacity of raw rabbit burgers.

Item	Formulation (F)				Storage Time (ST, Days)			*p*-Value			RMSE
	C	G	S	GS	D0	D4	D7	F	ST	F × ST	
pH	5.97 [a]	5.96 [a]	5.87 [b]	5.86 [b]	5.89 [y]	5.89 [y]	5.97 [x]	0.002	0.002	0.360	0.058
L*	57.24 [a]	57.65 [a]	51.31 [b]	51.04 [b]	55.38	53.80	53.75	<0.001	0.099	0.861	2.769
a*	5.86 [a]	4.62 [b]	5.53 [a]	4.09 [b]	5.57 [x]	5.34 [x]	3.97 [y]	<0.001	<0.001	0.343	0.759
b*	6.07 [b]	7.43 [a]	5.93 [b]	6.41 [ab]	7.04	5.97	6.37	0.013	0.066	0.252	0.938
C*	10.34 [a]	8.83 [ab]	8.29 [b]	7.71 [b]	10.19 [x]	8.10 [y]	8.07 [y]	<0.001	<0.001	0.234	1.209
h*	40.70 [b]	58.46 [a]	48.19 [b]	58.70 [a]	44.22 [y]	49.34 [y]	60.97 [x]	<0.001	<0.001	0.224	6.250
Drip loss%	0.63 [a]	0.77 [a]	0.40 [b]	0.44 [b]	0.00 [z]	0.73 [y]	0.95 [x]	0.016	<0.001	0.316	0.835
Cooking loss%	24.33 [a]	22.00 [a]	17.83 [b]	16.15 [b]	21.07 [x]	21.17 [x]	17.99 [y]	<0.001	0.032	0.828	0.024

C: control; G: control + 0.25% of garlic powder; S: control + 1% of salt; GS: control + 0.25% of garlic powder + 1% salt. [a,b] Different letters in the same row indicate significant differences at $p < 0.05$ for F. [x,y,z] Different letters in the same row indicate significant differences at $p < 0.05$ for ST.

During storage time a slight increase in pH was revealed at D7 ($p = 0.002$), that might be imputable to an alkalinisation of meat resulting from an increase in ammoniacal nitrogen levels and to the degradation of proteins and amino acids by Gram-negative bacteria [20–22]. Also the reduction of a* and C* and the increase of h* during storage time ($p < 0.001$ for all indexes) could be ascribed to bacteria metabolism actions and due to the formation of metmyoglobin produced by myoglobin oxidation [23,24].

Water holding capacity was affected principally from salt addition as both S and GS burgers showed lowest drip and cooking losses ($p = 0.016$ and $p < 0.001$, respectively), indeed salt due to its chemical properties contributes to water and fat binding in meat products and this property is enhanced by mincing processing [25–27]. As expected during storage time drip loss increased due to the natural water release, furthermore, the loss of water by the raw product during the storage lead to day 7 at more dry samples with a consequence of a lowest cooking loss.

Cooking flattered the formulation differences, with statistical evidence only on L* and h* parameters (Table 2). After cooking the presence of salt in the samples affected the L* value with lower values of S ang GS than C and G samples ($p < 0.001$), following the trend reported in raw samples. Even if a* and b* coordinates were not affected by F ($p = 0.065$ and $p = 0.278$, respectively) the h* index revealed a difference in colour between C and the other formulation that appeared lighter in colours ($p = 0.004$). During storage time a rise in b* value was highlighted ($p < 0.001$) as degradation of the pale pink colour of rabbit meat and the formation of yellowness complex. As consequence of b* value rise also both C* and h* increased their values meaning a lighter vivid yellowness samples at D7 than D0 ($p < 0.001$ and $p = 0.032$, respectively).

Table 2. pH and colour parameters of cooked rabbit burgers.

Item	Formulation (F)				Storage Time (ST, Days)			p-Value			RMSE
	C	G	S	GS	D0	D4	D7	F	ST	F × ST	
pH	6.13	6.11	6.07	6.06	6.10	6.07	6.11	0.218	0.053	0.893	0.102
L*	69.70 [a]	70.60 [a]	66.43 [b]	65.31 [b]	67.76	68.61	67.66	<0.001	0.452	0.201	2.278
a*	7.19	6.48	6.09	6.18	6.29	6.21	6.95	0.065	0.567	0.084	0.959
b*	14.77	14.27	15.23	16.14	14.09 [y]	14.58 [y]	16.64 [x]	0.278	<0.001	0.522	1.341
C*	16.43	15.70	16.42	17.29	15.44 [y]	15.86 [y]	18.09 [x]	0.567	<0.001	0.291	1.436
H*	64.09 [b]	65.98 [a]	68.02 [a]	68.97 [a]	65.90 [y]	66.97 [x]	67.42 [x]	0.004	0.032	0.102	2.563

C: control; G: control + 0.25% of garlic powder; S: control + 1% of salt; GS: control + 0.25% of garlic powder + 1% salt. [a,b] Different letters in the same row indicate significant differences at $p < 0.05$ for F. [x,y] Different letters in the same row indicate significant differences at $p < 0.05$ for ST.

Cooking might affect several chemical, physical, and even biological characteristics of the products. As burger require a cooking section to be eaten it is important how this final step is performed in order to maintain the chemical and nutritional properties added via formulation and physical properties related to sensory acceptance by consumers [28,29].

Colour differences (ΔEs) within F between ST are reported in Table 3; colour differences (ΔEs) within ST between F are reported in Table 4. All the F changed in a noticeable way the overall colour during the 7 days of storage (D0–D7). The addition of the sole garlic powder induced a strong variation colour after 4 days as reported by the ΔE value of D0–D4 period. That might be related to the rapid oxidation of the garlic compounds and the formation of a green-yellow hue that mitigate the pink rabbit meat colour. This modification in raw burgers affected also the ΔE between cooked burgers at D0 versus D4 as G formulation was the only one to reported values over the threshold of 2.3 points. On the contrary C, GS S cooked burgers showed higher variation in colour between D4 and D7.

Table 3. Colour differences (ΔE) within Formulation (F) between Storage Time (ST).

	Storage Time (ST, Days)								
	Raw Samples			Cooked Samples			Raw–Cooked Samples		
Formulation (F)	D0–D4	D4–D7	D0–D7	D0–D4	D4–D7	D0–D7	D0	D4	D7
C	2.21	3.09 *	5.18 *	0.97	3.38 *	4.32 *	13.65 *	14.93 *	17.42 *
G	4.66 *	2.51 *	3.57 *	3.69 *	6.96 *	3.58 *	12.63 *	18.39 *	14.51 *
GS	2.72 *	2.57 *	5.18 *	2.78 *	3.06 *	4.50 *	15.76 *	19.30 *	20.05 *
S	2.48 *	2.24	4.33 *	1.96	3.11 *	2.37 *	15.27 *	16.57 *	20.34 *

* Value over the threshold (2.3 points) with a noticeable difference in colour between the samples. C: control; G: control + 0.25% of garlic powder; S: control + 1% of salt; GS: control + 0.25% of garlic powder + 1% salt.

Table 4. Colour differences (ΔE) within Storage Time (ST) between Formulation (F).

	Storage Time (ST, Days)											
	D0				D4				D7			
Formulation (F)	C	G	S	GS	C	G	S	GS	C	G	S	GS
C		_2.81 *_	_8.21 *_	_7.21 *_		_4.84 *_	_7.29 *_	_7.39 *_		_4.94 *_	_5.37 *_	_4.94 *_
G	**0.67**		_8.27 *_	_6.64 *_	**4.03 ***		_6.15 *_	_5.27 *_	**0.51**		_7.02 *_	_8.10 *_
S	**5.77 ***	**5.40 ***		_1.73_	**2.67 ***	**6.26 ***		_3.70 *_	**2.84 ***	_2.41 *_		_1.15_
GS	**5.34 ***	**4.82 ***	**2.25**		**6.21 ***	**9.93 ***	**1.88**		**3.53 ***	**3.06 ***	**1.68**	

* Value over the threshold (2.3 points) with a noticeable difference in colour between the samples. C: control; G: control + 0.25% of garlic powder; S: control + 1% of salt; GS: control + 0.25% of garlic powder + 1% salt. In the columns of ΔEs within ST between F the values in italic or bold refer respectively to raw and cooked samples.

Colour differences at the same storage time between F highlighted that salt addition strongly changed the burgers' colour at D0 in both raw and cooked samples, reaching high ΔEs between S and GS in relation to C and G formulations. This tendency was maintained also at D4 even though also G burgers increase their colour distances from the control burgers. After cooking no noticeable differences was reported between S and GS burgers, countering the garlic effect on colour. After 7 days of storage the cooked burgers showed the lowest ΔE values between the formulations highlighting that the oxidation process occurred by the time affected all the samples inducing a general colour variation.

In Table 5 are reported the microbial loads of the raw burgers. No statistical differences were highlighted for the F main factor; thus, no effect of salt or garlic addition was evidenced on the microbial load. *Escherichia coli* and *Brochothrix thermosphacta* were not detected in the burgers both in relation to the formulation and the storage time. All the detected bacterial loads increased during storage time ($p < 0.001$ for all the parameters), mostly with differences between each fixed day of analysis. Only the total aerobic psychrotrophic bacteria showed to reach the highest value at D4 and to maintain it at D7.

Table 5. Microbial determinations on raw rabbit burgers.

Item	Formulation (F)				Storage Time (ST, Days)			*p*-Value			RMSE
	C	G	S	GS	D0	D4	D7	F	ST	F × ST	
Enterobacteriaceae	3.80	3.94	3.77	3.94	1.59 [z]	3.96 [y]	6.04 [x]	0.824	<0.001	0.691	0.445
Pseudomonas spp.	5.41	5.31	4.91	4.96	2.23 [z]	5.73 [y]	7.48 [x]	0.507	<0.001	0.984	0.748
Lactic acid bacteria	3.49	3.79	3.42	3.99	2.21 [z]	3.36 [y]	5.44 [x]	0.633	<0.001	0.969	0.923
Yeast and moulds	3.69	3.44	3.19	3.09	1.99 [z]	3.34 [y]	4.73 [x]	0.355	<0.001	0.665	0.686
Total aerobic mesophilic bacteria	6.62	6.49	6.05	6.21	4.22 [z]	6.73 [y]	8.08 [x]	0.293	<0.001	0.692	0.603
Total aerobic psychrotrophic bacteria	6.29	5.71	5.95	5.53	3.61 [y]	6.42 [x]	7.58 [x]	0.703	<0.001	0.794	1.299

C: control; G: control + 0.25% of garlic powder; S: control + 1% of salt; GS: control + 0.25% of garlic powder + 1% salt. [x,y,z] Different letters in the same row indicate significant differences at $p < 0.05$ for ST.

Spices and other ingredients added into minced meat products, such as burgers, might affect in different ways the growth of microorganisms. Turmeric and ginger powders showed a bacteriostatic effect against several different bacteria in rabbit burgers stored at 4 °C [20,23]. Similarly, pork burgers/patties supplemented with different plant products, such as passion fruit co-products and tea or grape extracts, showed a lower bacterial growth than the respective control treatment [30–32]. Ingredients' activity against microorganisms' growth might be also related to the physical form and technological transformations, to the employed concentrations or to the meat used. Indeed, Sallam et al. [33] reported activities against aerobic plate count of fresh garlic (30 g/kg) and garlic powder (9 g/kg) added in chicken sausages stored at 3 °C up to 21 days. Also Aydin et al. [34] reported activities of fresh garlic (10%) in ground beef refrigerated for 24 h in terms of total aerobic mesophilic bacteria and coliform bacteria.

4. Conclusions

The additions of garlic powder and salt to rabbit meat could bring several characteristics modifications to burgers along with culinary perceptions. Both garlic powder and salt also played a role in the colour changes in relation to the storage time. No effects on the microbial loads suggest that higher concentrations of garlic powder or salt are needed if their use is to be also intended as bacteriostatic additives. Different garlic products such as fresh minced or extracts could produce higher/lower beneficial effects, thus further studies are needed to better the potential application of this spice and how addition of salt and/or garlic could affect burgers' flavour and consumers' acceptance.

Author Contributions: Conceptualization, S.M. (Simone Mancini), R.N. and G.P.; methodology, S.M. (Simone Mancini) and G.P.; formal analysis, S.M. (Simone Mancini), R.N. and F.P.; writing—original draft preparation, S.M. (Simone Mancini); writing—review and editing, S.M. (Simone Mancini), S.M. (Simona Mattioli), R.N., F.P., A.D.B. and G.P.; supervision, G.P. All authors have read and agreed to the published version of the manuscript.

Funding: This research received no external funding.

Conflicts of Interest: The authors declare no conflict of interest.

References

1. Kallas, Z.; Gil, J.M. A dual response choice experiments (DRCE) design to assess rabbit meat preference in Catalonia: A heteroscedastic extreme-value model. *Br. Food J.* **2012**, *114*, 1394–1413. [CrossRef]
2. Petracci, M.; Soglia, F.; Baldi, G.; Balzani, L.; Mudalal, S.; Cavani, C. Technical note: Estimation of real rabbit meat consumption in Italy. *World Rabbit Sci.* **2018**, *26*, 91–96. [CrossRef]
3. González-Redondo, P.A.; Contreras-Chacón, G.M. Perceptions among university students in seville (SPAIN) of the rabbit as livestock and as a companion animal. *World Rabbit Sci.* **2012**, *20*, 155–162. [CrossRef]
4. Petracci, M.; Cavani, C. Rabbit meat processing: Historical perspective to future directions. *World Rabbit Sci.* **2013**, *21*, 217–226. [CrossRef]
5. Zotte, A.D. Rabbit farming for meat purposes. *Anim. Front.* **2014**, *4*, 62–67. [CrossRef]
6. Embuscado, M.E. Spices and herbs: Natural sources of antioxidants—A mini review. *J. Funct. Foods* **2015**, *18*, 811–819. [CrossRef]
7. Betts, G.; Everis, L.; Betts, R. Microbial issues in reducing salt in food products. In *Reducing Salt in Foods*; Elsevier: Amsterdam, The Netherlands, 2007; pp. 174–200, ISBN 9781845690182.
8. Casella, S.; Leonardi, M.; Melai, B.; Fratini, F.; Pistelli, L. The role of diallyl sulfides and dipropyl sulfides in the in vitro antimicrobial activity of the essential oil of garlic, *Allium sativum* L., and leek, *Allium porrum* L. *Phyther. Res.* **2013**, *27*, 380–383. [CrossRef]
9. Chung, L.Y. The antioxidant properties of garlic compounds: Allyl cysteine, alliin, allicin, and allyl disulfide. *J. Med. Food* **2006**, *9*, 205–213. [CrossRef]
10. Mancini, S.; Nuvoloni, R.; Pedonese, F.; Paci, G. Effects of garlic powder and salt additions in rabbit meat burgers: Preliminary evaluation. *J. Food Process. Preserv.* **2019**, *43*, e13894. [CrossRef]
11. Mancini, S.; Mattioli, S.; Nuvoloni, R.; Pedonese, F.; Dal Bosco, A.; Paci, G. Effects of garlic powder and salt additions on fatty acids profile, oxidative status, antioxidant potential and sensory properties of raw and cooked rabbit meat burgers. *Meat Sci.* **2020**, *169*, 108226. [CrossRef]
12. Commission Internationale de l'Eclairage. *Official Recommendations on Uniform Colour Spaces, Colour Differences Equations and Metric Colour Terms*; CIE: Paris, France, 1976.
13. Sharma, G.; Bala, R. *Digital Color Imaging Handbook*; Sharma, G., Ed.; Electrical Engineering & Applied Signal Processing Series; CRC Press: Boca Raton, FL, USA, 2002; Volume 11, ISBN 978-0-8493-0900-7.
14. Lundström, K.; Malmfors, G. Variation in light scattering and water-holding capacity along the porcine Longissimus dorsi muscle. *Meat Sci.* **1985**, *15*, 203–214. [CrossRef]
15. AMSA. *Research Guidelines for Cookery, Sensory Evaluation and Instrumental Tenderness Measurements of Fresh Meat*; National Live Stock and Meat Board: Chicago, IL, USA, 1995.
16. R Core Team. *R: A Language and Environment for Statistical Computing*; The R Foundation for Statistical Computing: Vienna, Austria, 2015.
17. Lee, E.-J.; Cho, J.-E.; Kim, J.-H.; Lee, S.-K. Green pigment in crushed garlic (*Allium sativum* L.) cloves: Purification and partial characterization. *Food Chem.* **2007**, *101*, 1677–1686. [CrossRef]
18. Wang, D.; Yang, X.; Wang, Z.; Hu, X.; Zhao, G. Isolation and identification of one kind of yellow pigments from model reaction systems related to garlic greening. *Food Chem.* **2009**, *117*, 296–301. [CrossRef]
19. Wang, D.; Li, X.; Zhao, X.; Zhang, C.; Ma, Y.; Zhao, G. Characterization of a new yellow pigment from model reaction system related to garlic greening. *Eur. Food Res. Technol.* **2010**, *230*, 973–979. [CrossRef]
20. Mancini, S.; Preziuso, G.; Dal Bosco, A.; Roscini, V.; Szendrő, Z.; Fratini, F.; Paci, G. Effect of turmeric powder (*Curcuma longa* L.) and ascorbic acid on physical characteristics and oxidative status of fresh and stored rabbit burgers. *Meat Sci.* **2015**, *110*, 93–100. [CrossRef]
21. Kılıç, B.; Şimşek, A.; Claus, J.R.; Atılgan, E. Encapsulated phosphates reduce lipid oxidation in both ground chicken and ground beef during raw and cooked meat storage with some influence on color, pH, and cooking loss. *Meat Sci.* **2014**, *97*, 93–103. [CrossRef] [PubMed]
22. Rodríguez-Calleja, J.M.; García-López, M.-L.; Santos, J.A.; Otero, A. Development of the aerobic spoilage flora of chilled rabbit meat. *Meat Sci.* **2005**, *70*, 389–394. [CrossRef]
23. Mancini, S.; Preziuso, G.; Fratini, F.; Torracca, B.; Nuvoloni, R.; Dal Bosco, A.; Paci, G. Qualitative improvement of rabbit burgers using *Zingiber officinale* Roscoe powder. *World Rabbit Sci.* **2017**, *25*, 367. [CrossRef]

24. Choe, J.-H.; Jang, A.; Lee, E.-S.; Choi, J.-H.; Choi, Y.-S.; Han, D.-J.; Kim, H.-Y.; Lee, M.-A.; Shim, S.-Y.; Kim, C.-J. Oxidative and color stability of cooked ground pork containing lotus leaf (*Nelumbo nucifera*) and barley leaf (*Hordeum vulgare*) powder during refrigerated storage. *Meat Sci.* **2011**, *87*, 12–18. [CrossRef]

25. Fernández-Ginés, J.M.; Fernández-Lopez, J.; Sayas-Barbera, E.; Perez-Alvarez, J.A. Meat Products as Functional Foods: A Review. *J. Food Sci.* **2005**, *70*, R37–R43. [CrossRef]

26. Mariutti, L.R.B.; Bragagnolo, N. Influence of salt on lipid oxidation in meat and seafood products: A review. *Food Res. Int.* **2017**, *94*, 90–100. [CrossRef] [PubMed]

27. Overholt, M.F.; Mancini, S.; Galloway, H.O.; Preziuso, G.; Dilger, A.C.; Boler, D.D. Effects of salt purity on lipid oxidation, sensory characteristics, and textural properties of fresh, ground pork patties. *LWT Food Sci. Technol.* **2016**, *65*, 890–896. [CrossRef]

28. King, N.J.; Whyte, R. Does it look cooked? A review of factors that influence cooked meat color. *J. Food Sci.* **2006**, *71*, R31–R40. [CrossRef]

29. Rodriguez-Estrada, M.T.; Penazzi, G.; Caboni, M.F.; Bertacco, G.; Lercker, G. Effect of different cooking methods on some lipid and protein components of hamburgers. *Meat Sci.* **1997**, *45*, 365–375. [CrossRef]

30. López-Vargas, J.H.; Fernández-López, J.; Pérez-Álvarez, J.Á.; Viuda-Martos, M. Quality characteristics of pork burger added with albedo-fiber powder obtained from yellow passion fruit (*Passiflora edulis* var. flavicarpa) co-products. *Meat Sci.* **2014**, *97*, 270–276. [CrossRef] [PubMed]

31. Lorenzo, J.M.; Sineiro, J.; Amado, I.R.; Franco, D. Influence of natural extracts on the shelf life of modified atmosphere-packaged pork patties. *Meat Sci.* **2014**, *96*, 526–534. [CrossRef]

32. Mancini, S.; Paci, G.; Fratini, F.; Torracca, B.; Nuvoloni, R.; Dal Bosco, A.; Roscini, V.; Preziuso, G. Improving pork burgers quality using *Zingiber officinale* Roscoe powder (ginger). *Meat Sci.* **2017**, *129*, 161–168. [CrossRef]

33. Sallam, K.I.; Ishioroshi, M.; Samejima, K. Antioxidant and antimicrobial effects of garlic in chicken sausage. *LWT Food Sci. Technol.* **2004**, *37*, 849–855. [CrossRef]

34. Aydin, A.; Bostan, K.; Erkan, M.E.; Bingöl, B. The Antimicrobial Effects of Chopped Garlic in Ground Beef and Raw Meatball (Çiğ Köfte). *J. Med. Food* **2007**, *10*, 203–207. [CrossRef]

Article

Impact of Nutritional Information on Consumers' Willingness to Pay for Meat Products in Traditional Wet Markets of Taiwan

Shang-Ho Yang [1], Ardiansyah Azhary Suhandoko [2] and Dennis Chen [3,*]

[1] Graduate Institute of Bio-Industry Management, National Chung Hsing University, No. 145 Xingda Rd., South Dist., Taichung City 40227, Taiwan; bruce.yang@nchu.edu.tw

[2] International Master Program of Agriculture, National Chung Hsing University, No. 145 Xingda Rd., South Dist., Taichung City 40227, Taiwan; ardiansyahazhary@gmail.com

[3] Massey College of Business, Belmont University, 1900 Belmont Boulevard, Nashville, TN 37212, USA

* Correspondence: dennis.chen@belmont.edu

Received: 6 July 2020; Accepted: 3 August 2020; Published: 9 August 2020

Abstract: The application of nutritional labels provides information regarding the health and nutritional value of products and allows consumers to engage in healthier dietary habits. However, not all types of retail markets provide full nutrition information for meat products. Since there is no nutritional information for fresh meat products in traditional wet markets, this study aimed to investigate consumer purchasing intention and willingness to pay (WTP) for this nutritional information in Taiwanese traditional wet markets. A total of 1420 valid respondents were examined using the random utility theory to explain consumer purchasing intention and WTP for nutritional information. Results showed that most (over 60%) consumers in traditional wet markets have positive purchasing intent for meat products with the nutrition information provided. Furthermore, the nutrition information in traditional wet markets significantly boosts consumers' purchasing intention and WTP when consumers have a personal health awareness on meat, have proficient experience in buying meat, and continuously receive information from health-related media. Specifically, consumers' shopping background and their level of health consciousness would be the key factors that would alter their WTP, if provided nutritional claims.

Keywords: traditional wet market; food product's label; nutritional information; willingness to pay

---

1. Introduction

A healthy lifestyle awareness regarding healthy eating habits among consumers is rising in these recent times. Nowadays, consumers are starting to think about the healthiness of food as one of the most important attributes and are starting to buy more products that positively relate to their health [1]. Additionally, unhealthy food selection leads to health issues such as diabetes, hypertension, and other non-communicable diseases (NCDs) [2–4]. Therefore, consumers are starting to be more selective when choosing their foods [5], and Taiwan had risen to be one of the top countries whose citizens think health is the most important factor when purchasing foods in the market [6].

Taiwanese consumers' behavior in selecting food is consistent regardless of buying food at restaurants or for cooking at home. Moreover, it is widely believed that making food at home is associated with healthier food [7] and better life quality [8]. Thus, on many occasions, Taiwanese consumers are still keen on cooking at home. The habits of the Taiwanese are conducive to the prevalence of them visiting traditional markets because home cooks often visit there to purchase the freshest ingredients. However, cooking at home may be decreasing due to many factors, one of which is a lack of access or limited information regarding choosing healthy food products [9], especially in Taiwan [10]. Accordingly, this correlates with the decline of people visiting Taiwan's traditional markets [11].

Taiwan's traditional markets are quite popular among food buyers. These markets are often referred to as traditional wet markets [10,11] because they are often damp as a result of melting ice at meat stands and a sprinkling of water at vegetable vendors [12]. Nevertheless, buyers are still willing to frequent them despite their modest environments. Traditional wet markets are often visited by consumers of varied professions to buy foods, ranging from the elderly to managers of hotels and restaurants, particularly when seeking fresh products [13–15]. Traditional wet markets have other strengths that outshine other types of markets. These strengths include freshness [16], quality, social benefits (personal trust with buyers, buying–selling dialogue, and personal connectivity) [17–20], and the bargaining experience that saves money [12,21–23]. There are more than 50,000 merchants located in roughly 650 traditional wet markets across Taiwan [24], and their sales account for over $5.4 billion (3.84%) [25]. On the other hand, the vast emergence of hypermarkets (e.g., Carrefour and Costco) and the trend of younger Taiwanese consumers (which comprise approximately one-third of the population) who prefer to eat out put traditional wet markets' continued popularity at stake [26–29]. Favorably, there are still distinctive products in traditional wet markets that can offset this worry and can possibly sustain their superiority relative to products in the more modern hypermarkets.

For example, meat products consistently attract consumers in Taiwanese traditional wet markets. People would rather buy meat in traditional wet markets because of its freshness, the flexibility in choosing particular meat parts, and lower costs [21]. Meat products in traditional wet markets draw more attention than other markets, resulting in the fact that up to 50% of the items sold in traditional wet markets are meat [16]. However, a significant risk remains that meats are still often connected with occurrences of non-communicable diseases [30,31]. Thus, consumers need tools to guide them to be healthy and to support their decision at the point of purchase.

Food labels might be able to help because food labels are believed to be a marketing tool and information strategy that eventually impacts consumers' perceptions of food quality [32,33]. Additionally, if food labels regarding quality arc applied to a product, they may create positive outcomes such as a willingness to pay significantly more or the ability to lure the consumers into becoming loyal buyers [34–40]. In addition, to address the healthier choice issue on consumers, nutritional information could help sellers target this better. Nutrition information is one of the helpful attributes of food labels that are used in many countries around the world, including Taiwan, to help consumers in deciding what to buy and to develop health-conscious food choices [41–44]. It has also been suggested by the Food and Drug Administration (FDA) and the United States Department of Agriculture (USDA) to apply this method. Food labels with nutrition information give easier information related to nutrition content and health in any food products corresponding to the food guide pyramid [45–49]. Furthermore, products with nutrition information should have a higher consumer willingness to pay (WTP) at the time of purchasing [50]. However, the benefits and utilization of nutrition information have not been realized among meat products in traditional wet markets.

Previous research studies involving traditional wet markets have been commonly focused on their management, marketing, and pricing strategies in comparison to other types of markets. Research regarding the attributes of consumers' behaviors in traditional wet markets has not been widely observed [51,52]. The positive impact of nutritional information on consumers' preferences about food products has been researched earlier in menus [53,54], supermarkets [55–57], grocery stores [58], cafés [59], and restaurants [60]. In addition, for specific products like meat, the effect of increasing purchasing power has been found to be similar in packaged meats in supermarkets [61,62], processed meats in hypermarkets [63,64], and meats at restaurants [60], which later increased buyers' preferences. Therefore, this study fills this gap in research regarding nutritional information, as it has not yet been applied to traditional wet markets.

This study used an open-ended contingent valuation method to elicit consumers' valuation on non-existing product attributes in the real market by asking them to what extent they are willing to pay extra in hypothetical markets [65]. Since nutritional labels are quite rare (or even not yet available) in the real traditional wet market environment, respondents evaluated the WTP of meat products with nutrition information like they do with non-market goods. Since this model follows an open-ended method, it is commonly used for this kind of research [66,67]. This research also adopts the random utility theory (RUT) to understand consumers' behavior based on their circumstances and their habits [68]. Thus, the main objectives of this study were to investigate the impact of consumers' preferences of nutritional information on meat products in traditional wet markets of Taiwan and to assess their WTP meat products bearing nutritional information. The first hypothesis suggested that consumers would give positive feedback differently based on their demographic background, even when applied in traditional wet markets. The second hypothesis suggested consumers would want to pay more for meat products that command premium prices. This paper also examined what kind of consumers prefer more or less of these types of meat products, as nutrition information's effect would be varied based on their demographic profile. In addition, an increased consumer purchasing tendency may be a result when compared to prior research that has been done in different market contexts [58,60,64].

2. Materials and Methods

Traditional wet market consumers are likely to be unfamiliar with nutrition information for fresh meat products, so this study aimed to investigate consumers' WTP if provided such nutrition information. This study adopted the contingent valuation method to explain consumer WTP for nutritional claims' impact on meat products in a hypothetical condition within a traditional wet market. The consumers were surveyed with an open-ended questionnaire. To observe each consumers' utility in buying meat products with nutrition information, the RUT was adopted in the research. Details of the methodology are explained further in the following section.

2.1. Participants and Survey Design

The study was conducted between July and August 2015 in cities across Taiwan. The respondents were interviewed at Taiwan traditional markets and train stations. The instructions given for the questionnaire were thinking about pork belly products that provided nutritional information. Pork belly meat products were chosen, because it is one of the most prominent animal parts bought in the Taiwan markets. Furthermore, pork belly's price was at approximately 110–150 NT\$/600 g in 2015 [69]. Thus, we acknowledge that based on the market study, the price for pork belly meat products across Taiwan (from south to north) were within this price range. We designed the survey based on the price that was currently available in the market. In the end, we took the 130 NT\$ as the middle price.

A total valid 1420 respondents were collected. The sortation of 1420 people then provided options of who was reported to know the market price or who did not know/not sure the market price. For the respondents who knew the market price, they were sorted in one of the following categories: (i) 110 NT\$, (ii) 130 NT\$, or (iii) 150 NT\$. People who chose (iv) do not know/not sure the market price were put in the situation of 130 NT\$ price. Randomization was automatically generated by SurveyMonkey. The final sample sizes were as follows: (i) 110 NT\$ group ($N = 467$), (ii) 130 NT\$ group ($N = 223$), (iii) 150 NT\$ group ($N = 102$), and (iv) the do not know the market price group ($N = 628$). An example of how participants are categorized in each of the three groups are shown in Table 1.

Table 1. The description of sample statistics of variables ($N = 1420$).

Variables		Mean	Description
Dependent Variables			
Positive WTP		0.61	BV = 1 if respondent is willing to pay any extra from 0 NT$ for nutrition information, 0 o/w
WTP for nutrition information		5.16	CV = Respondent's WTP for nutrition information
Independent Variables **Socio-Demographic**			
Female		0.66	BV = 1 if respondent is female, 0 o/w
Age		41.07	CV = Years of age
Family number		4.14	CV = Number of members at home
Family income		65.47	CV = Monthly average household or family income
Education		15.22	CV = Years of education
Housewife		0.13	BV = 1 if occupancy of the respondent is housewife, 0 o/w
Northern Taiwan		0.49	BV = 1 if respondent is from Northern Taiwan, 0 o/w
Central Taiwan		0.28	BV = 1 if respondent is from Central Taiwan, 0 o/w
Urban		0.64	BV = 1 if respondent is from urban area, 0 o/w
Shoppers' Customs			
Frequency cook at home		6.74	CV = Frequency to cook at home (the average number of times in one week)
Main-shopper (Always)		0.50	BV = 1 if respondent is always a major food shopper in the house, 0 o/w
Main-shopper (Sometimes)		0.32	BV = 1 if respondent is sometimes a major food shopper in the house, 0 o/w
Time spent (30–60 min)		0.50	BV = 1 if respondent spends 30–60 min to buy food in the traditional market, 0 o/w
Time spent (>1 h)		0.14	BV = 1 if respondent spends over 1 h to buy food in the traditional market, 0 o/w
Shopping time (5–11 a.m.)		0.43	BV = 1 if respondent purchases the food in the traditional market at 5–11 a.m., 0 o/w
Shopping time (11–5 p.m.)		0.22	BV = 1 if respondent purchases the food in the traditional market at 11 a.m.–5 p.m., 0 o/w
Nutrition-Related Information			
Safety certificate		0.73	BV = 1 if respondent examines safety certificate and meat safety are relevant, 0 o/w
Meat grade information	Fair	0.30	BV = 1 if respondent thinks the meat grade as a potential service provided by butcher is fairly important, 0 o/w
	Important	0.47	BV = 1 if respondent thinks the meat grade as a potential service provided by butcher is important, 0 o/w
	Very Important	0.15	BV = 1 if respondent thinks the meat grade as a potential service provided by butcher is very important, 0 o/w
Nutrition and calorie label		0.20	BV = 1 if respondent thinks the nutrition and calorie label item can increase the willingness to buy meat, 0 o/w
Fat and lean ratio information		0.36	BV = 1 if respondent thinks the fat and lean ratio item can increase the willingness to buy meat, 0 o/w
Health media concern		0.40	BV = 1 if respondent often watches the health-related content on TV or magazines, 0 o/w

Source: Grouped by this research. Note: (BV) and (CV) represent the binary and continuous variables, respectively, WTP represents the willingness to pay, and the o/w represents otherwise.

Before going further into the questions, the respondents were asked several screening questions, namely: (1) "Have you been to any traditional wet market in the past 12 months?" and (2) "Have you purchased any fresh meat products at a traditional wet market in the past 12 months?" If the respondents chose "No, I have not" or "No, I do not remember" and "No, I have not" or "No, I do not know," respectively, then they were be considered for further analysis. To know the WTP of nutrition information, these kinds of questions were necessary to reduce sampling bias.

As the WTP for nutrition information was treated as a dependent variable in Equation (5), there were questions based on many independent variables categorized into the following three groups.

The first independent variable group was socio-demographic, which consisted of (1) gender, (2) age, (3) family number, (4) family income, (5) education level, (6) housewife status, (7) location of survey (north, central, or south), and (8) respondents' origin (urban or rural). The second independent variable group was shoppers' customs, which consisted of (1) frequency of cooking at home, (2) main-shopper habits, (3) time spent for shopping, and (4) when visiting the market. The last independent variable group was nutrition-related information. This group consisted of (1) safety certificate, (2) meat grade information (whether it was fairly important, important, or very important to the respondents), (3) nutrition and calorie label, (4) fat and lean ratio information, and (5) health media concern. A detailed explanation for each variable's measurement is shown in Table 1.

2.2. Theoretical Model Used

In traditional wet markets, specific attributes may contribute to consumers' final purchase decisions, e.g., the cleanliness of the atmosphere, the bargaining situation, or other attributes that relate to their habits such as visiting time to a market. All of these attributes can affect their final decision. The RUT describes a consumers' utility given the alternatives of attributes. The RUT is the model for each individual's utility given the same situation of research with various ranges of each individual's behavior and individual story. In other words, the RUT can be used to capture personal mobility choices. It has a basic hypothetical thought that every consumer is a decision-maker and they can maximize their utility relative to his/her choices [68]. By using the RUT, the shopper's utility from one product can be comprised of the product's function of attributes [70]. Moreover, regarding the budget issue of shopper's perception, prior literature has suggested picking the set of attributes that might enlarge consumers' utilities [71].

In this research, consumers' personal choices were determined by the following three types of independent variables: socio-demographic, shoppers' customs, and nutrition-related information. Therefore, the derived RUT mathematical model was written in vector notation as:

$$U_{ij} = \beta_k X_{ijk} + \varepsilon_{ij} \tag{1}$$

where U_{ij} represents the utility of ith shopper for pork j with nutrition information, β represents a homogenous vector of coefficients in which located among consumers, X_{ijk} represents the kth attribute of pork j for the ith shopper, and ε_{ij} represents the random residual that is unknown deviation for the user i's utility.

2.3. Data Analysis

The consumers' decision on meat products bearing nutrition information does not solely rely on the meat's health perception, as it is also associated with the independent variables that consist of socio-demographic, shoppers' customs, and nutrition-related information. Thus, this study observed these factors predicting a probability regarding whether consumers would like to pay extra for meat products bearing nutrition information in traditional wet markets. Therefore, the data were analyzed using the logit model with the probability to give a positive WTP:

$$p = pr(y_i = 1|X_i) = F(X'\beta) = \frac{e^{X'\beta}}{1 + e^{X'\beta}} = \frac{\exp(X'\beta)}{1 + \exp(X'\beta)} \tag{2}$$

where $y_i = 1$ stands for the probability to give a positive WTP and X_i stand for independent variables such as socio-demographic, shoppers' customs, and nutrition-related information. Moreover, $\partial p/\partial x j = F'(x'\beta)\beta_j$ shows the calculation of the marginal effect in this logit model.

Furthermore, this study attempted to estimate how much WTP for nutritional information of meat products in traditional markets. Interval regression was utilized because it has a mathematical simplicity and asymptotic characteristics, which constrained the predicted probabilities to a range of 0–1 and forecast the probability of willingness to pay [72] for nutritional labeling on pork belly products. Regarding the known interval boundaries of WTP, the interval regression model set-up can be demonstrated as below:

$$y_i^* = x_i'\beta + u_i \tag{3}$$

$$\Pr\left[a_j < y^* \leq a_{j+1}\right] = \Pr\left[y^* \leq a_{j+1}\right] - \Pr\left[y^* \leq a_j\right] = F^*\left(a_{j+1}\right) - F^*\left(a_j\right) \tag{4}$$

where y_i^* is observed to be in the $(J+1)$ mutually exclusive intervals $(-\infty, a_1], (a_1, a_2], \ldots, (a_J, \infty)$. Given the answers individuals gave in the survey, y^* was found to lie in corresponding intervals, i.e., $y^* \leq 0, 1 < y^* \leq 3, 4 < y^* \leq 6, \ldots$, and $16 \leq y^*$. The empirical specification for the WTP for nutrition information on meat products in traditional wet markets is as follows:

$$\text{WTP for nutrition information} = y^* = \beta_0 + \beta_1 X_1 + \beta_2 X_2 + \ldots + \beta_{23} X_{23} + \varepsilon \tag{5}$$

where the WTP for nutrition information is explained by twenty-three independent variables (grouped into "socio-demographic" variables, "shoppers' customs" variables, and "nutrition-related information" variables) that are represented by X_s. Then, β_s represent the parameters to be estimated, and ε denotes the unobserved error term.

3. Results and Discussion

3.1. Sample Distribution

The distribution of the sample is presented in Table 1. According to Table 1, the willingness to pay for pork products with nutrition information in Taiwan traditional wet markets was found to be approximately 5.16 NT\$/600 g. The socio-demographic data showed that more than half of the shoppers were 41-year-old or older females who had at least an associate degree (education period at least 15.22 years). Due to this fact, the monthly household income averaged 65,470 NT\$ (or 785,640 NT\$ as an annual household income). Moreover, about 13% of total respondents are purely identified as a housewife at their family, and the average number of family members are about four people in a family. Finally, the respondents were distributed geographically as follows: close to half were from Northern Taiwan, and one quarter were from Central Taiwan. These demographic patterns were similar to the previous studies in the Taiwan market that most of the samples are occupied by women (50%), with respondents above 30 years of age, high school graduates or below in education, approximately 775,673 NT\$ annual household income, and housewives accounting for 25% of the sample [22,73–77]. Therefore, the sample means were very close to the population means in income and other measured categories. This revealed that our sample results may have been a good representative for the overall market conditions—though our sample means were representative of the population means, they had no effect on the WTP estimation. However, being representative of the population means may imply that our WTP estimations were close to the market condition as well.

Shoppers' customs suggested that they are likely to cook their meals daily. The data also suggested that half of the respondents were the primary buyers of groceries for the family, while the rest were just casual buyers. Furthermore, the schedule and the duration of the shopping indicated that roughly half of the shoppers liked to shop in the morning between 5 and 11 a.m. and liked to take 30–60 min to buy food. It can be presumed that these respondents went for this period because the composition of them consisted of females and housewives who would have had time in the morning and the evening. However, roughly 10% of the buyers are the people who like to spend over 1 h wandering around the traditional wet market. Lastly, nutrition-related information suggested that 73% of the respondents thought that meat safety is relevant when supported by a safety certificate. Half of them also reported that the meat grade as provided by the butcher as a potential service is important (47%) or very important (17%). The reason for this was that almost half of the respondents were eager to get health-related information. Lastly, less than 40% of the people believed the nutrition and calorie label items, as well as fat and lean ratio information on meat products, would increase their willingness to pay.

3.2. The Probability of WTP an Extra

According to Table 2, the logit regression model fit well with these 23 independent variables, based on the indication of Wald χ^2 test. Female shoppers showed a significant difference in wanting to pay extra if given additional nutrition information as compared to the males in this study. As shown in Table 1, the largest distribution of the sample was from the northern part of Taiwan. However, from Table 2, it can be seen that people from Central Taiwan had a higher WTP if provided nutrition information than the people from Southern or Northern Taiwan. This was supported by a previous researcher who mentioned that the penetration of the traditional market in the central area was stronger than other parts of Taiwan [78].

Table 2. The summary of logit model and marginal effect results for WTP an extra for nutrition information ($N = 1420$).

Independent \ Dependent	Positive WTP	
	Coefficient	M.E.
Socio-Demographic		
Female	0.23 *	0.05 *
Age	0.00	0.00
Family number	0.04	0.01
Family income	−0.04	−0.01
Education	0.00	0.00
Housewife	−0.23	−0.05
Northern Taiwan	0.18	0.04
Central Taiwan	0.30 *	0.07 **
Urban	0.06	0.01
Shoppers' Customs		
Frequency cook at home	0.00	0.00
Main-shopper (Always)	0.03	0.01
Main-shopper (Sometimes)	0.26	0.06
Time consumed (30–60 min)	−0.08	−0.02
Time consumed (>1 h)	−0.08	−0.02
Morning shopping time (5–11 a.m.)	0.08	0.02
Evening shopping time (11–5 p.m.)	0.04	0.01
Nutrition-Related Information		
Safety certificate	0.15	0.03
Meat grade information — Fairly Important	0.47 **	0.10 **
Meat grade information — Important	0.47 **	0.11 **
Meat grade information — Very important	0.63 ***	0.11 ***
Nutrition and calorie label	0.59 ***	0.13 ***
Fat and lean ratio information	0.09	0.02
Health media concern	0.20 *	0.05 *
Constant	−0.40	
Log-Likelihood	−923.51	−923.51
Wald χ^2	51.28	
Pseudo R^2	0.03	0.03

Source: Calculated by this research. Note: (***), (**), and (*) denote statistical significance at the 1%, 5%, and 10% significance, respectively.

As seen in Table 2, the variables in shoppers' customs did not indicate any significance in the results of estimated coefficients and marginal-effects likelihood. On the contrary, the variables within nutrition-related information were observed to show positive answers towards the importance of the potential service items provided by the butcher, namely meat grade information. The classification of meat grades—whether fairly important, just important, or very important—has a positive chance to add money to a consumer's WTP. Nevertheless, the people who think meat grade information is very important, have the highest willingness to pay for nutrition label when compared to the other categories.

Moreover, from Table 2, it is seen that only nutrition and calorie labels would add to consumers' WTP to purchase meat in the Taiwan traditional wet market, while fat and lean ratio information would not. These findings might link to health media concerns. Whereas, it is indicated that the respondents who frequently gain health information for themselves from mass media would give a more positive effect on WTP for nutrition information than the people who occasionally or never pay their attention to those platforms [79]. It can be said that females who think additional nutritional information can increase WTP are the groups of people who are aware of the healthy and food-borne disease. This finding was similar to the previous study that females tend to buy healthier meat than males in a market situation [80]. In short, the WTP for nutrition information is mostly affected by nutrition-related information that has an absolute impact.

3.3. The WTP for Nutrition Information on Pork

Based on Table 3, it can be seen that the respondents presented mostly as groups who knew the price. These groups represented 55% of the sample. However, 44% of consumers chose "do not know" or were not sure about the current meat price in Taiwan traditional markets. Since the groups who knew the price were based on the respondents' knowledge about prices, this could be considered a random sortation for each group. Therefore, the largest group being those who knew the price value might have indicated that more people go to the lowest priced market or that more traditional markets are adopting the lower price market strategy. Within the groups of those who knew the price, the group with the lowest price value (110 NT$) dominated the sample, with 467 observations. They were followed by the other groups of people who selected 130 NT$ and 150 NT$, who had sample sizes of 223 and 102, respectively. From Table 3, it appears that the consumers who shopped for the lowest meat price might have had the highest WTP if given additional nutrition information.

Table 3. The estimation of WTP for nutrition information.

Independent \ Dependent	Chosen 110 NT$	Chosen 130 NT$	Chosen 150 NT$	Chosen Do Not Know (130 NT$)
Socio-Demographic				
Female	0.25	−0.23	−1.88	1.37 **
Age	0.01	−0.02	−0.19	0.03
Family number	0.32	0.94 ***	−0.79	0.09
Family income	0.01	−0.03	0.00	0.01
Education	−0.39 **	−0.06	−0.39	−0.01
Housewife	−1.32	0.30	0.57	−1.86 **
Northern Taiwan	1.15	−0.18	−0.71	0.40
Central Taiwan	−0.01	−0.22	0.22	0.80
Urban	−0.28	1.46	−4.26	0.95 *
Shoppers' Customs				
Frequency cook at home	−0.01	0.05	0.06	−0.04
Main-shopper (Always)	−1.06	−1.36	10.11 **	−0.05
Main-shopper (Sometimes)	−0.23	−1.19	12.61 ***	0.95
Time consumed (30–60 min)	0.31	1.04	−2.86	−0.78
Time consumed (>1 h)	0.37	1.12	−6.54 *	0.03
Morning shopping (5–11 a.m.)	−1.21	−0.60	5.49 **	0.67
Evening shopping (11–5 p.m.)	−0.43	0.95	3.73	0.36

Table 3. *Cont.*

Independent \ Dependent		Chosen 110 NT$	Chosen 130 NT$	Chosen 150 NT$	Chosen Do Not Know (130 NT$)
Nutrition-Related Information					
Safety certificate		0.31	0.49	−0.80	2.00 ***
Meat grade information	Fairly Important	2.00	1.31	7.31	2.89 **
	Important	2.47 *	0.72	7.54	2.63 **
	Very important	4.05 ***	4.14**	11.45	4.63 ***
Nutrition and calorie label		2.38 **	1.29	3.43	1.94 ***
Fat and lean ratio information		0.83	−0.36	−3.97	0.01
Health media concern		0.43	2.00 **	6.51 ***	0.85
Constant		*9.94 ***	*−2.89*	*−2.66*	*−7.311 **
Observations (*n*)		467	223	102	628
Log-Likelihood		−1014.52	−380.42	−127.37	−1110.35
Wald X^2		43.67	42.94	42.65	59.49
AIC		2079.04	810.84	304.73	2270.70

Source: Calculated by this research. Note: (***), (**), and (*) denote statistical significance at the 1%, 5%, and 10% significance, respectively.

3.3.1. Ordinary Buyers

The people who chose the 110 NT$ option had shorter education histories and could be assessed as ordinary buyers since they selected the lowest price (110 NT$) when they knew the price. They were regular people who would go to Taiwan traditional markets and might utilize additional information, such as nutrition and calorie labels and meat grade classification, as their decision-making factors. Though this was different than results from a prior study that stated that higher education people would pay attention to nutrition information more [81], this study showed that these Taiwanese people with less formal education would pay attention to nutrition information on meat products in traditional wet markets. When this group of people thought that this item was a very important potential service item provided by the butcher, they said that they would have a 4 NT$ more WTP for meat grade classification information. This group of ordinary buyers also said that they would increase their WTP by up to 2 NT$ if nutrition and calorie labels were added. Thus, this group's dependability and WTP level on nutrition information were significant because they reported wanting the meat grade classification and nutrition and calorie label.

3.3.2. Nutrition Information-Oriented Buyers

Ordinary buyers' dependency on nutrition-related information was similar to the group who did not know the meat price. This group said that when they were provided with nutrition-related information attributes, they increased their WTP. This group could be considered nutrition information-oriented buyers since almost all the entire nutrition-related information factors showed significant and positive WTP. In addition, female consumers in this category reported that they would add about 1.3 NT$ to their WTP if provided nutrition labels. As described, females in this group did not know the price, so they relied on other information given by the butcher in nutrition information. This conclusion was supported by similar research found that that women are willing to pay an extra price for higher nutritional content in meat [80,81] since a woman's decisions in the market often depend on external impulses [75]. Additionally, other findings have shown that this group raises its WTP by roughly 1 NT$ if the constituents are from an urban area and not a housewife. This finding might suggest these women are career women who live in big cities because they do not shop as often as housewives do. This is similar to prior research on career consumers that found that if products are given additional information, they might buy them more [82]. However, this result does not cover the fact that more career women exist than housewives nowadays. As stated before, the number of Taiwanese who are likely to eat out is growing [28,29], so these consumers tend to rely more on nutrition information when they do visit the market.

Moreover, nutrition information-oriented buyers who scored the meat grade classification as very important showed an approximately 0.60 NT$ higher WTP compared to ordinary buyers. This was consistent with previous studies that have shown that when nutrition information-oriented buyers care the most about meat grade, they are willing to pay more for the meat because they care greatly about quality, food safety, and health benefits [83–85]. This cluster also showed that they paid more attention to a safety certificate than the clusters who chose 110 NT$, 130 NT$ and 150 NT$. Thus, even though they did not know the price, if the meat product was given this kind of certificate, the WTP for nutrition information would rise by roughly 2 NT$. However, these consumers showed less WTP for nutrition and calorie labels than ordinary buyers, which means that they most likely perceived that safety and nutrition information were more important for their food choice. This is also supported by other prior studies that have stated nutrition information-oriented buyers' preferences on nutrition information are based on their particular and personal nutrition utility satisfaction regarding meat products [62,86]. In general, nutrition information-oriented buyers are highly dependent on nutrition information, as they show a positive WTP in many of the nutrition-related information independent variables.

3.3.3. Family-Oriented Buyers

The respondents who selected 130 NT$ as the meat price considered how many people lived in their house. The more family members they had, the more they were willing to pay at about 1 NT$. Additionally, this cluster was aware of the health media content through many mass media outlets since their WTP elicits approximately 2 NT$ more than the groups who do not know the price or chose 110 NT$. In addition, because they thought that meat grade was very important, they added an additional 4 NT$ to their WTP for this information. Thus, we can say that the shoppers who chose 130 NT$ were family-oriented buyers because they buy meat based on quality. Additionally, the management of food stock in their house was found to relate to their family member numbers. This finding aligned with similar prior research that found that a family with four or more members prefers to buy foods in traditional markets [77] and have a higher WTP if given additional nutrition information [87]. This is also supported by their knowledge about nutrition information from the mass media they access, so they know about the relevance of nutrition information for a healthy diet. Based on prior research, health media is one of the keys to increasing consumers' preferences regarding nutrition information [79]. However, because they are family-oriented, these consumers may choose a standard price (130 NT$) because they need to manage their money for the welfare of the whole family. Therefore, this group was found to have a relatively higher dependability and WTP for nutrition information than ordinary buyers.

3.3.4. Experienced and Proficient Buyers

The buyers who chose 150 NT$ were the people who relied on their previous buying experience in traditional wet markets. The effect of health content publication on mass media caused a rise of the WTP of the respondents in this group. It was shown that the health-related content, which they often obtained through mass media, increased their WTP by 6.5 NT$. Due to this habit, these consumers were seen as the only group who significantly relied on shoppers' customs variables, such as their custom as the family's shopper, their time management, and visiting time selection at the market. First of all, this group reported to prefer to be the primary shopper for their household, whether always or sometimes, and were willing to pay 10–12 NT$ more if they were given the nutrition information. Through these findings, it can certainly be observed that these people were more experienced than other categories and had more knowledge about the nutrition label, so they would prefer the highest price (150 NT$). For this reason, they were considered experienced and proficient buyers. This was aligned with a previous study that found that well-informed consumers show a high WTP on products bearing nutrition information [63]. Because of this reason, they would rather go to Taiwan traditional wet markets in the morning rather than the evening or night because they already know the best time to buy the best quality products [12,88]. They also seemed more efficient than other clusters because they did not prefer to stay in the market for more than 15 min since they already knew about the personal indicators that would affect their choices. It can be said that this category was more experienced about healthy diets because of their purchasing experiences including their knowledge from mass media [89]; thus, although they are a smart buyer, they rely on nutrition information and are willing to pay more for it.

3.4. The Estimation of Additional WTP for Nutrition Information

The results in this section describe the estimation of an additional WTP for nutrition information. These results were calculated by multiplying the average of each variable with the WTP in different consumers' groups that showed significant results. Based on Table 4, the highest additional WTP of the socio-demographic group was found with consumers who had a higher family number in the household. These people were willing to pay up to roughly 134 NT$ when they chose 130 NT$. The next category of female shoppers, career shoppers, and urban shoppers who chose "do not know" showed a willingness to pay up to approximately 132 NT$. It can be said that people who chose 110 NT$ (who were previously considered ordinary buyers) suggested that the less education they have in their life, the more they need the nutrition level and the more additional money they add to the price. However, the overall WTP estimation was still negative for education. These results might indicate that consumers with a higher education would have a much less WTP than those who have a lower education. One reason that nutrition information-oriented buyers and family-oriented buyers were found to only have 2–4 NT$ additional budget might be that they think this 2–4 NT$ is a reasonable supplementary value for pork with nutritional information in traditional markets, if purchased at a middle price (130 NT$).

Table 4. The estimation of additional WTP for nutrition information.

Independent \ Dependent	Average	Chosen 110 NT$	Chosen 130 NT$	Chosen 150 NT$	Chosen Do Not Know (130 NT$)
Socio-Demographic					
Female	1	-	-	-	1.37
Family number	4.14	-	3.91	-	-
Education	15.22	−5.88	-	-	-
Housewife	1	-	-	-	−1.86
Urban	1	-	-	-	0.95
Shoppers' Customs					
Main-shopper (Always)	1	-	-	10.11	-
Main-shopper (Sometimes)	1	-	-	12.61	-
Time consumed (>1 h)	1	-	-	−6.54	-
Morning shopping (5–11 a.m.)	1	-	-	5.49	-
Nutrition-Related Information					
Safety certificate	-	-	-	-	2.00
Meat grade information — Fairly Important	-	-	-	-	2.89
Meat grade information — Important	1	2.47	-	-	2.63
Meat grade information — Very important	1	4.05	4.14	-	4.63
Nutrition and calorie label	1	2.38	-	-	1.94
Health media concern	-	-	2.00	6.51	-
Constant	1	9.94	-	-	−7.23
Total of Additional WTP		12.96	10.04	28.19	7.23

Source: Calculated by this research.

Following shoppers' customs in Table 4, the top two additional WTPs were found not only from their behavior as a major buyer in the household but also their decision to choose a high-ranking price of 150 NT$. These experienced and proficient buyers who were the main shoppers were eager to buy pork with nutrition information until a price of 163 NT$. However, people in this group who were early bird shoppers and liked to spend less than one-hour shopping in traditional markets elicited a WTP of around 156 NT$. These experienced buyers (who were also primary shoppers) have the highest additional WTP. That is probably because they were either sometimes or always the decision-makers in their house for buying food, so nutrition information was important for them. Thus, when it came to the price, they did not hesitate to pay a premium, with an added price of up to 13 NT$. However, early-bird shoppers and speedy shoppers might have some personal preferences related to the time consumed and time visiting that interfere their decision in traditional markets. Thus, when facing pork with nutrition information, they were willing to pay a premium price, but only by about 7 NT$—about half of the added price of main shoppers' attribute.

As for the nutrition-related information, the inferior additional WTP came from consumers who value safety certificate attributes, meat grade information (fairly important and important) attributes, and nutrition and calorie label attributes. These people were spread across consumers who chose 110 NT$ and 130 NT$ and did not know price options. Generally, they desired to pay roughly only an additional 2 NT$ for pork with nutrition information. This was consistent with a previous study that discussed meat product with a safety certificate and found that if consumers are provided with safety certificate information, they, in turn, have a higher positive WTP [90]. The middle place was the consumers who presumed that meat grade was very important for the supplementary vendor's service. All buyers from all groups, except those who chose 150 NT$, said that they would pay an additional 4 NT$. Moreover, the elevated additional WTP was found on people from experienced and proficient buyers (chose 150 NT$), who are often concerned about health material learned from mass media. When they chose 150 NT$, they wanted to pay up to 157 NT$ for pork with nutrition information. The high additional WTP, up to 157 NT$, for nutrition-related information might be the result of their health knowledge gained from various platforms of mass or social media. The other groups with other nutrition-related information variables did not have a concern for health-related content, so their additional WTP only allowed them to upgrade by around 2–4 NT$. It can be stated that the more consumers have this access, the more they will increase their WTP, regardless of the attributes of nutrition-related information or price options, since they are considered health-conscious individuals [91].

The total for each additional WTP for nutrition information is discussed further in the following visualization of Figures 1 and 2. As seen in Figure 1, the additional WTP for each consumer group was combined based on three variable groups. Regarding the group who chose 110 NT$, it can be seen that nutrition-related information variables were important to them when deciding whether to buy pork with nutrition information. The people who chose 110 NT$ have significantly higher WTP for nutrition-related information (approximately 8 NT$) than the groups who chose 130 NT$ or 150 NT$. However, the total WTP of socio-demographic variables were negative among higher-education buyers for those who chose 110 NT$. This marked a significantly higher negative value than that of socio-demographic variables in the nutrition information-oriented buyers and family-oriented buyers' groups. However, regarding shoppers' customs variables, only the people who chose 150 NT$ showed a positive WTP, while the rest just revealed zero. From this result, it can be interpreted that ordinary buyers care more about nutrition-related information attributes when they visit traditional wet markets and buy pork with nutrition information. Those factors help them to better finalize their decision than the other two groups of independent variables. Nevertheless, as can be seen in Figure 1, family-oriented buyers' nutrition-related information variables were not significantly higher than experienced and proficient buyers', but these variables seemed to be the most important to them. Even though their socio-demographic variables were lower than the ordinary buyers', they were higher than those of the nutrition information-oriented buyers and experienced and proficient buyers. This showed that when buying pork with nutrition information in a traditional market, both factors could impact a consumer's additional WTP.

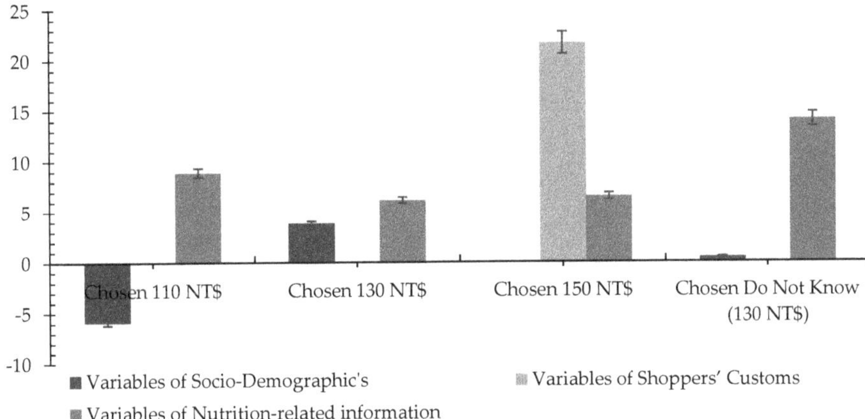

Figure 1. The estimation of WTP for nutrition information. Note: The error bars denote statistical significance at 5% significance. Source: Illustrated by this research.

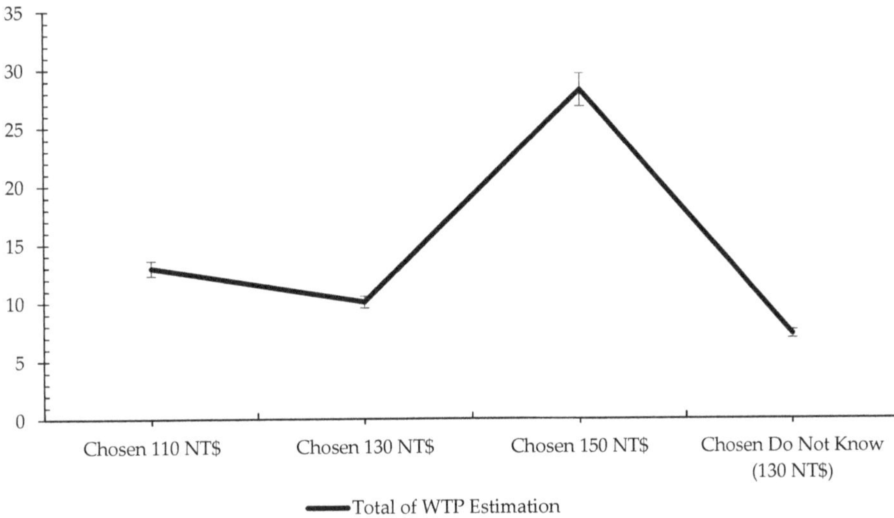

Figure 2. The total estimation of WTP for nutrition information. Note: The error bars denote statistical significance at 5% significance. Source: Illustrated by this research.

Moreover, in Figure 1, when combining the additional WTP of nutrition-information-oriented buyers' (did not know the price) socio-demographic and shoppers' customs variables, they did not have as much of an impact as much as nutrition-related information variables. It is because they both showed a value of almost zero. while nutrition-related information variables rose up to 15 NT$ and revealed significantly higher than ordinary buyers' (8 NT$). It can be said that whenever people go to traditional markets not knowing the price and preferring nutrition attributes, they increase their WTP for pork with nutrition information. The reason for this is that they need more information to buy or simply because they prefer products with more nutrition-related information. Lastly, from Figure 1, we can also infer that experienced and proficient buyers' shoppers' customs variables showed significantly higher impacts than for the other price option groups, accounting for up to 22 NT$ when others accounted for nothing. This result could be seen as proof that when buyers have more experience and knowledge, their behavior on buying things in traditional markets impacts their preferences, especially on their additional WTP. Thus, providing pork with nutrition information is a good strategy to gain these people's attention and money.

In addition, Figure 2 visualizes the total of all variables in every consumer group. According to Figure 2, the sum of the total variables showed significant differences across the four groups. This figure also shows a similar pattern in additional WTP, but one obvious result was that only three groups had the same 3 NT$ interval of the total. They were ordinary buyers, family-oriented buyers, and nutrition information-oriented buyers. Though they were significantly different from each other, their gap was still considerably smaller than each of their gaps with experienced and proficient buyers. When people chose 130 NT$ or did not know the price, their WTP rose until 140 NT$ and 137 NT$, which means that the people in these groups had an averagely similar thought pattern when facing pork with nutrition information. However, when they chose either the lowest or premium price, their WTP was added up to more than 10 NT$. Like ordinary buyers, their WTP went up to around 123 NT$. It can be said that when they chose lower price and were presented with pork with nutrition information, their WTP could reach middle or regular prices, which means that nutrition information gives them a chance to choose 130 NT$ in the future. However, for the experienced and proficient buyers, the addition of nutritional information can also capture a new possibility for pork belly's price at around 175 NT$/600 g. We can conclude that these experienced and proficient buyers are likely to appreciate more nutrition information on meat products in traditional markets. Ultimately, a market vendor could enact a more optimal strategy by placing a premium price, since these experienced and proficient buyers are predisposed to pay an even higher price for nutrition information. This finding is similar to that of previous study that also concluded that additional improved-quality information for meat products made buyers willing to pay a premium price [63,71,91], even if it is over the current price.

4. Conclusions

Since many options, yet little health information, exist in traditional wet markets in Taiwan, consumers need guidance to decide which healthy products to choose. Nutrition information in traditional wet markets seems to be a tool in this modern era to help people pay more attention to healthy eating habits. Through this study, it was found that the nutrition information generally gives a positive impact throughout the independent variables and across several types of traditional wet market buyers. In conclusion, female shoppers, career woman shoppers, primary shoppers, morning shoppers, and big family shoppers reported their ability to increase their WTP for nutrition information on meat products in traditional wet markets. On the other hand, a less-educated buyer with longer shopping time in a market was found to not prefer nutrition information. Generally, variables in nutrition-related information were important in all of the groups. The variables in socio-demographic and shoppers' customs might adjust to the consumers' preferences. This adjustment is based on the influence of health-related-content (from mass media) and on their personal health preferences of meat. Thus, it can be said that when people think that the healthiness of food is one of the most important attributes, they are willing to pay attention to the additional labels.

Secondly, nutrition information showed a positive impact on the family-oriented buyers and experienced and proficient buyers since these two groups might watch and access the health-related content in mass media so they were varied from medium to high levels in terms of their reliability on nutrition information and WTP enhancement for the price selection of 130 NT$ and 150 NT$, respectively. The nutrition information-oriented buyers showed a more positive impact for nutrition-related information variables, because they were not equipped enough for the knowledge of nutrition labels and price (do not know the price). Thus, they reported requiring more instructions to finalize their decision. However, ordinary buyers showed a low reliability and WTP regarding nutrition information on meat products since they come from lower educational backgrounds and prefer to pay the lowest price (110 NT$).

Since there has not been much research observing traditional wet market consumers' WTP concerning meat products bearing nutrition information, this research shows that the choice of applying nutrition information will bring positive feedback from the buyers and particular groups of consumers are even willing to pay more for this nutrition information. Thus, these findings will benefit traditional wet marketers in Taiwan if they apply these lessons to their marketing strategies. Lastly, these findings might be useful for the government to promote nutrition labels and other health-related content using mass media. Thus, the awareness of people to buy healthy food will be improved, including when they visit traditional wet markets, most of which currently do not provide additional nutrition labels. A limitation of this study was that the WTP estimation was done in hypothetical situations, and what a consumer intends to do (stated preferences in contingent valuation surveys) [92] and what they do in the real market could be different. Future research on this topic might approach different types of nutrition information such as nutrition claims, health claims, and traffic light labels, so meat consumers' choices can be further described to help Taiwan traditional wet marketers build their marketing strategies.

Author Contributions: The following authors contributed to this article in the following ways: Conceptualization, S.-H.Y. and A.A.S.; methodology, S.-H.Y. and A.A.S.; software, S.-H.Y.; validation, S.-H.Y. and A.A.S.; formal analysis, S.-H.Y. and A.A.S.; investigation, S.-H.Y. and A.A.S.; resources, S.-H.Y. and A.A.S.; data curation, S.-H.Y. and A.A.S.; writing—original draft preparation, S.-H.Y., A.A.S., and D.C.; writing—review and editing, S.-H.Y., A.A.S., and D.C.; visualization, S.-H.Y., A.A.S., and D.C.; supervision, S.-H.Y.; project administration, S.-H.Y.; funding acquisition, S.-H.Y. All authors have read and agreed to the published version of the manuscript.

Funding: This research is supported in part by the funding from the Ministry of Science and Technology (MOST) of Taiwan (two funding numbers MOST-105–2410-H-005–003 and MOST-103–2410-H-5–5-SSS).

Conflicts of Interest: The authors declare no conflict of interest.

References

1. Lin, S. Relationship between Nutritional Labeling and Consumer Purchase Intention. Master's Thesis, Wenzao Ursuline University of Languages, Kaohsiung, Taiwan, 2014.
2. World Health Organization. Obesity and Overweight. 2020. Available online: https://www.who.int/en/news-room/fact-sheets/detail/obesity-and-overweight (accessed on 2 July 2020).
3. McFerran, B.; Mukhopadhyay, A. Lay theories of obesity predict actual body mass. *Psychol. Sci.* **2013**, *24*, 1428–1436. [CrossRef] [PubMed]
4. Gracia, A.; Loureiro, M.L.; Nayga, R.M. Consumers' valuation of nutritional information: A choice experiment study. *Food Qual. Prefer.* **2009**, *20*, 463–471. [CrossRef]
5. Moorman, C.; Matulich, E. A model of consumers' preventive health behaviors: The role of health motivation and health ability. *J. Consum. Res.* **1993**, *20*, 208–228. [CrossRef]
6. Prescott, J.; Young, O.; O'Neill, L.; Yau, N.J.N.; Stevens, R. Motives for food choice: A comparison of consumers from Japan, Taiwan, Malaysia and New Zealand. *Food Qual. Prefer.* **2002**, *13*, 489–495. [CrossRef]
7. Wolfson, J.A.; Bleich, S.N. Is cooking at home associated with better diet quality or weight-loss intention? *Public Health Nutr.* **2015**, *18*, 1397–1406. [CrossRef] [PubMed]
8. Chen, R.C.Y.; Lee, M.S.; Chang, Y.H.; Wahlqvist, M.L. Cooking frequency may enhance survival in Taiwanese elderly. *Public Health Nutr.* **2012**, *15*, 1142–1149. [CrossRef]
9. Caraher, M.; Dixon, P.; Carr-Hill, R.; Lang, T. The state of cooking in England: The relationship of cooking skills to food choice. *Br. Food J.* **1999**, *101*, 590–609. [CrossRef]
10. Trappey, C.; Lai, M.K. Differences in factors attracting consumers to Taiwan's supermarkets and traditional wet markets. *J. Fam. Econ. Issues* **1997**, *18*, 211–224. [CrossRef]
11. Huang, C.T.; Tsai, K.H.; Chen, Y.C. How do wet markets still survive in Taiwan? *Br. Food J.* **2015**, *117*, 234–256. [CrossRef]
12. Trappey, C.V. Are Wet Markets Drying Up? Taiwan Review, Taiwan Today. 1997. Available online: https://taiwantoday.tw/news.php?post=12589&unit=8,29,32,45 (accessed on 2 May 2020).
13. Frederick, C.; Fu, C. *Taiwan Food Service: Hotel Restaurant Institutional*; USDA Foreign Agriculture Service: Washington, DC, USA, 2018.
14. Goldman, A.; Krider, R.; Ramaswami, S. The persistent competitive advantage of traditional food retailers in Asia: Wet markets' continued dominance in Hong Kong. *J. Macromarketing* **1999**, *19*, 126–139. [CrossRef]

15. Ho, S.C. Evolution versus tradition in marketing systems: The Hong Kong food-retailing experience. *J. Public Policy Mark.* **2005**, *24*, 90–99. [CrossRef]
16. Gorton, M.; Sauer, J.; Supatpongkul, P. Wet markets, supermarkets and the 'big middle' for food retailing in developing countries: Evidence from Thailand. *World Dev.* **2011**, *39*, 1624–1637. [CrossRef]
17. Kniazeva, M.; Belk, R.W. Supermarkets as libraries of postmodern mythology. *J. Bus. Res.* **2010**, *63*, 748–753. [CrossRef]
18. Paswan, A.; Pineda, M.D.L.D.S.; Ramirez, F.C.S. Small versus large retail stores in an emerging market-Mexico. *J. Bus. Res.* **2010**, *63*, 667–672. [CrossRef]
19. Jones, M.A.; Reynolds, K.E.; Mothersbaugh, D.L.; Beatty, S.E. The positive and negative effects of switching costs on relational outcomes. *J. Serv. Res.* **2007**, *9*, 335–355. [CrossRef]
20. Malhotra, N.K.; Ulgado, F.M.; Wu, L.; Agarwal, J.; Shainesh, G. Dimensions of service quality in developed and developing economies: Multi-country cross-cultural comparisons. *Int. Mark. Rev.* **2005**, *22*, 256–278. [CrossRef]
21. Goldman, A.; Ramaswami, S.; Krider, R.E. Barriers to the advancement of modern food retail formats: Theory and measurement. *J. Retail.* **2002**, *78*, 281–295. [CrossRef]
22. Varela, O.; Huang, W.; Sanyang, S.E. Consumer behavior and preference in the fruit markets of Taiwan. *Agric. Econ. Mark. J.* **2009**, *2*, 19–28.
23. Ackerman, D.; Tellis, G. Can culture affect prices? A cross-cultural study of shopping and retail prices. *J. Retail.* **2001**, *77*, 57–82. [CrossRef]
24. Ford, M.; Chang, C. *Taiwan Retail Foods Annual*; USDA Foreign Agriculture Service: Washington, DC, USA, 2014.
25. Frederick, C.; Chang, C. *Taiwan Retail Foods Annual*; USDA Foreign Agriculture Service: Washington, DC, USA, 2019.
26. Jang, Y.J.; Kim, W.G.; Bonn, M.A. Generation Y consumers' selection attributes and behavioral intentions concerning green restaurants. *Int. J. Hosp. Manag.* **2011**, *30*, 803–811. [CrossRef]
27. Kueh, K.; Voon, B.H. Culture and service quality expectations: Evidence from generation Y consumers in Malaysia. *Manag. Serv. Qual.* **2007**, *17*, 656–680. [CrossRef]
28. Department of Household Registration Affairs. Ministry of the Interior, Republic of China (Taiwan), History. 2018. Available online: https://www.ris.gov.tw/app/en/3911 (accessed on 5 May 2020).
29. Yang, E.C.L.; Khoo-Lattimore, C. Food and the perception of eating: The case of young Taiwanese consumers. *Asia Pac. J. Tour. Res.* **2015**, *20*, 1545–1564. [CrossRef]
30. Ranucci, D.; Miraglia, D.; Branciari, R.; Morganti, G.; Roila, R.; Zhou, K.; Jiang, H.; Braconi, P. Frankfurters made with pork meat, emmer wheat (*Triticum dicoccum* Schübler) and almonds nut (*Prunus dulcis* Mill.): Evaluation during storage of a novel food from an ancient recipe. *Meat Sci.* **2018**, *145*, 440–446. [CrossRef] [PubMed]
31. Wolfer, T.L.; Acevedo, N.C.; Prusa, K.J.; Sebranek, J.G.; Tarté, R. Replacement of pork fat in Frankfurter-type sausages by soybean oil oleogels structured with rice bran wax. *Meat Sci.* **2018**, *145*, 352–362. [CrossRef]
32. Kumar, N.; Kapoor, S. Do labels influence purchase decisions of food products? Study of young consumers of an emerging market. *Br. Food J.* **2017**, *119*, 218–229. [CrossRef]
33. Font-i-Furnols, M.; Guerrero, L. Consumer preference, behavior and perception about meat and meat products: An overview. *Meat Sci.* **2014**, *98*, 361–371. [CrossRef]
34. Chrysochou, P.; Krystallis, A.; Giraud, G. Quality assurance labels as drivers of customer loyalty in the case of traditional food products. *Food Qual. Prefer.* **2012**, *25*, 156–162. [CrossRef]
35. Czine, P.; Török, Á.; Pet, K.; Horváth, P.; Balogh, P. The impact of the food labeling and other factors on consumer preferences using discrete choice modeling—The example of traditional pork sausage. *Nutrients* **2020**, *12*, 1768. [CrossRef]
36. Bernués, A.; Olaizola, A.; Corcoran, K. Labelling information demanded by European consumers and relationships with purchasing motives, quality and safety of meat. *Meat Sci.* **2003**, *65*, 1095–1106. [CrossRef]
37. Bernués, A.; Ripoll, G.; Panea, B. Consumer segmentation based on convenience orientation and attitudes towards quality attributes of lamb meat. *Food Qual. Prefer.* **2012**, *26*, 211–220. [CrossRef]
38. Bacarella, S.; Altamore, L.; Valdesi, V.; Chironi, S.; Ingrassia, M. Importance of food labeling as a means of information and traceability according to consumers. *Adv. Hortic. Sci.* **2015**, *29*, 145–151. [CrossRef]

39. Sirieix, L.; Delanchy, M.; Remaud, H.; Zepeda, L.; Gurviez, P. Consumers' perceptions of individual and combined sustainable food labels: An UK pilot investigation. *Int. J. Consum. Stud.* **2013**, *37*, 143–151. [CrossRef]

40. Bialkova, S.; van Trijp, H. What determines consumer attention to nutrition labels? *Food Qual. Prefer.* **2010**, *21*, 1042–1051. [CrossRef]

41. Sacks, F.M.; Svetkey, L.P.; Vollmer, W.M.; Appel, L.J.; Bray, G.A.; Harsha, D.; Obarzanek, E.; Conlin, P.R.; Miller, E.R.; Simons-Morton, D.G.; et al. DASH-Sodium Collaborative Research Group. Effects on blood pressure of reduced dietary sodium and the dietary approaches to stop hypertension (DASH) diet. *N. Engl. J. Med.* **2001**, *344*, 3–10. [CrossRef]

42. Balasubramanian, S.K.; Cole, C. Consumers' search and use of nutrition information: The challenge and promise of the nutrition labeling and education act. *J. Mark.* **2002**, *66*, 112–127. [CrossRef]

43. Hawley, K.L.; Roberto, C.A.; Bragg, M.A.; Liu, P.J.; Schwartz, M.B.; Brownell, K.D. The science on front-of-package food labels. *Public Health Nutr.* **2013**, *16*, 430–439. [CrossRef]

44. American Diabetes Association. Evidence-based nutrition principles and recommendations for the treatment and prevention of diabetes and related complications. *Diabetes Care* **2002**, *25*, 202–212. [CrossRef]

45. Institute of Medicine. *Front-of-Package Nutrition Rating Systems and Symbols: Phase I Report*; The National Academies Press: Washington, DC, USA, 2010. [CrossRef]

46. Asian Productivity Organization. *Food Standards and Labeling Systems in Asia and the Pacific*; APO Seminar on Food Standards and Labeling System: Tokyo, Japan, 2002.

47. Perez-Escamilla, R.; Haldeman, L. Food label use modifies association of income with dietary quality. *J. Nut.* **2002**, *132*, 768–772. [CrossRef]

48. Cioffi, C.E.; Levitsky, D.A.; Pacanowski, C.R.; Bertz, F. A nudge in a healthy direction: The effect of nutrition labels on food purchasing behaviors in university dining facilities. *Appetite* **2015**, *92*, 7–14. [CrossRef]

49. Miller, L.M.S.; Cassady, D.L. The effects of nutrition knowledge on food label use: A review of the literature. *Appetite* **2015**, *92*, 207–216. [CrossRef]

50. Fonseca, M.D.C.P.D.; Salay, E. Beef, chicken and pork consumption and consumer safety and nutritional concerns in the city of Campinas, Brazil. *Food Control* **2008**, *19*, 1051–1058. [CrossRef]

51. Minten, B.; Reardon, T. Food prices, quality, and quality's pricing in supermarkets versus traditional markets in developing countries. *Rev. Agric. Econ.* **2008**, *30*, 480–490. [CrossRef]

52. Schipmann, C.; Qaim, M. Modern food retailers and traditional markets in developing countries: Comparing quality, prices, and competition strategies in Thailand. *Appl. Econ. Perspect. Policy* **2011**, *33*, 345–362. [CrossRef]

53. Sinclair, S.E.; Cooper, M.; Mansfield, E.D. The influence of menu labeling on calories selected or consumed: A systematic review and meta-analysis. *J. Acad. Nutr. Diet.* **2014**, *114*, 1375–1388. [CrossRef] [PubMed]

54. Roberto, C.A.; Larsen, P.D.; Agnew, H.; Baik, J.; Brownell, K.D. Evaluating the impact of menu labeling on food choices and intake. *Am. J. Public Health* **2010**, *100*, 312–318. [CrossRef] [PubMed]

55. 55 Glanz, K.; Hewitt, A.M.; Rudd, J. Consumer behavior and nutrition education: An integrative review. *J. Nutr. Educ.* **1992**, *24*, 267–277. [CrossRef]

56. Bládic, A.; O'Reilly, S.; O'Sullivan, K. Consumer attitudes to nutrition labelling. *Br. Food J.* **1997**, *99*, 283–289. [CrossRef]

57. Dubois, P.; Albuquerque, P.; Allais, O.; Bonnet, C.; Bertail, P.; Combris, P.; Lahlou, S.; Rigal, N.; Ruffieux, B.; Chandon, P. Effects of Front-of-pack labels on the nutritional quality of supermarket food purchases: Evidence from a large-scale randomized controlled trial. *J. Acad. Mark. Sci.* **2020**. [CrossRef]

58. Berning, J.P.; Chouinard, H.H.; Manning, K.C.; McCluskey, J.J.; Sprott, D.E. Identifying consumer preferences for nutrition information on grocery store shelf labels. *Food Policy* **2010**, *35*, 429–436. [CrossRef]

59. Lowe, M.R.; Tappe, K.A.; Butryn, M.L.; Annunziato, R.A.; Coletta, M.C.; Ochner, C.N.; Rolls, B.J. An intervention study targeting energy and nutrient intake in worksite cafeterias. *Eat. Behav.* **2010**, *11*, 144–151. [CrossRef]

60. Goodman, S.; Vanderlee, L.; White, C.M.; Hammond, D. A quasi-experimental study of a mandatory calorie-labelling policy in restaurants: Impact on use of nutrition information among youth and young adults in Canada. *Prev. Med. (Baltim.)* **2018**, *116*, 166–172. [CrossRef] [PubMed]

61. Piedra, M.A.; Schupp, A.R.; Montgomery, D.E. Consumer use of nutrition labels on packaged meats. *J. Food Distrib. Res.* **1996**, *27*, 42–47. [CrossRef]

62. Schupp, A.; Gillespie, J.; Reed, D. Consumer awareness and use of nutrition labels on packaged fresh meats: A pilot study. *J. Food Distrib. Res.* **1998**, *29*, 24–30. [CrossRef]
63. Barreiro-Hurlé, J.; Gracia, A.; De-Magistris, T. Market implications of new regulations: Impact of health and nutrition information on consumer choice. *Span. J. Agric. Res.* **2009**, *7*, 257. [CrossRef]
64. Schnettler, B.; Ares, G.; Sepúlveda, N.; Bravo, S.; Villalobos, B.; Hueche, C.; Lobos, G. Are consumers willing to pay more for reformulated processed meat products in the context of the implementation of nutritional warnings? Case study with frankfurters in Chile. *Meat Sci.* **2019**, *152*, 104–108. [CrossRef]
65. Owusu, V.; Anifori, M.O. Consumer willingness to pay a premium for organic fruit and vegetable in Ghana. *Int. Food Agribus. Manag. Rev.* **2013**, *16*, 67–86. [CrossRef]
66. Hanemann, M.; Loomis, J.; Kanninen, B. Statistical efficiency of double-bounded dichotomous choice contingent valuation. *Am. J. Agric. Econ.* **1991**, *73*, 1255–1263. [CrossRef]
67. Hanemann, W.M. Valuing the environment through contingent valuation. *J. Econ. Perspect.* **1994**, *8*, 19–43. [CrossRef]
68. Cascetta, E. *Transportation Systems Engineering: Theory and Methods*; Pardalos, P.M., Hearn, D., Eds.; Kluwer Academic Publishers: Napoli, Italy, 2001; Volume 49, pp. 95–169.
69. Liu, Y.; Sheu, Y. Study on pork consumption and CAS sign recognize for different urban degree. *Chin. J. Agribus. Manag.* **2002**, *8*, 223–245, (In Traditional Chinese).
70. Lancaster, K.J. A new approach to consumer theory author. *J. Polit. Econ.* **1996**, *74*, 132–157. [CrossRef]
71. Viegas, I.; Nunes, L.C.; Madureira, L.; Fontes, M.A.; Santos, J.L. Beef credence attributes: Implications of substitution effects on consumers' WTP. *J. Agric. Econ.* **2014**, *65*, 600–615. [CrossRef]
72. Prathiraja, P.; Ariyawardana, A. Impact of nutritional labeling on consumer buying behavior. *Sri Lankan J. Agric. Econ.* **2003**, *5*, 35–46. [CrossRef]
73. Cheng, H.W.; Chien, C.L.; Woodburne, P.R. The impact of cause-related marketing on store switching: An analysis of the chain convenience stores in Taiwan. *Corp. Manag. Rev.* **2017**, *37*, 87–126. [CrossRef]
74. Shiu, E.C.; Dawson, J.A. Demographic segmentation of shoppers at traditional markets and supermarkets in Taiwan. *J. Segm. Mark.* **2001**, *4*, 69–85. [CrossRef]
75. Li, H.; Houston, J.E. Factors affecting consumer preferences for major food markets in Taiwan. *J. Food Distrib. Res.* **2001**, *32*, 97–109. [CrossRef]
76. Chern, W.S.; Chang, C.Y. Benefit evaluation of the country of origin labeling in Taiwan: Results from an auction experiment. *Food Policy* **2012**, *37*, 511–519. [CrossRef]
77. Hsu, J.L.; Chang, W.H. Market segmentation of fresh meat shoppers in Taiwan. *Int. Rev. Retail. Distrib. Consum. Res.* **2002**, *12*, 423–436. [CrossRef]
78. Ho, C. Consumer behaviour and habit in 1995. *Retail Mark.* **1996**, *240*, 13–17.
79. Mills, S.; Brown, H.; Wrieden, W.; White, M.; Adams, J. Frequency of eating home cooked meals and potential benefits for diet and health: Cross-sectional analysis of a population-based cohort study. *Int. J. Behav. Nutr. Phys. Act.* **2017**, *14*, 1–11. [CrossRef]
80. Rimal, A.P.; Fletcher, S.M. Understanding consumers' attitude toward meat labels and meat consumption pattern. In Proceedings of the Southern Agricultural Economics Association Annual Meeting, Mobile, Alabama, AL, USA, 1–5 February 2003; no. 770. pp. 1–19. [CrossRef]
81. Mozaffarian, D.; Afshin, A.; Benowitz, N.L.; Bittner, V.; Daniels, S.R.; Franch, H.A.; Jacobs, D.R., Jr.; Kraus, W.E.; Kris-Etherton, P.M.; Krummel, D.A.; et al. Population approaches to improve diet, physical activity, and smoking habits: A scientific statement from the American heart association. *Circulation* **2012**, *126*, 1514–1563. [CrossRef]
82. Warden, C.A.; Stanworth, J.; Chen, J.F.; Huang, S.C.T. Strangers in strange lands: Hypermarkets and chinese consumer culture misalignment. *Int. J. Mark. Res.* **2012**, *54*, 799–820. [CrossRef]
83. Makweya, F.L.; Oluwatayo, I.B. Consumers preference and willingness to pay for graded beef in Polokwane municipality, South Africa. *Ital. J. Food Saf.* **2019**, *8*, 46–54. [CrossRef] [PubMed]
84. Polkinghorne, R.J.; Thompson, J.M. Meat standards and grading. A world view. *Meat Sci.* **2010**, *86*, 227–235. [CrossRef] [PubMed]
85. Lyford, C.P.; Thompson, J.M.; Polkinghorne, R.; Miller, M.F.; Nishimura, T.; Neath, K.; Allen, P.; Belasco, E.J. Is willingness to pay (WTP) for beef quality grades affected by consumer demographics and meat consumption preferences? *Australas. Agribus. Rev.* **2010**, *18*, 1–17. [CrossRef]

86. McArthur, L.; Chamberlain, V.; Howard, A.B. Behaviors, attitudes, and knowledge of low-income consumers regarding nutrition labels. *J. Health Care Poor Underserved* **2001**, *12*, 415–428. [CrossRef] [PubMed]

87. Cannoosamy, K.; Pugo-Gunsam, P.; Jeewon, R. Consumer knowledge and attitudes toward nutritional labels. *J. Nutr. Educ. Behav.* **2014**, *46*, 334–340. [CrossRef]

88. Maruyama, M.; Wu, L. Quantifying barriers impeding the diffusion of supermarkets in China: The role of shopping habits. *J. Retail. Consum. Serv.* **2014**, *21*, 383–393. [CrossRef]

89. Wang, R.; Liaukonyte, J.; Kaiser, H.M. Does advertising content matter? Impacts of healthy eating and anti-obesity advertising on willingness to pay by consumer body mass index. *Agric. Resour. Econ. Rev.* **2018**, *47*, 1–31. [CrossRef]

90. Berges, M.; Casellas, K.; Rodríguez, R.; Errea, D. Willingness to Pay for Quality Attributes of Fresh Beef. In Implications on the Retail Marketing. In Proceedings of the International Conference of Agriculture Economists, Milan, Italy, 8–14 August 2015. [CrossRef]

91. Flowers, S.; McFadden, B.R.; Carr, C.C.; Mateescu, R.G. Consumer preferences for beef with improved nutrient profile. *J. Anim. Sci.* **2019**, *97*, 4699–4709. [CrossRef]

92. Ajzen, I.; Brown, T.C.; Carvajal, F. Explaining the discrepancy between intentions and actions: The case of hypothetical bias in contingent valuation. *Personal. Soc. Psychol. Bull.* **2004**, *30*, 1108–1121. [CrossRef]

Article

Authentication of Meat and Meat Products Using Triacylglycerols Profiling and by DNA Analysis

Vojtech Hrbek [1], Kamila Zdenkova [2,*], Diliara Jilkova [2,3], Eliska Cermakova [2], Monika Jiru [1], Katerina Demnerova [2], Jana Pulkrabova [1] and Jana Hajslova [1]

[1]   Department of Food Analysis and Nutrition, University of Chemistry and Technology, Technicka 5, 166 28 Prague 6, Czech Republic; Vojtech.Hrbek@vscht.cz (V.H.); Monika.Jiru@vscht.cz (M.J.); Jana.Pulkrabova@vscht.cz (J.P.); Jana.Hajslova@vscht.cz (J.H.)

[2]   Department of Biochemistry and Microbiology, University of Chemistry and Technology, Technicka 5, 166 28 Prague 6, Czech Republic; Eliska2.Cermakova@vscht.cz (E.C.); Katerina.Demnerova@vscht.cz (K.D.)

[3]   Research Department, Food Research Institute Prague, Radiová 1285/7, 10231 Prague 10, Czech Republic; Diliara.Akhatova@vupp.cz

*   Correspondence: Kamila.Zdenkova@vscht.cz

Received: 4 July 2020; Accepted: 8 September 2020; Published: 10 September 2020

Abstract: Two alternative, complementary analytical strategies were successfully used to identify the most common meat species—beef, pork and chicken—in meat products. The first innovative high-throughput approach was based on triacylglycerols fingerprinting by direct analysis in real time coupled with high-resolution mass spectrometry (DART–HRMS). The second was the classic commonly used DNA analysis based on the use of nuclear or mitochondrial DNA in multiplex polymerase chain reaction (mPCR). The DART–HRMS method represents a rapid, high throughput screening method and was shown to have a good potential for the authentication of meat products. Nevertheless, it should be noted that due to a limited number of samples in this pilot study, we present here a proof of concept. More samples must be analyzed by DART–HRMS to build a robust classification model applicable for reliable authentication. To verify the DART–HRMS results, all samples were analyzed by PCRs. Good compliance in samples classification was documented. In routine practice under these conditions, screening based on DART–HRMS could be used for identification of suspect samples, which could be then examined and validated by accurate PCRs. In this way, saving of both labor and cost could be achieved. In the final phase, commercially available meat products from the Czech market were tested using this new strategy. Canned meats—typical Czech sausages and luncheon meats, all with declared content of beef, pork and chicken meat—were used. Compliance with the label declaration was confirmed and no adulteration was found.

Keywords: meat; authentication; triacylglycerols; ambient mass spectrometry; DNA; PCR

1. Introduction

The adulteration of food is a current socioeconomic worldwide problem. Consumer demand for correct and understandable food labeling is growing. The informed choice of the products that they want to buy is issue of high concern.

One relatively common fraudulent practice is the replacement or dilution of a highly valuable commodity by a cheaper one. This problem may be encountered in meat products—specifically in minced ones. Pork, beef or chicken meat are among the most popular and nutritionally valuable food commodities. Nevertheless, as chicken meat is the cheapest, it may be used to substitute for expensive beef. In addition to economic fraud, hazard for consumers suffering of allergic reaction to certain meats (e.g., chicken protein allergy) must be considered. Moreover, religious aspects may be of concern, as eating pork is not acceptable in Muslim populations.

With regards to the above facts, analytical strategies that can quickly, affordably and reliably detect such unfair practices are urgently needed. A number of approaches have been employed for authentication purposes. The most common of them is target analysis of specific markers such as nucleic acids, peptides/proteins or metabolites. Wide variety of techniques including chromatographic, electrophoretic, spectroscopic immunochemical and molecular–biologic methods have been used for meat authentication [1–5]. In routine practice also immunological methods based on the interaction between an antigen and antibody (mainly ELISA) are used for authentication of animal species in meat products.

Various molecular techniques used for food authentication have been developed and reviewed [6,7]. These techniques are based on the DNA polymorphism between species and are classified into three types. (1) polymerase chain reaction (PCR)-based techniques, (2) hybridization-based techniques and (3) sequencing-based techniques such as DNA barcoding to analyze short standard DNA sequences and forensically informative nucleotide sequencing FINS [5,8–14]. Various PCR methods have been described: endpoint PCR, real-time qPCR and digital dPCR that allow the amplification of a chosen region of genomic or mitochondrial DNA. In animal cells, mitochondrial DNA (mtDNA) is present in many copies, while genomic DNA is mostly present in one copy. Genomic DNA is an appropriate target for quantification, while a small addition of undeclared species to the sample can be demonstrated by the detection of mtDNA [12].

Regarding specific markers screening by chromatographic methods, in our recent study, we demonstrated the possibility to recognize addition of 2% chicken meet addition to pork by assessing specific ratios of amino acids 1-methylhisitidine/3-methylhistidine [15]. However, this approach is rather time-consuming. The same applies e.g., for proteomics-based procedures in which meat species authenticity is performed by means of well-defined proteogenomic annotation, carefully chosen surrogate tryptic peptides ad analysis using a high resolution mass spectrometry HRMS [16]. This technique, thanks to high spectral resolution, enables skipping over chromatographic separation. The use of ambient ionization method such as direct analysis in real time (DART) allows a great simplification and increase in the speed of mass spectrometry-based measurements. In particular case, the sample investigation can be performed in the open environment of the laboratory by its introduction into the ionization region, where it is exposed to a stream of ionizing medium [17]. The attractive features of DART have made this technique, apart of other applications, a challenging tool rapid characterization of food composition and/or assessment of its authenticity based on metabolomic fingerprinting [18]. For processing of data generated by DART–HRMS advanced statistical methods represented by principal component analysis (PCA) followed by discriminant analysis, e.g., partial least squares discriminant analysis (PLS-DA) are commonly used.

The objective of this pilot study was to demonstrate the applicability of a new authentication strategy for large sets of selected meat products enabling both labor and cost saving. In the first phase, samples with suspect triacylglycerols (TAGs) profile were rapidly identified by DART–HRMS technique, in the next step, the confirmation was performed by validated PCRs. To our knowledge, this is the first study that analyzes meat samples by both mentioned methods and discusses their discrimination potential.

2. Materials and Methods

2.1. Samples

Samples were purchased through the retail network of the Czech Republic. In total, 36 samples were available: pork ($n = 3$), chicken ($n = 3$), beef ($n = 3$) and heat-treated meat products: ham ($n = 3$), sausages ($n = 11$), luncheon meat ($n = 7$) and meat in its natural juices ($n = 6$). Approximately 200-g of each sample was homogenized by means of an electric grinder (Grindomix GM 200, Retsch, Düsseldorf, Germany) and stored in a plastic container at −20 °C until analysis. The fat content and meat composition of the commercial meat products are shown in Table 1. All samples were tested using both analytical approaches, DART–HRMS–represents a rapid screening method and PCR–verification method. DART–HRMS models for data evaluation were created from nine samples of meat (pork, chicken and beef), 27 samples (meat products) were used for comparison of use in practice. It must be mentioned, that due to the very low number of analyzed samples, this innovative strategy was being tested to determine the potential of this method for the purpose of authentication of meat and meat products. The purpose of this pilot study was to present the suitability of this combination; it is proof of the concept.

Table 1. Declared composition of analyzed samples.

Sample	Product	Pork Meat/Lard (%)	Chicken Meat (%)	Beef Meat (%)	Max Fat Content (%)
1	Chicken ham	–	92	–	1.5
2	Poultry ham	–	60	–	1.5
3	Pork ham	95	–	–	10
4	Sausage	16/Y	–	35	40
5	Sausage	40/Y	–	10	40
6	Sausage	62/Y	–	23	40
7	Sausage	40/25	–	10	34
8	Sausage	54/Y	–	26	44
9	Luncheon meat	N/Y	–	Y	40
10	Sausage	16/Y	–	35	45
11	Sausage	17/Y	–	26	N
12	Sausage	17.5/Y	–	38.5	45
13	Meat in natural juices	70	–	–	33
14	Luncheon meat	79	–	–	30
15	Meat in natural juices	92	–	–	N
16	Meat in natural juices	70	–	–	40
17	Meat in natural juices	70	–	–	30
18	Sausage	33/30	–	22	N
19	Sausage	71/Y	–	16	45
20	Sausage	43/30	–	17	45
21	Luncheon meat	48	Y	–	40
22	Meat in natural juices	30 + MSM/Y		–	N
23	Luncheon meat	18	32	–	30
24	Luncheon meat	35	30	–	25
25	Luncheon meat	71		–	40
26	Luncheon meat	31/Y	39	–	26
27	Meat in natural juices	Y	Y	70	27

Y—label on packaging indicates the usage, but the percent content is not stated; N—not labeled on the packaging; MSM—mechanically separated meat.

2.2. Direct Analysis in Real Time Coupled with High-Resolution Mass Spectrometry (DART–HRMS) Analysis

2.2.1. Sample Preparation for Instrumental Analysis

6 mL of hexane was added to the homogenized sample (2 g) in a 15-mL plastic tube. The sample in the plastic tube was extracted for 1 min using a Turrax instrument (T 10 basic ULTRA-TURRAX®, IKA, Staufen, Germany). After centrifugation (5 min, 20 °C, 10,000 rpm), the extract was transferred to a glass vial and ready for DART–HRMS analysis. (QC) 100 µL of each pure meat hexane extract (beef, chicken and pork) was mixed as a quality control sample.

2.2.2. Conditions of DART–HRMS Analysis

For the analysis using ambient mass spectrometry, the DART ion source (DART-SVP) was fitted with a 12Dip-It^{TM} tip scanner autosampler (IonSense, Saugus, MA, USA) coupled to an Exactive^{TM} benchtop (Thermo Fisher Scientific, Bremen, Germany). A Vapur^{TM} interface (IonSense, Saugus, MA, USA) was employed to couple the ion source to the mass spectrometer and low vacuum in the interface chamber was maintained with a membrane pump (Vacuubrand, Wertheim, Germany). The distance between the exit of the DART gun and the ceramic transfer tube of the Vapur was set to 10 mm, the gap between the ceramic tube and the inlet to the heated capillary of the Exactive was 2 mm.

The DART and MS instruments were operated in positive ionization mode and the optimized settings were as follows: helium pressure: 5.5 bar; gas temperature: 450 °C; discharge needle voltage: 1000 V; grid electrode: 250 V. For mass spectrometric detection, the settings were as follows: capillary voltage: 60 V; tube lens voltage: 120 V; capillary temperature: 250 °C. The sheath, auxiliary, and sweep gases were disabled during DART–MS analysis.

The mass spectrometer was operated at a mass resolving power of 50,000 FWHM calculated for m/z 200. The mass spectra acquisition rate was 2 spectra s^{-1}. Liquid samples were delivered into the DART ionization region with the use of a 12 Dip-It tip scanner autosampler. Dip-It^{TM} tips (IonSense, Saugus, MA, USA) were inserted into a holder. μL of hexane extract of each sample were individually placed on the end of glass tips. The Dip-It holder was mounted on the body of the autosampler and the Dip-It tips were automatically moved at a constant speed of 0.5 mms^{-1} through the helium gas between the exit of the DART gun and the inlet of the Vapur interface.

Standard external mass calibration of the MS system in the range of 50–1000 m/z was performed in positive mode prior to the measurement of every sample set (sequence) according to the manufacturer's instructions. Moreover, an adjusted mass calibration for ESI(+) in the mass range of m/z 50–1000 using collision-induced dissociation (CID) at 25 eV was subsequently performed to cover the lower masses.

2.2.3. Data Analysis

Chemometric analysis included multivariate data analysis using unsupervised and supervised models. Principal component analysis (PCA) and partial least squares discriminant analysis (PLS-DA) were employed based on SIMCA software (v. 13.0, 2011, Umetrics, Umea, Sweden; www.umetrics.com).

In the first stage, data processing and data pretreatment must be carried out to capture the bulk of the variation between different datasets. In this way, raw data generated by meat samples analysis employing the DART–HRMS technique (TAGs signals in positive ionization mode) in the form of absolute peak intensities were preprocessed using a constant row sum, that is, each variable was divided in the sum of all variables for each sample; this procedure transformed all the data to a uniform range of variability. In other words, the intensities of the variables obtained from the profile of the analyzed sample were summed and then each specific variable was divided by the sum thus obtained, thus avoiding the different intensities of the individual ions which would be caused by the measurement itself and not by the differences of analyzed samples. DART–HRMS data were initially processed with the software Xcalibur 2.2 and copied to MS Excel 2010. The macro function was used in the following step to create the final tables which were exported to the SIMCA software.

Subsequently, Pareto scaling was applied prior to PCA and PLS-DA [19]. Then, PCA analysis enabled the transformation of the original variables (normalized intensities of ions) to the new uncorrelated variables (principal components). In this way, the reduced dimensionality of the data were obtained while still preserving information from the original data set. Additionally, PLS-DA was subsequently applied to identify and reveal the most significant TAGs. PLS-DA was performed to provide a better distribution of samples and enable the creation of a statistical model and validation.

The quality of the models was evaluated by the goodness-of-hit parameter (R^2X), the proportion of the variance of the response variable that is explained by the model (R^2Y) and the predictive ability parameter (Q^2), which was calculated by a k-fold internal cross-validation of the data using a default option of the SIMCA software. In general terms, the value of R^2 must be higher than Q^2 and an acceptable value of Q^2 is more than 0.5 [20]. In addition, the models were also evaluated in terms of their recognition and prediction abilities. Recognition ability represents the percentage of samples in the training set that were correctly classified. Prediction ability is the percentage of samples in the test set that are correctly classified by using the model developed during the training step. For this purpose, seven-fold internal cross-validation was used [21]. For the control of the Q^2 values, if they were stable and relevant (correctly calculated), the permutation test was used [22].

S-plot illustrating the distribution of the detected features involved in the statistical evaluation was used as a tool for 'marker' selection. Features at the extremes of the S-plot, the outermost ions can be considered as ´markers´ with the highest importance for sample separation. For sorting the ´markers´ according their importance, a VIP (variable importance in projection) plot that explains X and correlates to Y can be used. The most important variables in a given model are those with VIP score >1. The other tool for explaining/confirming ions as markers is a variable line plot, which illustrates the variability among the top ions across the sample sets.

The tentative identification of compounds behind the marker ions was based on the estimation/calculation of the elemental formula (accurate mass and mass error for respective m/z values in MS^1 and isotopic pattern were considered). The estimated molecular structure of the markers was compared with online databases such as ChemSpider (www.chemspider.com) or Metlin (www.metlin.scripps.edu/index.php).

2.2.4. Quality Control

To verify the absence of carryover effects and to control the stability of the recorded fingerprints, blank and quality control (QC) matrix samples were analyzed within the DART–HRMS sequence. It should be noted that the order of the tested samples within the sequence was random (established based on random number generation) to avoid any possible time-dependent changes during DART–HRMS analysis, which could result in false clustering. To check the overall performance of the instrumental system, QC samples were inserted into the sequence, always after a set of ten tested samples and analyzed under the same conditions. The QC sample was a pool of all meat sample extracts. In this way, the repeatability of sample fingerprints could be monitored. The good instrument performance was documented by a tight clustering of these QC samples (i.e., the similarity of their fingerprints) in the PCA plot.

2.3. Analysis by Polymerase Chain Reaction (PCR)

Multiplex mPCRs were used for the authentication of meat origin. The design of this study was as follows: after homogenization of the meat or whole meat product, the isolation of the DNA was performed, followed by PCRs. The mPCR based on mitochondrial cytochrome b gene amplification was used for qualitative analysis. Two mqPCRs (triplex and duplex) were used, based on the amplification of a single copy of chromosomally encoded gene sequences. Single-copy chromosomal genes were analyzed, such as cyclic phosphodiesterase for cattle, beta actin for pigs; interleukin-2 (Il-2) for chickens and the myostatin gene for mammals and poultry.

2.3.1. DNA Isolation

DNA was isolated from 200-mg homogenized samples (Section 2.1) using a cetyltrimethylammonium bromide (CTAB) method [23]. The quality of the isolated DNA was verified by 1% horizontal agarose electrophoresis in Tris/Borate/EDTA buffer (Bio-Rad, Hercules, CA, USA), DNA concentration and purity was determined spectrophotometrically with a nanophotometer (Implen, Munich, Germany).

2.3.2. Primers and Probes

The primers and probes used are shown in Table 2 and were synthetized by East Port (Prague, Czech Republic).

Table 2. Primers and probes used in this study.

Meat Species	Name of Primer	Target Sequence	Sequence of Primer [5′-3′]	Product [bp]	References
Universal F	SIM		GACCTCCCAGCTCCATCAAACATCTCATCTTGATGAAA		[12]
Beef R	B		CTAGAAAAGTGTAAGACCCGTAATATAAG	274	[12]
Pork R	P	Cytochrome b (mtDNA)	GCTGATAGTAGATTTGTGATGACCGTA	398	[12]
Chicken, turkey R	C		CGTATTGTACGTTCCGGCAAG	169	[24]
Horse R	H		CTCAGATTCACTCGACGAGGGTAGTA	439	[12]
Beef	Bos-PDE-f	Cyclic-GMP-phospho-diesterase (gDNA)	ACTCCTACCCATCATGCAGAT	104	[11,25]
	Bos-PDE-r		TGTTTTTAAATATTTCAGCTAAGAAAAA		
	Bos-PDE-p		TexasRed: AACATCAGGATTTTTGCTGCATTTGC:BHQ2		
Pork	Sus1-F	Beta-actin (gDNA)	CGAGAGGCTGCCGTAAAGG	107	[11,26]
	Sus1-R		TGCAAGGAACACGGCTAAGTG		
	Sus1-p		HEX:TCTGACGTGACTCCCCGACCTGG:BHQ1		
Mammals and poultry	MY-F	Myostatin (gDNA)	TTGTGCAAATCCTGAGACTCAT	97	[11,27]
	MY-R		ATACCAGTGCCTGGGTTCAT		
	My-p		FAM:CCCATGAAAGACGGTACAAGGTATACTG:BHQ1		
Chicken	ChIn-F	Interleukin-2 (gDNA)	TGTTACCTGGGAGAAGTGGTTACT	135	[25]
	ChIn-R		CTGACCATAAAGAATACCTACCG		[24]
	ChIn-p		TAMRA:TGAAGAAAGAAACTGAAGATGACACTGAAATTAAAG:BHQ2		[25]

2.3.3. Multiplex mPCR

For this method, primers complementary to mitochondrial DNA cytochrome b were used [12,24]. mPCR amplification was conducted in 15 μL 1.5-mM $MgCl_2$, 0.2-mM dNTP mix (Promega, Madison, WI, USA), primer mix (Metabion International AG, Planegg, Germany), 100 ng template DNA and 0.4 U Platinum^{TM} DNA polymerase (Thermo Fisher Scientific, Waltham, MA, USA). The primers were mixed in the ratio of 1:0.6:0.6:1.5:1.5 for SIM:B:P:C:H and used together to mPCR (ratio 1 means concentration 0.4 μmol·L^{-1}). Amplifications were performed in a Biometra T-Gradient (Whatman Biometra, Göttingen, Germany) as follows: initial denaturation at 94 °C for 2 min, 40 cycles of denaturation at 94 °C for 30 s, annealing at 53 °C for 30 s and extension at 72 °C for 30 s, final polymerization was for 5 min at 72 °C. Visualization and detection of amplicons were done on 2.5% agarose gel.

2.3.4. Multiplex mqPCRs

Primers and probes used for mqPCR were complementary to single-copy chromosomally encoded gene sequences. The reaction conditions for pork and beef were adopted from Iwobi et al. [11], chicken mqPCR from Zdenkova et al. [24]. The analyses were performed in an ABI 7500 (Applied Biosystems™, Foster City, CA, USA), the 7500 Software v2.0.6 was employed for data analysis. Four fluorescence channels were analyzed separately.

The result of the triplex qPCR was amplification of the 104-bp-long bovine gDNA segment with the Texas Red fluorescence curve; 107 bp from pork gDNA with a fluorescence curve of the HEX fluorophore and a 97-bp-long amplicon from the gDNA of mammals and poultry with the FAM fluorophore. The duplex qPCR amplified the 135-bp-long amplicon of chicken gDNA with the TAMRA fluorophore together with a 97-bp-long amplicon from the gDNA of mammals and poultry with the FAM fluorophore.

2.3.5. Data Analysis

To separate the PCR amplicons, a 2.5% agarose gel was used. The confirmation of amplicon size was based on comparing the length of the amplicons obtained from the samples with the length of the marker fragments (100-bp DNA ladder, New England Biolabs, Ipswich, MA, USA) and the positive control during the reaction (target DNA) which, together with the non-template control, was included in each amplification reaction.

The qPCR data analyses are based on evaluating the fluorescence curves of the amplification cycle. If the fluorescence value of the sample exceeds the base fluorescence value, the amplification is positively evaluated, and the sample thus contains the target segment. If the sample does not contain a target section or its sample content is lower than the detection limit of the method used, the fluorescence reading does not exceed the fluorescence baseline.

2.3.6. Quality Control

PCR controls were performed for each reaction; a positive control containing the target DNA and no-template control without any DNA added.

3. Results and Discussion

In this study, TAGs and DNA were used for identification of the meat origin. The workflow was as follows: In the first step, the potential of using metabolomics profiling, focusing on the analysis of TAGs, employing DART–HRMS to differentiate pork, beef and chicken samples was investigated. The aim was to design the conditions of the analysis, which would allow obtaining separate groups for different meats on the PCA and PLS-DA plots. If such conditions would not have been found, the authentication of meat cannot be done with this approach. As the samples were differentiated according to the type of meat, the same strategy for the analysis of commercial meat products was used. At the same time, the results obtained by the DART–HRMS method were confirmed by the established validated mPCRs. A theoretical comparison of the two methods used for our analyses was also performed. A possible combination of these methods that can facilitate the routine analysis of meat and meat products was suggested based on both theoretical and practical comparisons. See the chapters below for more details.

3.1. Results of DART–HRMS Analysis

Hexane was chosen as an extracting agent for sample preparation, since it is a nonpolar solvent suitable for the isolation of TAGs. Other solvents (e.g., methanol) were tested. However, when compiling the statistical models, the assumption was confirmed that, for pure meat, methanol and hexane can be used for the extraction and analysis with excellent classification of different types of meats. However, the classification is no longer satisfactory with methanol when analyzing meat products containing other ingredients (besides meat, e.g., spices). The reason for this is the co-isolation of many other substances that come from other ingredients used in the production of meat products and may be different for different products. Therefore, the only possible way to use the metabolomic approach for the authentication of meat in a meat product is TAGs analysis. The profiles of TAGs associated with various types of meat should be the same in the original meat as in the final meat product, as shown below.

3.1.1. DART–HRMS Fingerprints of Different Meat Types

Figure 1 shows the characteristic fingerprints, TAGs profiles, associated with all three types of analyzed meat (beef, pork and chicken). TAGs form $[M + NH_4]^+$ ions, which are in the m/z range 800–1000. Apparent differences in TAGs profiles of individual meat species are evident, especially in the ratios of relative intensities of the individual TAG ions present in the profiles. Different TAGs are dominant in various types of meat, for example, TAGs with m/z 848.7682, 874.7837 and 900.7996 are predominantly found in chicken meat, while TAGs with ions with 850.7836, 876.7995 and 902.8148 are predominantly found in pork and ions with m/z 850.7836, 876.7995 and 904.8306 in beef. The mentioned differences are mainly in terms of the ratios of ions relative intensities, as shown in Figure 1. This fact is particularly important from the point of view to reveal the economically motivated adulteration, i.e., to reveal the addition of undeclared cheap chicken meat, more often to pork meat products, but also to beef products. Due to the differences in the TAGs present and their ratios, in particular meat profiles, the employed DART–HRMS strategy indicates a good potential for detecting adulteration.

The fingerprint was converted to an ion list according to m/z and information about the intensities of the ions. The total number of detected ions related to TAGs in the profiles with signal intensity higher than 1000 cps was 40. However, it means 40 different summary formulas, but thanks to the isomers, the number of TAGs is probably higher. Data were transferred to MS Excel and TAG ions with VIP values higher than 1 (see Section 2.2.3) with their intensities were selected using the MAKRO function. The total number of selected ions was 15 (see Table 3). Furthermore, SIMCA v13.0 software was used for chemometric analysis. First of all, PCA was performed, followed by partial least squares discriminant analysis (PLS-DA). For the statistical analysis, 15 selected ions corresponding to TAGs were used. The obtained PCA and PLS-DA plots are shown in Figure 2.

As for the processing of DART–HRMS data, PCA clearly separated the meat varieties based on their TAGs profiles. In Figure 2 presenting the positive ionization data, PC1 and PC2 together described 79% of the sample set variability (64% and 15% for PC1 and PC2, respectively). Considering the fact that the first five PCs explain 99% (ESI+) of the total variance, The PC1/PC2 plot seemed to be a good starting point for sample clustering according to meat variety.

In the next step (following PCA analysis), PLS-DA was used, see Figure 2. As expected, efficient separation of samples into groups was achieved, and the mathematical model ($R^2X = 0.994$, $R^2Y = 0.989$, $Q^2 = 0.973$) obtained in this way reliably enabled the correct classification of an unknown sample; recognition ability (100%) and prediction ability (100%) were excellent.

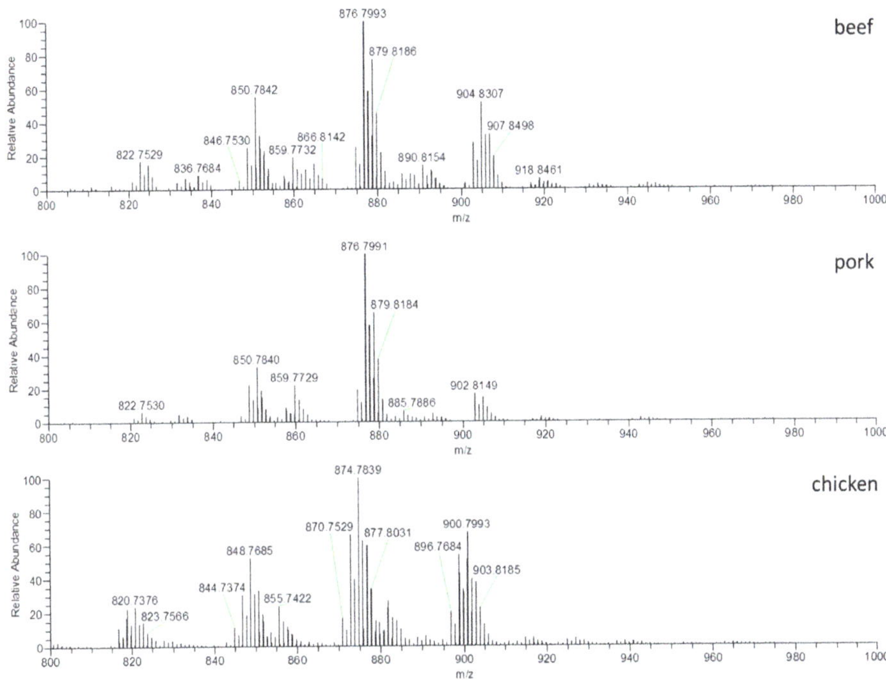

Figure 1. DART–HRMS profiles of triacylglycerols in hexane extract of beef, pork and chicken meat, *m/z* range 800–1000, positive ionization.

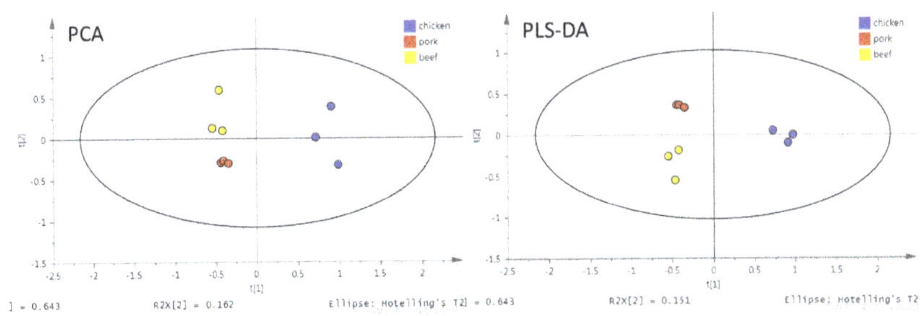

Figure 2. Principal component analysis (PCA) and partial least squares discriminant analysis (PLS-DA) plot created from ions related to triacylglycerols present in hexane extracts of chicken meat (blue), pork (red) and beef (yellow).

Although the model was created from a small number of samples, the differences between the samples of each species were noticeable. There is a certain risk of the model overfitting when using a small number of samples, for these reasons the model was control using a permutation test. It ought to be mentioned, that a larger number of samples would make the model probably more precise. In case of DART–HRMS the accuracy is secured by exact mass measurement and low mass differences (Δppm). Furthermore, the specificity is based on exact mass measurement and selection of appropriate *m/z* values. Estimation of sensitivity was based on the lowest intensity of selected ions (1000 cps). This corresponds to addition of approximately 3% of different meat into the other meat species. However, in the case of this study, the DART–HRMS method was designed and tested for screening purposes. The obtained results suggested that our DART–HRMS protocol could be used for rapid testing of a large number of samples. For precise authentication of suspicious or unclear samples, the use of validated PCR method is recommended.

Table 3. Characteristic ions for pork, beef and chicken.

m/z	Δppm	Formula	Identification	Significant Ions for
846.7535	1.259	$C_{53}H_{100}NO_6$	C 50:3	chicken
848.7682	1.339	$C_{53}H_{102}NO_6$	C 50:2	chicken/pork
850.7836	1.559	$C_{53}H_{104}NO_6$	C 50:1	chicken/pork
852.7975	4.820	$C_{53}H_{106}NO_6$	C 50:0	chicken/pork
872.7683	2.104	$C_{55}H_{102}NO_6$	C 52:4	chicken
874.7837	1.562	$C_{55}H_{104}NO_6$	C 52:3	chicken
876.7995	1.672	$C_{55}H_{106}NO_6$	C 52:2	pork/beef
878.8130	2.522	$C_{55}H_{108}NO_6$	C 52:1	pork/beef
880.8246	4.112	$C_{55}H_{110}NO_6$	C 52:0	pork/beef
896.7689	1.412	$C_{57}H_{102}NO_6$	C 54:6	chicken
898.7841	1.898	$C_{57}H_{104}NO_6$	C 54:5	chicken
900.7996	2.050	$C_{57}H_{106}NO_6$	C 54:4	chicken
902.8148	3.636	$C_{57}H_{108}NO_6$	C 54:3	pork/beef
904.8306	1.709	$C_{57}H_{110}NO_6$	C 54:2	beef
906.8457	2.974	$C_{57}H_{112}NO_6$	C 54:1	beef

3.1.2. DART–HRMS Analysis of Real Meat Products

The next step was application of suggested process for real meat products control. As representative meat products, sausages (as a traditional meat product consumed in the Czech Republic and other countries) and luncheon meat (as a product with high meat content), were selected for the verification of the declared meat type on packing. In addition, samples of pork and chicken ham-meat products containing almost no other ingredients (only pure meat) were analyzed as references. From all of these products, the hexane extracts, including lipophilic fraction with TAGs, were prepared and to obtain the TAGs profiles DART–HRMS was used, Figure 3.

The sample preparation chosen for the purposes of this study was affected as little as possible by the composition (ingredients) of the analyzed meat products. The DART–HRMS method was based on the analysis of TAGs, which are isolated from samples by a very nonpolar solvent (hexane). Such a sample preparation procedure naturally discriminates against the influence of inorganic salts in particular, which are rather polar in nature. Nonpolar substances could be a problem, but as this is not a quantification of individual substances, this is not a significant problem. Even if there is a possible suppression of the signal due to the lack of separation, all monitored TAGs would be suppressed similarly. Due to the use of data normalization during data processing, this effect should be compensated.

The PCA and PLS-DA chemometric analyses for data processing were used in the same way as it was mentioned in previous Section 3.1.1 (see Figure 4). As input data, the same TAGs ($n = 15$, see Table 3) were used, as in the case of chemometric analysis of data related to pure meat samples extracts. The PCA analysis focused on PC1 and PC2, which accounted for 81% of the sample set variability (64% and 17% for PC1 and PC2, respectively). Since the first five PCs explain 94% of the total variance, The PC1/PC2 plot seemed to be again a good starting point for sample classification.

The PLS-DA plot shows the classification of all three types of meat and meat products together. Most meat products samples were very close to the group of samples corresponding to pork meat. The values of the coefficients of recognition and prediction had a low value ($R^2X = 0.806$, $R^2Y = 0.434$, $Q^2 = 0.329$), because the plot includes four groups of samples. However, the classification of meat samples into groups by type of meat was perfect, and the attribution of meat product samples to pork meat samples matched the declaration on the packaging. Two samples (chicken ham), marked with the number 1 (Figure 4), were separated from the rest of the meat products. They were very close to chicken meat samples, which again reflected the declaration on the packaging. The sample marked with number

2 (meat in natural juices, sample number 27)—near the group of beef samples (Figure 4)—contained 70% beef, which again corresponds with the meat declaration on the packaging.

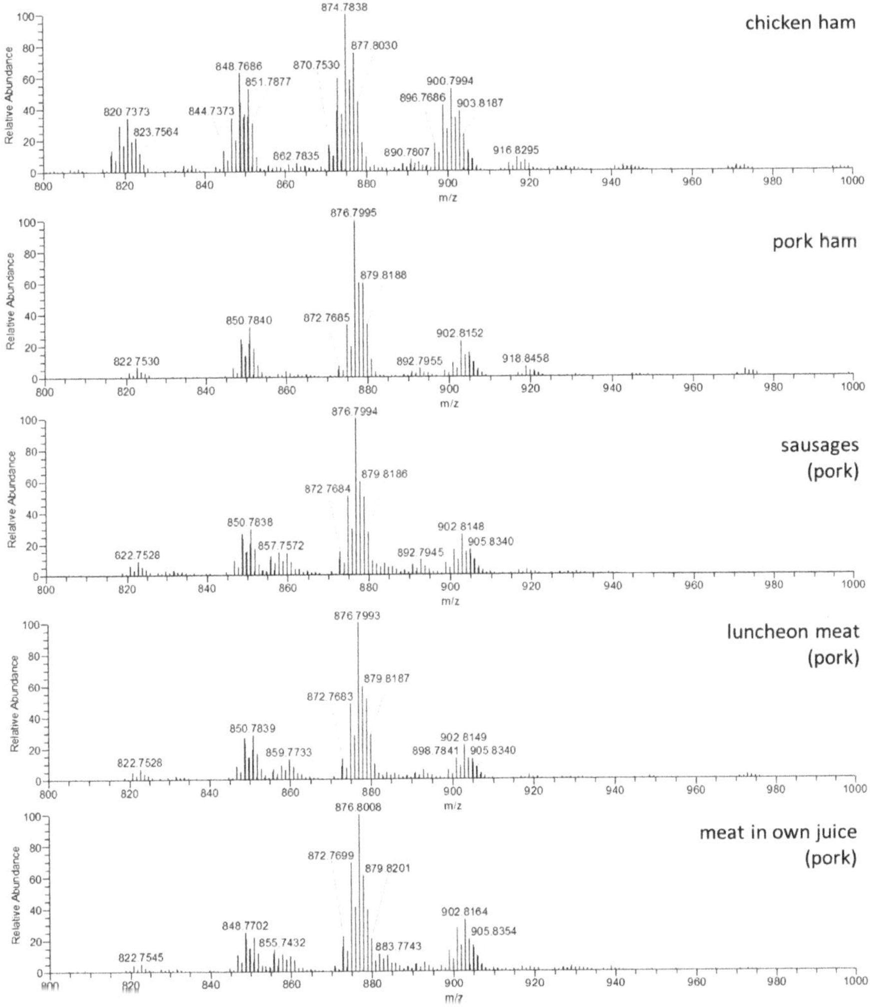

Figure 3. DART–HRMS profiles of triacylglycerol and hexane extracts of chicken ham, pork ham, sausages, luncheon meat and meat in natural juices, *m/z* range 800–1000, positive ionization.

The strategy presented in this article was based on the use of the rapid screening method DART–HRMS, which after data processing and visualization using a PCA or PLS-DA model will show suspicious samples in terms of authentication. Sample number 2 was an example of a "suspicious sample". The sample was detected from among a group of authentic samples (reference samples) and thus became suspicious. It was necessary to verify the authenticity using the exact PCR method. In our case, the content of pork and beef was indicated on the packaging of the product. The position of the sample in the model corresponded to its composition, and thus the functionality of the model was verified. For an unknown sample, said PCR analysis could be additionally performed.

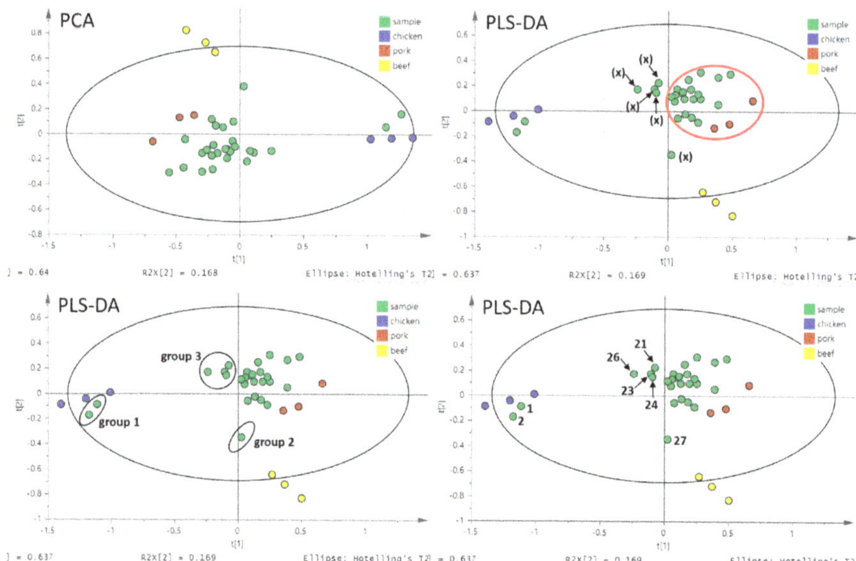

Figure 4. PCA and PLS-DA plots created from ions related to triacylglycerols present in hexane extracts of meat products (green), chicken (blue), pork (red) and beef (yellow). Group 1—chicken ham samples; group 2—meat products containing 70% of beef; group 3—pork meat products containing mechanically separated chicken meat and the rest of the green points represent meat products based on pork meat.

It can be very difficult to identify a sample as suspicious based only on a visual assessment. Therefore, the PCR method was used to confirm the results of DART–HRMS analyses. All samples were analyzed by PCRs. Thanks to these analyses, it was possible to determine the boundary (bold red line) in the PLS-DA model behind which suspicious samples are located, and it is necessary to subject these samples to confirmatory PCR analysis. Other samples that are located from the border (bold red line) towards the reference samples are considered as authentic samples containing only pork, as it demonstrated in Figure 4.

It is worth noting that four pork meat products (marked (x) that contained mechanically separated chicken meat tended to separate from the meat product made only from pork meat. They were situated towards the chicken meat samples in the PCA and PLS-DA plot, marked with the number 3 (see Figure 4). The other analyzed meat products mainly contained pork (without beef or chicken meat addition) were assigned to a group of authentic pork meat samples. From the analyses carried out and the obtained PLS-DA model, it is clear that this procedure correctly evaluates and assigns samples into groups according to the meat composition, the type of meat used in the production of meat products (sausages, luncheon meat, meat in natural juices). TAGs profiling by DART–HRMS could be used as a screening method to verify the composition of meat in meat products and to detect of adulteration by chicken meat.

To better demonstrate/simulate pork adulteration by chicken meat (the situation is similar for beef adulterated by chicken meat), PLS-DA plot (Figure 5A) was created using only the data for samples of chicken and pork and meat products composed of them. The plots show an excellent separation of chicken and pork samples ($R^2X = 0.784$, $R^2Y = 0.929$, $Q^2 = 0.894$, recognition ability = 100% and prediction ability = 100%). In case of adulteration, in the sense of mislabeling, i.e., a meat product labeled as a pork product contains (undeclared) chicken meat, the sample would be assigned more to the chicken meat group. For better understanding, one of the chicken samples was marked as a pork sample. The model did not place it among the other pork samples (red dots), but correctly assigned/kept it in the chicken meat samples group (red dot among the blue dots group—chicken meat samples), as shown in Figure 5B.

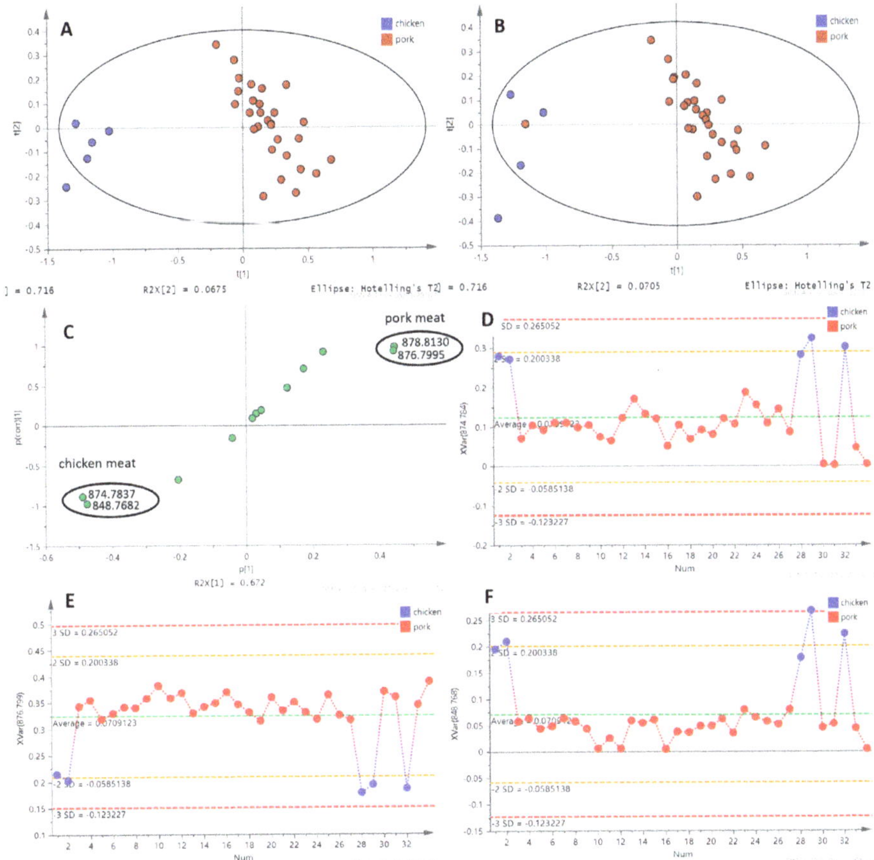

Figure 5. Examples of chemometric analysis of DART–HRMS data obtained from measurement of hexane extract of chicken and pork samples (meat and meat products). (**A**) PLS-DA plot; (**B**) PLS-DA plot illustrating the situation where the sample is adulterated; (**C**) S-plot and important chicken and pork markers; (**D**) variable line plot (trend plot) for ion *m/z* 874.7837—chicken meat marker; (**E**) variable line plot ion *m/z* 876.7995—pork meat marker; (**F**) variable line plot ion *m/z* 848.7682—chicken meat marker. Group of samples marked by number 1—four pork meat products which contained mechanically separated chicken meat.

To visualize ions that can be considered as 'markers' with the highest importance for sample separation, several plots were created from the acquired data. Figure 5 shows an example of an S-plot (Figure 5C) created from the data obtained from pork and chicken meat (TAGs ions, $[M + NH_4]^+$), which were analyzed by DART–HRMS in positive ionization mode. Four the most remote ions visualized in the S-plot (highlighted with black circles), which also had the highest values in the VIP-plot, were selected as 'markers' . In fingerprinting-based authentication strategies, the identification of the detected metabolome components is not essential for sample separation. On the other hand, under some conditions, e.g., when only marker ions are considered for sample profiling, then the identification of unique markers may be of interest. The identification was based on comparing their estimated elemental composition, mass difference (Δppm) and isotopic profiles with the data available in online libraries and scientific papers [28,29]. The variable line plot for the ions at *m/z* 874.7837, 848.7682 (markers for chicken meat, Figure 5D,F) and *m/z* 876.7995 (marker for pork meat, Figure 5E) illustrates the changes in the content of the respective TAGs in the pork and chicken meat. The estimated elemental composition of these ions together with their tentative identification are shown in Table 3, where the markers also for beef are included. These ions are not markers in the true sense of the word, because they are not unique to one group (meat type). However, due to the fact that statistical data processing takes into account not only the presence and intensity of the ions, but also their relative ratio, these selected ions are essential for authentication. For beef, pork and chicken, significant (very abundant) ions were selected for each type of meat, which are determinants and key for verifying an undeclared addition of—for example—chicken meat into a meat product or determining which type of meat was used.

It could be concluded that the usage of supervised PLS-DA model could lead to a distortion of the separation (overfitting). This model was used for completeness and, also for better visualization of the obtained data and explanation of the context. The PLS-DA model is very similar as the PCA model in terms of sample separation. Permutation tests were also performed during data processing and confirmed its applicability. Due to the fact, that DART–HRMS analysis is intended primarily for rapid screening of large number of samples and identification of only suspicious samples for an additional accurate PCR method, it is possible to use the PCA model for this purpose.

3.1.3. Confirmation of Isolated DNA Quality and Quantity

DNA was isolated using a CTAB method according to ČSN EN ISO 21,571 (ISO 21571, 2005). Both the yield and quality of the DNA obtained from pure musculature of beef, pork, chicken, turkey and horse were higher than 90 ng·μL^{-1} for all tested, commercially important meat species. More than 100 ng·μL^{-1} DNA was obtained from chicken, pork and beef samples. More than 50 ng·μL^{-1} DNA was isolated from 26 real meat products; the highest yield (316 ng·μL^{-1}) was obtained from meat in its natural juices (sample number 22). The lowest yield of DNA was isolated from one sausage (sample number 6), 15 ng·μL^{-1} was isolated with the ratio of absorbance (A 260 nm/A 280 nm) equal to 1.6 corresponding to nucleic acids with the presence of proteins. DNA of appropriate quality as well as quantity for subsequent amplifications was isolated from all samples tested.

3.1.4. DNA Analysis of Meat and Meat Products

PCRs for identification of beef, pork, horse and poultry (chicken, turkey) meat were designed and experimentally verified in our previous work; used PCRs were selective and allowed the detection of 30 copies of the haploid pig genome, 26 copies of the haploid beef-cattle genome and 11 copies of haploid chicken genome [24]. The limit of quantification for the mqPCR system was 12.5 ng of DNA per reaction. Four types of meat products were analyzed in this work: ham, Czech sausage, meat in its natural juices and luncheon meat. The results of mPCR and mqPCR analysis are shown in Table 4, examples of primary results in Figure 6.

Table 4. Results of meat sample by DNA analysis.

	Product	Declared Composition			mPCR			mqPCR		
		Pork	Chicken	Beef	Pork	Chicken/Turkey	Beef	Pork	Chicken	Beef
1	Chicken ham	-	+	-	-	+	-	-	+	-
2	Poultry ham	-	+	-	-	+	-	-	+	-
3	Pork ham	+	-	-	+	-	-	+	-	-
4	Sausage	+	-	+	+	-	+	+	-	+
5	Sausage	+	-	+	+	-	+	+	-	+

Table 4. Cont.

	Product	Declared Composition			mPCR			mqPCR		
		Pork	Chicken	Beef	Pork	Chicken/Turkey	Beef	Pork	Chicken	Beef
6	Sausage	+	-	+	+	-	+	+	-	+
7	Sausage	+	-	+	+	-	+	+	-	+
8	Sausage	+	-	+	+	-	+	+	-	+
9	Luncheon meat	+	-	+	+	-	+	+	-	+
10	Sausage	+	-	+	+	-	+	+	-	+
11	Sausage	+	-	+	+	-	+	+	-	+
12	Sausage	+	-	+	+	-	+	+	-	+
13	Meat in natural juices	+	-	-	+	-	-	+	-	-
14	Luncheon meat	+	-	-	+	-	-	+	-	-
15	Meat in natural juices	+	-	-	+	-	-	+	-	-
16	Meat in natural juices	+	-	-	+	-	-	+	-	-
17	Meat in natural juices	+	-	-	+	-	-	+	-	-
18	Sausage	+	-	+	+	-	+	+	-	+
19	Sausage	+	-	+	+	-	+	+	-	+
20	Sausage	+	-	+	+	-	+	+	-	+
21	Luncheon meat	+	+	-	-[1]	+	-	+	+	-
22	Meat in natural juices	+	-	-	+	-	-	+	-	-
23	Luncheon meat	+	+	-	+	+	+[1]	+	+	-
24	Luncheon meat	+	+	-	-[1]	+	-	+	+	-
25	Luncheon meat	+	-	-	+	-	-	+	-	-
26	Luncheon meat	+	+	-	+	+	+[1]	+	+	-
27	Meat in natural juices	+	+	+	+	+	-[1]	+	+	+

Legend: + amplicon present; - amplicon not detected; [1] - difference in results of mPCR and mqPCR analysis.

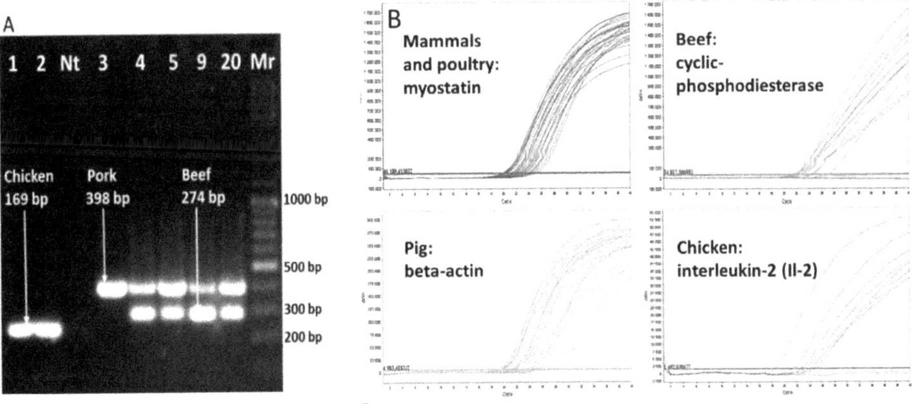

Figure 6. Examples of DNA analysis. Electropherogram of (**A**) mPCR amplicons and (**B**) fluorescent curves of mqPCR. Nt—no template control; x-axis—number of PCR cycles; y-axis—delta Rn (fluorescence).

Analyses of hams by quadruplex PCR of mtDNA and multiplex qPCR of genomic DNA gave the same results, the presence of the declared animal species (i.e., chicken or pig) was established. Eleven samples of typical Czech sausages were analyzed by mPCR, with identical results obtained for both target DNA (mtDNA and gDNA). The presence of the animal species declared in the label (i.e., beef and/or pork) was proved. Six cans of meat in natural juices were analyzed, five were made mainly from pork and one was made mainly from beef. Results that matched the labels were obtained for both target DNAs (mtDNA and gDNA) for all five pork cans. Beef meat in natural juices (sample 27) also contained mechanically separated poultry meat and pork skin, all three animal´s DNA was detected by mqPCR with gDNA as a target. It worked despite the heat treatment. Seven different Luncheon meats were analyzed, all the results obtained by mqPCR with gDNA as a target corresponded to the declaration on the label. A discrepancy between the results of mtDNA and gDNA analysis was observed in this type of meat product (luncheon meat). White adipose tissue (univacuolar adipocytes) contains a low number of mitochondria. The use of mqPCR amplified gDNA, which enables the better detection of a low amount of target DNA, is recommended for the analysis of samples containing large quantities of fat (e.g., luncheon meat).

Results obtained by PCR analyses of mtDNA and gDNA were the same for ham and sausages; for highly processed samples (cans) only PCRs targeting the gDNA appeared to be appropriate. Four samples of luncheon meat (21, 23, 24 and 26) and one meat in natural juices (27) showed differences in result of mtDNA and gDNA analyses, these samples were also categorized as a "suspicious sample" by DART–HRMS as explained in Section 3.1.2. The DART–HRMS analysis is a quick, cheap and high-throughput compare to moderate time-consuming PCRs (See Table 5). Therefore, DART–HRMS was proposed to be used as a screening method followed by through, a highly accurate, more laborious and more expensive amplification of gDNA by mqPCR only of suspicious samples. The advantages of such combinations are time, money and laboratory personnel-capacity saving.

Table 5. Comparison of DART–HRMS and PCR.

Parameters	DART–HRMS	PCR
Target molecule	Triacylglycerols	DNA
Preparation step	Hexane extract–lipophilic fraction containing triacylglycerols	DNA isolation—many methods available
Capacity of the machine	High-throughput method (+++)	Mainly 96 reactions in one run (++)
Cost of the analysis (only the retail price of chemicals is included)	Very low (+++)	Low (++)
Duration	Extraction: moderate (+++) Analysis: quick (+++) Evaluation: long	Extraction: moderate (++) Analysis: moderate (++) Evaluation: quick (+++)
Price of the required instrumentation	High (-)	Low for classical PCR instrument, moderate for qPCR device (+)
Feasibility for analysis of raw products	Yes (+++)	Yes, reliable (+++)
Feasibility for analysis of heat-treated meat products	Yes (++)	Yes, reliable (+++)
Feasibility for analysis of mixtures	Yes (++)	Yes, reliable (+++)
Feasibility for analysis of products containing high amounts of fat	Yes (+++)	Yes, gDNA is recommended as a target (+++)
Conduction of the experiment	Laboratory device (-), performing the analysis (+++)	Laboratory device (++), performing the analysis (+)
Claims for evaluation of results	Demanding for evaluation–experience is needed, because of the statistical analysis.	Simple to evaluate (+++). The PCR amplicon or fluorescence curve is or is not there, which is clearly visible from the primary results
Usage	Screening method	Confirmatory method

The results of mqPCR analyses were identical to the declarations by the manufacturer for the product labels. This fact is quite encouraging, considering the relatively common occurrence of fraudulent food practices, including for meat products. One possible explanation is that we focused on analyzing regional sausages and luncheon meats. The traditional (original) recipe for sausage and luncheon meat does not allow the addition of chicken meat (it is forbidden) and it can be said that the manufacturers did not follow these rules when preparing these regional sausages. With luncheon meat, almost half of the products analyzed were mislabeled.

The requirements are set out in Act No. 326/2001 Coll., issued by the Czech Ministry of Agriculture, which imposes requirements for meat labeling. Ham is made from the musculature of pork, poultry or beef, a heat-treated product. As for sausages, in this study, typical Czech sausages called "Špekáčky" were analyzed. These are a kind of sausage made from a finely minced mixture of pork and beef with smoked bacon pieces that gave the product its name. The true sausage according to the original recipe should consist of 50% beef, 20% pork cut out of the skin and 30% chopped bacon pieces. The basic raw materials according to current regulations are beef, pork or veal; a minimum of 40% meat and a maximum of 45% fat are required, and it does not allow the use of mechanically separated meat and poultry. The last two products were canned, i.e., products hermetically sealed in packaging and sterilized. According to the current regulation, canned food called "pork in natural juices" must contain at least 70% meat. The water content must not exceed 70% and the fat may be up to 40%. Canned food called "beef in natural juices" must contain at least 70% meat, the water content must not exceed 80%, and the fat may be up to 20%. The sum of all parameters exceeds 100% due to the fact that much of the fat and water in the product comes from the meat used. Luncheon meat should be made from pork and beef. The law specifies pork luncheon meat, which has the same limits on meat, fat and water content as "pork in natural juices".

3.1.5. Comparison of Methods

Commonly used molecular genetic methods such as PCR are based on the isolation and analysis of different DNA types, while the new innovative DART–HRMS analytical method is focused on the analysis of TAGs profiles. PCRs use DNA markers, which are unique, stable, well-known and independent of the type of animal breeding. Detection/quantification of these DNA markers can be used also for the investigation of mixed animal meat products. The quality and quantity of isolated DNA is the crucial parameter for the successful use of molecular genetic methods. There are many different procedures for isolating and analyzing DNA, which differ in the yield, cost and time requirements of the procedure. In PCR analysis, the whole procedure usually takes hours. Sample preparation for DART–HRMS is very simple and fast; the DART–HRMS analysis itself requires only a few seconds. On the other hand, the evaluation of the obtained data are significantly simplified with PCR analysis compared to the use of advanced chemometric analysis in the DART–HRMS evaluation procedure. With the DART–HRMS process, up to 130 samples per hour can be measured, while these 130 samples can take at least one working day to be analyzed by mPCR. On the other hand, PCR provides accurate results and is considered an arbitrary method, whereas the DART–HRMS method is primarily a screening method for testing a large number of samples and selecting only suspect specimens, which are then confirmed by DNA analysis (Table 5).

4. Conclusions

In the presented work the multiplex mPCR analysis of DNA and DART–HRMS analysis of the TAGs profile of meat (pork, beef, chicken) in combination with multivariate statistical analysis were compared theoretically and practically. In both cases, comparable results were obtained. DART–HRMS is a quick method with a great potential for screening a large number of samples. PCRs analysis can be used for precise animal species identification; however, it takes more time to get results.

Our results suggest that DART–HRMS could be used primarily as a screening method and suspected samples could be subsequently analyzed by PCR method. This combination of both approaches has potential for meat type verification or detection of adulteration, respectively.

Author Contributions: The following authors contributed to this article in the following ways: Conceptualization and methodology, K.Z. and V.H.; software, V.H. and M.J.; validation, J.P., E.C. and D.J.; formal analysis, K.Z., V.H., J.P., E.C. and D.J.; investigation and data curation, K.Z. and V.H.; resources, K.D., J.P. and J.H.; writing—original draft preparation, all authors; writing—review and editing, K.Z., V.H., J.H. and K.D.; visualization, D.J.; supervision, project administration and funding acquisition, K.D., J.P. and J.H. All authors have read and agreed to the published version of the manuscript.

Funding: This research was funded by a grant from the Ministry of Agriculture of the Czech Republic: National Agency for Agricultural Research project number QJ1530272: Complex strategies for the effective detection of food fraud in the production-consumer chain and the projects ERDF CZ.2.16/3.1.00/21537 and CZ.2.16/3.1.00/24503 and METROFOOD-CZ research infrastructure project (MEYS Grant No.: LM2018100) including access to its facilities.

Conflicts of Interest: The authors declare no conflict of interest.

References

1. Hsieh, Y.P.; Woodward, B.B.; Ho, S.H. Detection of Species Substitution in Raw and Cooked Meats Using Immunoassays. *J. Food Prot.* **1995**, *58*, 555–559. [CrossRef] [PubMed]
2. Ballin, N.Z.; Vogensen, F.K.; Karlsson, A.H. Species determination—Can we detect and quantify meat adulteration? *Meat Sci.* **2009**, *83*, 165–174. [CrossRef] [PubMed]
3. Fajardo, V.; González, I.; Rojas, M.; García, T.; Martín, R. A review of current PCR-based methodologies for the authentication of meats from game animal species. *Trends Food Sci. Technol.* **2010**, *21*, 408–421. [CrossRef]
4. Köppel, R.; Daniels, M.; Felderer, N.; Brünen-Nieweler, C. Multiplex real-time PCR for the detection and quantification of DNA from duck, goose, chicken, turkey and pork. *Eur. Food Res. Technol.* **2013**, *236*, 1093–1098. [CrossRef]
5. Iwobi, A.; Sebah, D.; Spielmann, G.; Maggipinto, M.; Schrempp, M.; Kraemer, I.; Gerdes, L.; Busch, U.; Huber, I. A multiplex real-time PCR method for the quantitative determination of equine (horse) fractions in meat products. *Food Control* **2017**, *74*, 89–97. [CrossRef]
6. Griffiths, A.M.; Sotelo, C.G.; Mendes, R.; Pérez-Martín, R.I.; Schröder, U.; Shorten, M.; Silva, H.A.; Verrez-Bagnis, V.; Mariani, S. Current methods for seafood authenticity testing in Europe: Is there a need for harmonisation? *Food Control* **2014**, *45*, 95–100. [CrossRef]
7. Murugaiah, C.; Al-Talib, H.; Radu, S. Forensics: Food Authentication Using MtDNA. *J. Nutr. Health Food Sci.* **2015**, *3*, 1–10. [CrossRef]
8. Mozola, M.A.; Peng, X.; Wendorf, M.; Artiga, L. Evaluation of the GeneQuence® DNA hybridization method in conjunction with 24-h enrichment protocols for detection of Salmonella spp. in select foods: Collaborative study. *J. AOAC Int.* **2007**, *90*, 738–755. [CrossRef]
9. Tichoniuk, M.; Ligaj, M.; Filipiak, M. Application of DNA hybridization biosensor as a screening method for the detection of genetically modified food components. *Sensors* **2008**, *8*, 2118–2135. [CrossRef]
10. Chapela, M.J.; Sotelo, C.G.; Calo-Mata, P.; Pérez-Martín, R.I.; Rehbein, H.; Hold, G.L.; Quinteiro, J.; Rey-Méndez, M.; Rosa, C.; Santos, A.T. Identification of Cephalopod Species (Ommastrephidae and Loliginidae) in Seafood Products by Forensically Informative Nucleotide Sequencing (FINS). *J. Food Sci.* **2002**, *67*, 1672–1676. [CrossRef]
11. Iwobi, A.; Sebah, D.; Kraemer, I.; Losher, C.; Fischer, G.; Busch, U.; Huber, I. A multiplex real-time PCR method for the quantification of beef and pork fractions in minced meat. *Food Chem.* **2015**, *169*, 305–313. [CrossRef] [PubMed]
12. Matsunaga, T.; Chikuni, K.; Tanabe, R.; Muroya, S.; Shibata, K.; Yamada, J.; Shinmura, Y. A quick and simple method for the identification of meat species and meat products by PCR assay. *Meat Sci.* **1999**, *51*, 143–148. [CrossRef]
13. Barbuto, M.; Galimberti, A.; Ferri, E.; Labra, M.; Malandra, R.; Galli, P.; Casiraghi, M. DNA barcoding reveals fraudulent substitutions in shark seafood products: The Italian case of "palombo" (*Mustelus* spp.). *Food Res. Int.* **2010**, *43*, 376–381. [CrossRef]
14. Thakur, M.; Singh, S.; Shukla, M.; Sharma, L.; Agarwal, N.; Goyal, S.; Sambandam, S. Identification of Galliformes through Forensically Informative Nucleotide Sequencing (FINS) and its Implication in Wildlife Forensics. *J. Forensic Res.* **2012**, *4*. [CrossRef]
15. Jírů, M.; Stranska, M.; Kocourek, V.; Krmela, A.; Tomaniová, M.; Rosmus, J.; Hajslova, J. Authentication of meat species and net muscle proteins: Updating of an old concept. *Czech J. Food Sci.* **2019**, *37*, 205–211. [CrossRef]
16. Ruiz Orduna, A.; Husby, E.; Yang, C.T.; Ghosh, D.; Beaudry, F. Assessment of meat authenticity using bioinformatics, targeted peptide biomarkers and high-resolution mass spectrometry. *Food Addit. Contam. Part A* **2015**, *32*, 1709–1717. [CrossRef]

17. Cody, R.B.; Laramée, J.A.; Durst, H.D. Versatile new ion source for the analysis of materials in open air under ambient conditions. *Anal. Chem.* **2005**, *77*, 2297–2302. [CrossRef]

18. Hajslova, J.; Cajka, T.; Vaclavik, L. Challenging applications offered by direct analysis in real time (DART) in food-quality and safety analysis. *TrAC Trends Anal. Chem.* **2011**, *30*, 204–218. [CrossRef]

19. Worley, B.; Powers, R. Multivariate Analysis in Metabolomics. *Curr. Metab.* **2013**, *1*, 92–107. [CrossRef]

20. Blasco, H.; Błaszczyński, J.; Billaut, J.C.; Nadal-Desbarats, L.; Pradat, P.F.; Devos, D.; Moreau, C.; Andres, C.R.; Emond, P.; Corcia, P.; et al. Comparative analysis of targeted metabolomics: Dominance-based rough set approach versus orthogonal partial least square-discriminant analysis. *J. Biomed. Inform.* **2015**, *53*, 291–299. [CrossRef]

21. Berrueta, L.A.; Alonso-Salces, R.M.; Héberger, K. Supervised pattern recognition in food analysis. *J. Chromatogr. A* **2007**, *1158*, 196–214. [CrossRef]

22. Triba, M.N.; Le Moyec, L.; Amathieu, R.; Goossens, C.; Bouchemal, N.; Nahon, P.; Rutledge, D.N.; Savarin, P. PLS/OPLS models in metabolomics: The impact of permutation of dataset rows on the K-fold cross-validation quality parameters. *Mol. Biosyst.* **2015**, *11*, 13–19. [CrossRef]

23. ISO. *Foodstuffs-Methods of Analysis for the Detection of Genetically Modified Organisms and Derived Products-Nucleic Acid Extraction*; ANSI: New York, NY, USA, 2005.

24. Zdeňková, K.; Akhatova, D.; Fialová, E.; Krupa, O.; Kubica, L.; Lencová, S.; Demnerová, K. Detection of meat adulteration: Use of efficient and routine-suited multiplex polymerase chain reaction-based methods for species authentication and quantification in meat products. *J. Food Nutr. Res.* **2018**, *57*, 351–362.

25. Laube, I.; Zagon, J.; Spiegelberg, A.; Butschke, A.; Kroh, L.W.; Broll, H. Development and design of a 'ready-to-use' reaction plate for a PCR-based simultaneous detection of animal species used in foods. *Int. J. Food Sci. Technol.* **2007**, *42*, 9–17. [CrossRef]

26. Köppel, R.; Ruf, J.; Zimmerli, F.; Breitenmoser, A. Multiplex real-time PCR for the detection and quantification of DNA from beef, pork, chicken and turkey. *Eur. Food Res. Technol.* **2008**, *227*, 1199–1203. [CrossRef]

27. Laube, I.; Spiegelberg, A.; Butschke, A.; Zagon, J.; Schauzu, M.; Kroh, L.; Broll, H. Methods for the detection of beef and pork in foods using real-time polymerase chain reaction. *Int. J. Food Sci. Technol.* **2003**, *38*, 111–118. [CrossRef]

28. Hurkova, K.; Uttl, L.; Rubert, J.; Navratilova, K.; Kocourek, V.; Stranska-Zachariasova, M.; Paprstein, F.; Hajslova, J. Cranberries versus lingonberries: A challenging authentication of similar Vaccinium fruit. *Food Chem.* **2019**, *284*, 162–170. [CrossRef] [PubMed]

29. Hrbek, V.; Rektorisova, M.; Chmelarova, H.; Ovesna, J.; Hajslova, J. Authenticity assessment of garlic using a metabolomic approach based on high resolution mass spectrometry. *J. Food Compos. Anal.* **2018**, *67*, 19–28. [CrossRef]

MDPI

St. Alban-Anlage 66

4052 Basel

Switzerland

Tel. +41 61 683 77 34

Fax +41 61 302 89 18

www.mdpi.com

Foods Editorial Office

E-mail: foods@mdpi.com

www.mdpi.com/journal/foods